Advanced Solutions for the Efficient Integration of Electric Vehicles in Electricity Grids

Advanced Solutions for the Efficient Integration of Electric Vehicles in Electricity Grids

Editors

Evangelos Karfopoulos
Michele Roccotelli
Ioannis Karakitsios

MDPI

Basel • Beijing • Wuhan • Barcelona • Belgrade • Novi Sad • Cluj • Manchester

Editors

Evangelos Karfopoulos
National Technical University
of Athens
Athens
Greece

Michele Roccotelli
Politecnico di Bari
Bari
Italy

Ioannis Karakitsios
National Technical University
of Athens
Athens
Greece

Editorial Office
MDPI AG
Grosspeteranlage 5
4052 Basel, Switzerland

This is a reprint of articles from the Special Issue published online in the open access journal *Energies* (ISSN 1996-1073) (available at: https://www.mdpi.com/journal/energies/special_issues/4K0I7952ZA).

For citation purposes, cite each article independently as indicated on the article page online and as indicated below:

Lastname, A.A.; Lastname, B.B. Article Title. *Journal Name* **Year**, *Volume Number*, Page Range.

ISBN 978-3-7258-2617-9 (Hbk)
ISBN 978-3-7258-2618-6 (PDF)
doi.org/10.3390/books978-3-7258-2618-6

Contents

energies

MDPI

Article

Development and Validation of V2G Technology for Electric Vehicle Chargers Using Combo CCS Type 2 Connector Standards

Shahid Jaman [1,2], Boud Verbrugge [1,2], Oscar Hernandez Garcia [3], Mohamed Abdel-Monem [3], Blum Oliver [4], Thomas Geury [1,2] and Omar Hegazy [1,2,*]

1 MOBI-EPOWERS Research Group, ETEC Department, Vrije Universiteit Brussel (VUB), Pleinlaan 2, 1050 Brussel, Belgium
2 Flanders Make, Gaston Geenslaan 8, 3001 Heverlee, Belgium
3 Products Innovation & Systems Verification (PI&SV) Team, Powerdale (PWD), Witte Patersstraat 4, 1040 Brussel, Belgium
4 CR/AES3, Robert-Bosch-Campus 1, 71272 Renningen, Germany
* Correspondence: omar.hegazy@vub.be

Abstract: Vehicle-to-Grid (V2G) technology is viewed as a viable solution to offer auxiliary power system services. Currently, V2G operation is only possible through DC chargers using the CHAdeMO connector with the necessary communication protocol. However, in Europe, for high-power DC charging (>50 kW), the Combined Charging Service (CCS) Type 2 is preferred over CHAdeMO. Therefore, this work presents the development of a V2G testing system with a Combo CCSType 2 charger including communication via the ISO 15118-2 protocol. The BOSCH passenger car with a 400 V battery pack is used to test and validate the technical feasibility of V2G charging via a Combo CCS Type 2 connector standard. The V2G feature is characterized in terms of efficiency, signal delay, response proportionality, magnitude accuracy and noise precision. A data driven V2G charger simulation model based on the real-time data is also developed in MATLAB/Simulink. The performance under various operating settings is presented in the outcomes, emphasizing the need for appropriate hardware calibration, and understanding while delivering standard-compliant grid control services using V2G technology. Finally, the results of the simulation model are compared with the real hardware results in terms of error, noise level and data magnitude accuracy.

Keywords: Vehicle-to-Grid (V2G); testing; Combo CCS Type 2; ISO 15118-2; electric vehicle DC charger; electric vehicle charging analyzer; communication control unit (CCU)

Citation: Jaman, S.; Verbrugge, B.; Garcia, O.H.; Abdel-Monem, M.; Oliver, B.; Geury, T.; Hegazy, O. Development and Validation of V2G Technology for Electric Vehicle Chargers Using Combo CCS Type 2 Connector Standards. *Energies* **2022**, *15*, 7364. https://doi.org/10.3390/en15197364

Academic Editor: G.M. Shafiullah

Received: 14 September 2022
Accepted: 4 October 2022
Published: 7 October 2022

1. Introduction

Electric vehicles (EVs) are becoming a popular means of transportation to ensure a clean environment. The majority of nations are making efforts to electrify their transportation systems, which are currently one of the main causes of the world's carbon footprint [1]. For EVs, the electric charge of the battery is the main fuel to drive the car. To refuel the EV, we have to connect the EV to a charging station. For the EV charging pile, there are two options are available: (a) AC charging pile and (b) DC charging pile. They are differing from each other based on the charging voltage and current characteristics. In DC or direct current, the electrons flow in one direction steadily with zero frequency. In contrast, the electron changes its direction of mobility after a certain period periodically with 50 Hz frequency. The advantages of AC charging and DC charging are briefly discussed below [2]:

Benefits of DC charging over AC charging

- DC chargers are bigger and faster than AC chargers.
- All power conversion (AC/DC and/or DC/DC) occurs outside of the car. So, there are no power converters for DC charging inside the car. The charging current goes directly to the battery from the DC charger.

- The electric vehicle that accepts DC charging can fast recharge while traveling a long distance.
- Working with AC at high voltages is more dangerous than working with DC.

As the power converter is built within the DC charger itself, unlike AC chargers, electricity may be sent straight to the car's battery without needing to be converted via the onboard charger. Because it needs a power source or transformer (a lot of electricity, such as 125A from the grid), the cost of charging is greater.

Benefits of AC charging over DC charging

- Most parking spaces where a car will be parked for a longer period of time are ideal for AC charging.
- The installation of an AC charging station is less expensive than DC.
- An inductor or conductor can be used to lower the current without experiencing a significant energy loss because the current magnitude is not constant.
- Rectifiers make it simple to convert AC electricity to DC.
- A transformer may be used to produce a broad range of voltages.
- When AC is delivered at higher voltages over larger city distances, the line losses are less in comparison to a DC transmission.

The main disadvantage of AC chargers is the speed of charging. Since the on-board charger is used for AC charging, the charging speed is depending on the design and technologies used during the on-board charger design. The majority of the EVs are left in parking lots for a sizable portion of the day, which means that the energy in their batteries is not being used. To transfer this underutilized EV battery power to the grid, Vehicle-to-Grid (V2G) technology was developed [3].

The V2G technology was first conceptualized by Kempton and Letendre in 1997 [4]. The power is usually transmitted from the generation unit to the customer (home, aggregator, etc.) in a conventional power system. An EV charging station functions as a consumer in normal operation, connecting the EV battery to the grid to charge it via the power connection. The EVs with V2G technology may return energy to the grid when the car is parked. Additionally, the V2G features enable more advantages such as peak shaving and labeling the demand, ancillary service to avoid the shortage. On the contrary, V2G has some barriers to using it commercially. The EV battery lifetime is reduced by frequently discharging for V2G service. Additionally, the public's primary concerns with this strategy are the battery's level of charge, local charging stations and infrastructure, as well as the initial high investment prices. Therefore, the rate of V2G system participation may be impacted, but this issue may be handled by putting in place a well-organized and motivating plan to assure the public of this system's effectiveness [5]. The most significant barrier is to change the hardware modification inside the car. Moreover, establishing communication amongst the many stakeholders is the biggest obstacle to successfully using the V2G technology. To transfer the required instructions between the EVs, the electric vehicle supply equipment (EVSE), and third-party operators, V2G systems require communication protocols, both front-end and back-end [6]. The charging topologies (on-board/off-board), type of charging (conductive/inductive), safety, charging connections, communication, and cybersecurity criteria are all provided by the front-end protocols. They explain the relationship between EVs and EVSEs as well [7]. For instance, front-end protocols include communication standards such as IEC 61851, ISO 15118, SAE J2847, and CHAdeMO [8,9]. The back-end protocols describe criteria for communication and cybersecurity and provide links between EVSEs and third-party operators such as charge point operators (CPOs) [9,10]. A few examples of back-end protocols include the open charge point protocol (OCPP), IEC 63110, the open automated demand response (ADR), and EEBUS described in [11]. More information on these standards can be found in [12]. Due to the advancement in communication and devices, the V2G technology has attracted the researchers' and OEMs' attention to integrating EVs into the power grid with grid-supporting capabilities.

As a result of the increased attention being paid to V2G technology, approximately 50 V2G projects have been launched globally with the goal of identifying the best business models for all parties involved, including EV owners, chargers developers, utilities, and automotive OEMs [13]. Nissan is now collaborating with the German power company E.ON on the V2G services, distributed energy generation, and renewable energy systems business models [14]. Similarly, Volkswagen sees enormous potential in V2G [15–17]. The residential V2G charger from the Australian business EV-NRG has been released. It can control EV power depending on data from the vehicle and a series of sensors that track household loads and grid supply [11]. Additionally, V2G technology is developing via different projects around the globe. For instance, the ACES project tested V2G service in Bornholm, and Denmark tested V2G technology using the CHAdeMO standards to assess the feasibility of the provision of grid services [18]. The Parker project also demonstrated the V2G technology in 2016 with a CHAdeMO charging port to assess the grid support service to renewable power systems [19]. The real-time benefits of advanced vehicle-grid integration are assessed with the IEC 62196-2 standard on DC charging port in the INVENT project [20]. Moreover, this project enables UC San Diego to manage when and how electricity is drawn from EVs plugged into its charging stations to help power university buildings and facilities, saving both money and energy. The Grid Motion project has also investigated the V2G service feasibility in smart grid applications [21]. The summary of the state of the art of European V2G projects with associate charging standards is shown in Table 1.

Table 1. European V2G test projects with corresponding connector standards [22].

Projects	Country	Connector Standard	Project Duration
Parker	Denmark	CHAdeMO	2016–2018
Redispatch V2G	Germany	CHAdeMO	2018–2021
CITY-ZEN	Netherland	CHAdeMO	2014–2019
Smart Solar Charging	Netherland	CHAdeMO/CCS	2014–2019
Grid Motion	France	CHAdeMO/CCS	2017–2019
ACES	Denmark	CHAdeMO	2016-2018
Network Impact of Grid-Integrated Vehicle	UK	CHAdeMO/CCS	2017–2020
Porto Santo	Portugal	IEC 62196-2/CHAdeMO	2015–2020
ZEMtoALL	Spain	CHAdeMO	2012-2017
Vehicle-to-Coffee	Germany	IEC 62196-2/CHAdeMO	2015–2018
NewMotion V2G	Netherland	CHAdeMO	2017
Amsterdam V2G	Netherland	CHAdeMO	2013–2017
SEEV4-City	5 Cities in EU	CHAdeMO	2016–2020
SHAR-Q	Greece	IEC 62196-2/CHAdeMO	2016–2019
Denmark V2G	Denmark	CHAdeMO	2016–Ongoing
Genoa Pilot	Italy	CHAdeMO	2017–Ongoing
UK Vehicle-2-Grid (V2G)	UK	CHAdeMO	2016–Ongoing
GrowSmarter	Spain	CHAdeMO	2016–2019

The ChaoJi or GB/T connectors do not have a standard communication protocols whereas the CHAdeMO publishes the communication protocol for V2G in 2014 [23]. Currently, the V2G technology is only commercially possible through CHAdeMO DC chargers using the accompanying communication protocol, which also means that only cars using CHAdeMO, such as Nissan and Mitsubishi, can be used to test the V2G functionality.

However, in Europe, most EVs use the Combined Charging Service (CCS) type 2 for DC charging together with the ISO 15118 communication protocol. There are three separate versions of the ISO 15118 standard: 15118-1, 15118-2, and 15118-3. The broad specifications and use case descriptions for V2G communication are provided in ISO 15118-1. It offers to charge confirmation, service certification, and start and stop charging instructions for electric vehicles. In order to satisfy the use cases established in 15118-1 for the network and application layer needs for V2G communication, ISO 15118-2 defines the message types and formats between the EV Communication Controller (EVCC) and Supply Equipment Communication Controller (SECC) [24–26]. However, the ISO 15118-2 protocol still needs some modifications to ensure V2G services. Moreover, for the real-time test of V2G technology with Combo CCS type 2, a simple communication unit needs to be added to the EV to regulate the power flow between the EVs and the grid. Designing this communication technology is expensive [9]. Therefore, most (European) EVs are not equipped with this unit, and they require a setup modification or adaptation to successfully establish a bidirectional power flow. Accordingly, the majority of the EVs on the European market today is not yet ready for V2G technology except for a few EVs manufactured by companies such as Nissan and Mitsubishi.

This paper attempts to enable V2G technology for Combo CCS Type 2 chargers by developing and validating a V2G testing system using the modified ISO 15118-2 communication protocol. This research also investigated the performance of DC chargers to the provision of V2G-based services. Additionally, a thorough modeling of the system in MATLAB/Simulink 2019b software environment, which takes the results of the tests into account, is developed to complete the research. The low-fidelity accurate charger model can be used as an effective tool for power system simulation studies for the grid integration impact assessment of the EVs. The investigated modeling technique of the real V2G system hardware can be able to contribute during demanding smart grid services such as frequency containment reserves (FCR). The remainder of the paper is structured as follows. Section 2 outlines the background information on EV charging connectors and communication standards. Section 3 outlines the V2G test system development and modification strategy. Section 4 presents a comprehensive discussion of the testing results including statistical analysis. Section 5 described the modeling of the V2G charger low-fidelity model in MATLAB/Simulink based on the real-time testing data. Finally, the conclusions are presented in Section 6.

2. EV Charging Connector and Communication Standard Background

International organizations have been working for many years to enact their connection standards regarding the high-power charging of EVs. The commonly used technologies are: (a) AC on-board charging and (b) DC off-board charging. In previous years, European EV manufacturers have rather produced EVs with on-board chargers which are compatible with a high-power AC supply. However, since the company TEPCO introduced a standard Japanese socket for DC connection, namely CHAdeMO, DC off-board charging is becoming more common. CHAdeMO is now available in several European countries, and although it is not internationally standardized yet, it is widely used in the US and Japan [27]. The SAE also has been developing its own connector standard called Combo. In SAE J1772, the Combo coupler features extra pins compared to the standard version [28].

2.1. AC Charging Connector

The IEC 62196-2 standard defines three types of connectors for AC supply power in Europe. These types of connectors are described in Table 2.

Table 2. IEC 62196-2 Connector Standard for AC charging [29].

IEC 62196-2	Type 1	Type 2	Type 3
Coupler	1-phase	1-phase and 3-phase	1-phase and 3-phase with shutters
Related standard	SAE J1772 Type 1	VDE-AR-E 26232-2	-
Maximum current	32 A (80 A at the US)	70 A (1-phase) 63 A (3-phase)	16 A (1-phase) 32 A (3-phase)/63 A(3-phase)
Maximum voltage	250 V	480 V (3-phase)	250 V/400 V
Maximum power	19 kW	43.5 kW	22 kW (3-phase)
Pin and interlock	5 pins, mechanic lock	7 pins, electronic lock	4 pins or 5 pins
Control pin	Two short pins	One short, one long pin	-
Communication	PWM over CP	PWM over CO	-
		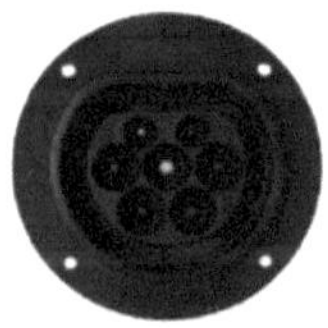	

The IEC 62196-2 Type 1 is not different compared to the SAE J1772 which is widely used in the US and Japan. In China, the IEC 62196-2 Type 1 is also known as GB/T 20234.2-2015 connector. Notice that China has modified it by changing from an electronic lock into a mechanic lock (with the option of electronic lock) and the control pilot is using two short pins.

2.2. DC Charging Connector

The most common use of fast charging technologies is to supply the direct current to the EV battery. This requires an off-board charging station to supply high levels of current and to transform the AC of the distribution grid to DC for the charging of the EV's battery. CHAdeMO was the first fast-charging method in the world, specified by the Japan Electric Vehicle Standard (JEVS) G105-1993 from the JARI (Japan Automobile Research Institute), and it has been developed by the Japanese company TEPCO. CHAdeMO is an abbreviation of "CHArge de MOve" or "move by charge" [30]. The connector includes two large pins for DC power, plus other pins to carry CAN-BUS connections. The CHAdeMO transmits the information by CAN-BUS and analog control lines. This hybrid communication protocol has more advantages compared to pure digital control, since it can let the system double-check the digital control system. Once the analog signal is lost, an immediate charging operation shutdown can be performed [28]. The CHAdeMO specifications comply with the international standard IEC 62196-3. The characteristics of CHAdeMO are shown in Table 3.

Table 3. CHAdeMO connector characteristics [31].

Parameter Description	Rated Values
Maximum current	120 A
Maximum voltage	500 V DC
Maximum power	50 kW
Maximum current (control system)	7 A
Maximum voltage (control system)	12 V DC
Level of charging	DC level 3
Control pin	7 pins
Communication protocol	CHAdeMO (CAN communication)

The configuration of CHAdeMO only specifies the necessary and effective parts for connection. A summary and pinout of CHAdeMO is shown in Table 4.

Table 4. CHAdeMO pinout functionalities [32].

Pin ID	Wire Cross Section (mm^2)	Description
1	0.75	Ground
2	0.75	Start/stop charging 1
3		None
4	0.75	Permission/prohibition charging
5	22 or 40	DC supply negative
6	22 or 40	DC supply positive
7	0.75	Verification of the connector connection
8	0.75	CAN High
9	0.75	CAN Low
10	0.75	Start/stop charging 2

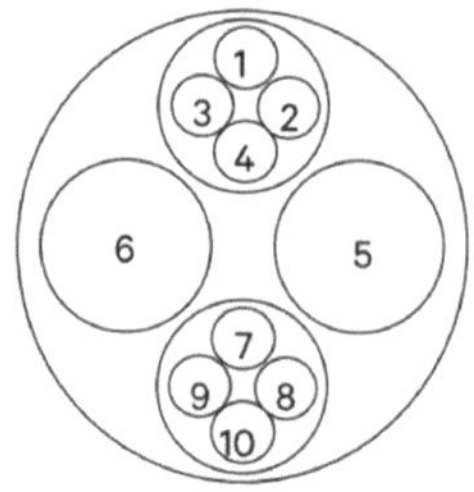

2.3. *Combined Charger Connector (J1772-2009 Combo)*

In parallel to CHAdeMO, the US and some European EV manufacturers such as Audi, BMW, Daimler, Ford, General Motors, Porsche, or Volkswagen have been developing a new system for fast charging: the Combined Charging System (CCS), i.e., J1772-2009 Combo. The main goal of this standard is to allow the EV to charge in both AC (slow/medium charging) and DC (fast charging), which is different from CHAdeMO as it only allows DC and needs an extra socket for AC charging.

There are two types of Combo CCS currently available. The Combo Type 1 connector for the US has been developed based on the AC Type 1 connector (specified by the standards SAE J1772/UNE EN 62196-2). The Combo Type 2 for Europe integrates a Type 2 connector as defined in the standard UNE EN 62196-2 [33]. The main characteristics of Combo connectors Type 1 and Type 2 are summarized in Table 5.

Table 5. Main characteristics of Combo connectors Type 1 and type 2 [31].

Parameters	Combo Type 1 (US)	Combo Type 2 (EU)

	DC Charging	
Maximum current	150 A	200 A
Maximum voltage	600 V	850 V
Charging mode	4	4
Maximum power	90 kW	170 kW
Connector type	Combo 1 (IEC 62196-3)	Combo 2 (IEC 62196-3)
	AC Charging	
Nominal current	32 A	70 A (1-phase)/63 A (3-phase)

Table 5. *Cont.*

Parameters	Combo Type 1 (US)	Combo Type 2 (EU)
Nominal voltage	250 V	230 V (1-phase)/400 V (3-phase)
Charging mode	3	3
Maximum power	13 kW	44 kW
Connector type	Type 1 (IEC 62196-2, SAE J1772)	Type 2 (IEC 62196-2)

The Combo Type 1 and Type 2 have a different pinout configuration based on their different functionalities. Table 6 shows the configuration and the functions of the pins in the Combo Type 1 and Type 2.

Table 6. Pin configuration and functionality of Combo connectors Type 1 and Type 2 [32].

Pin	Functions	Comments	Combo Type 1
PP	Communication/Charging process control	Proximity inlet	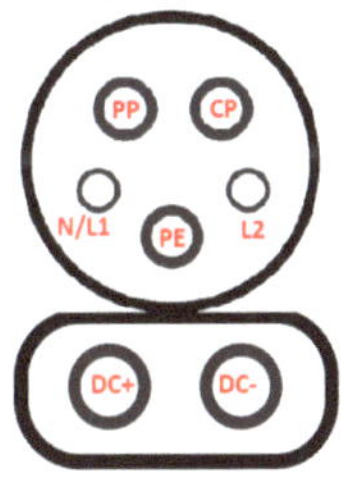
CP		Control pilot	
PE	Earth ground	EV to earth ground	
L1/N	AC 1-phase charging	Phase 1/Neutral	
L2		Phase 2	
DC+	DC charging	DC positive terminal	
DC-		DC negative terminal	
Pin	**Functions**	**Comments**	**Combo Type 2**
PP	Communication/Charging process control	Proximity inlet	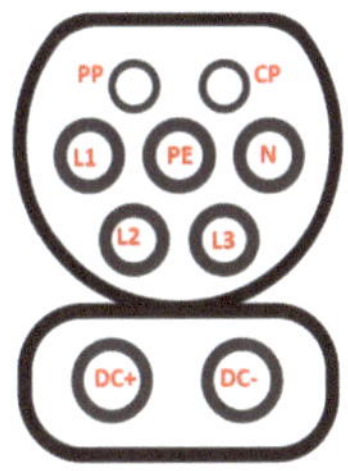
CP		Control pilot	
PE	Earth ground	EV to earth ground	
L1	AC 3-phase Charging	Phase 1	
L2		Phase 2	
L3		Phase 3	
N		Neutral	
DC+	DC Charging	DC positive terminal	
DC-		DC negative terminal	

Currently, GB/T was giving just 237.5 kW at 950 V and 250 A; thus, this will be about four times more competent in terms of power with other DC charging connectors. Power will be more than twice as high as the new 400 kW CHAdeMO and 350 kW CCS Combo specs, too. However, the V2G features is not supported by GB/T connector standard. The feature comparison among the DC charging connector standards is listed Table 7 below:

Table 7. Specification comparison of different DC charging connector standards [33].

Specification	New GB/T	GB/T	CHAdeMO	CCS Type 1	Tesla
	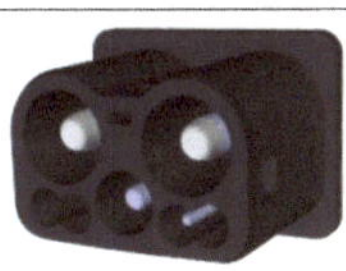				
Max Power	900 kW	237.5 kW	400 kW	400 kW	135 kW
No. of Control Pilot	2	0	3	1	1

Table 7. *Cont.*

Specification	New GB/T	GB/T	CHAdeMO	CCS Type 1	Tesla
Communication	CAN (SAE J1939)	CAN (SAE J1939)	CAN (ISO 11898)	PLC (ISO 15118)	CAN (SAE J2411)
+12 V Power Supply	Yes	Optional	Yes	No	No
V2G Compatible	Unknown	Under R&D	Yes	Under R&D	No
Coupler Lock	Inlet	Connector	Connector	Inlet	Inlet
Availability	China	China, India	Global	EU, US, A	Global
Related Standards	IEC 61851-23-1 IEC 61851-23-2	IEC 61851-23-1	IEC 61851-23-1 IEC 61851-23-2 IEEE 2030.1	IEC 61851-23-1 SAE J1772	None
Cooling Technique	Liquid-Cooled Cable Under Development	Liquid-Cooled Cable Not Available	Liquid-Cooled Cable Under Development	Liquid-Cooled Cable Under Development	Liquid-Cooled Cable Discontinued

2.4. EV Charging and V2G Communication Standards

The design of an EV charging system and the relevant communication standards are shown in Figure 1. The EV and charging supply unit are the main key systems for EV charging. The EV is connected to the charging supply unit using a charging cable including communication wires. They usually use IEC 61851 for application-level communication standards and ISO/IEC 15118 for CAN communication standards. This high-level communication is specifically carried out by the EVCC in the EV and the SECC to regulate the whole charging process [34,35]. Third-party actors provide supporting responsibilities on the other side of the EVSE, such as billing and power management.

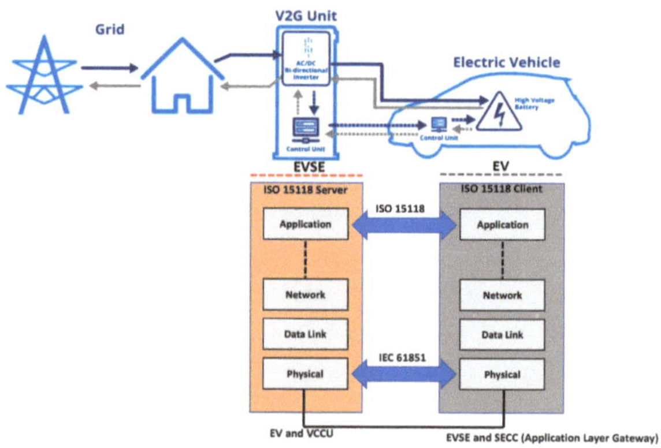

Figure 1. EV charging and V2G architecture with relevant communication standards [36].

In this work, a high-level communication based on ISO 15118-2 is implemented, where the bidirectional feature is defined for the Combo CCS Type 2 connector. The IEC 61850 standard is also used for the operator and EVSE to exchange values relating to electricity. The messages produced by ISO/IEC 15118 and IEC 61850, however, are incompatible. The IEC 61850-90-8 was suggested to bridge communication between ISO/IEC 15118 and IEC 61850 in [36].

3. Test System Development

The DC charging test system architecture is shown in Figure 2. The novelty of this test system is the incorporation of the Combo CCS Type 2 charging port standard during V2G testing. Additionally, the BOSCH car is used as a baseline vehicle during V2G with necessary hardware and software modifications. The system modification details are described in Sections 3.4 and 3.5. The test system consists of an Electric Vehicle Charging Analyzer (EVCA) and a battery emulator through which the system has been tested to ensure a safe operation in accordance with the current standards and charger interfaces. The required equipment to measure and verify communication, protection, safety, and load circuit—on the standard—conformity over the complete duration of charging and captures all deviations is used. In this way, it is possible to identify the non-conformity of charging and obtain the reasons for charge interruptions. The overall test system components are described in the next subsections.

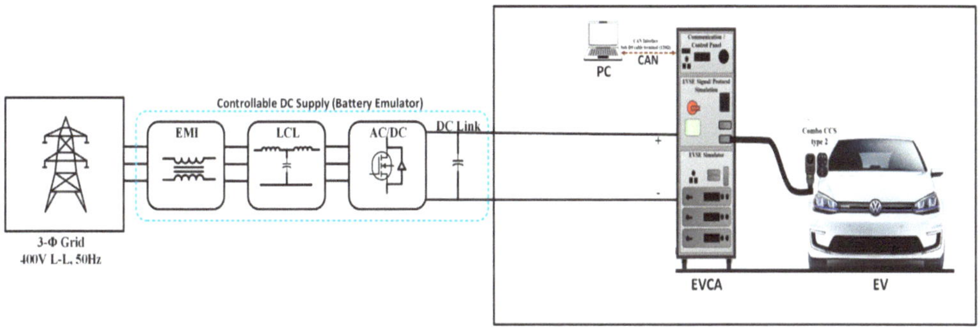

Figure 2. DC charging test system architecture with Combo CCS Type 2 connector.

3.1. *Controllable DC Supply/Load (Battery Emulator)*

DC programmable loads (80 kW): these devices can be used as battery simulators to get into the charge loop. The main features follow:

- Per module: max. 1000 V with energy recovery.
- Power up to 30 kW, and 150 kW by connecting two modules in parallel.
- CAN interface.
- Power monitoring module with redundant power contactors.

3.2. *Electric Vehicle Charging Analyzer (EVCA)*

An EVCA instrument measures and monitors both the load and the communication circuit for the entire charging procedure. By detecting and logging deviations from the standards, the conformity of all relevant parameters can be assessed when evaluating the measurement files. Results can be transferred by CAN to external data processing equipment. It can be used for testing all types of chargers (AC and DC chargers). The EVCA can operate in three modes: (1) EV emulator, (2) EVSE emulator (AC and DC), and (3) Man-in-the-middle. This unit is designed for testing the conductive charging of electric vehicles. The main features are:

- Data recording and controlling on CAN interface as "Remote Control".
- Isolated banana sockets for control pilot of EVSE for comparison with an oscilloscope.
- Integrated electronics for measuring Voltage up to 800 V and Current up to 200 A.
- Integrated high-power resistors for isolation test with ca. 100 mA leakage current at 500 V DC.

3.3. *Electric Vehicle DUT (BOSCH)*

ROBERT BOSCH GmbH supplied the passenger EV shown in Figure 3 to perform the charging test. The company technical team replaced the CHAdeMO port with a Combo CCS Type 2 connector to perform AC and DC charging tests compatible with the above-mentioned test infrastructure. Additionally, a vehicle control unit was installed to enable DC V2G with a Combo CCS Type 2 connector and the ISO 15118-2 communication protocol.

Figure 3. Passenger car to test the V2G using Combo CCS Type 2.

3.4. Additional EV Hardware Modification for V2G

The EV has been equipped with an additional vehicle control unit (VCU) as shown in Figure 4 to enhance the V2G setup. The VCU is usually used for vehicle state management, communication with the backend, coordination, etc. In this test setup, the additional VCU has also direct interfaces to connect debugging tools for developing/modifying the software and monitoring all vehicle internal bus messages as well as the ISO 15118-2 communication. The automotive-compliant changeover relay for switching the 12 V supply voltage between the series VCU and the additional VCU is shown in Figure 5. A dedicated relay terminal is also switched over with the same changeover relay. This means that only one VCU can be active at the same time.

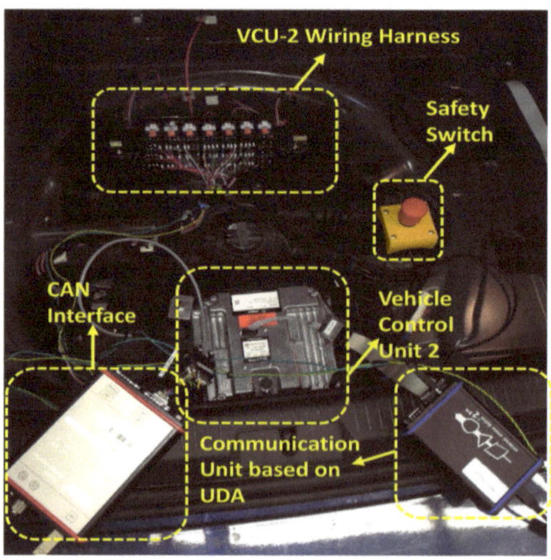

Figure 4. Additional VCU installation with necessary wiring and communication interfaces.

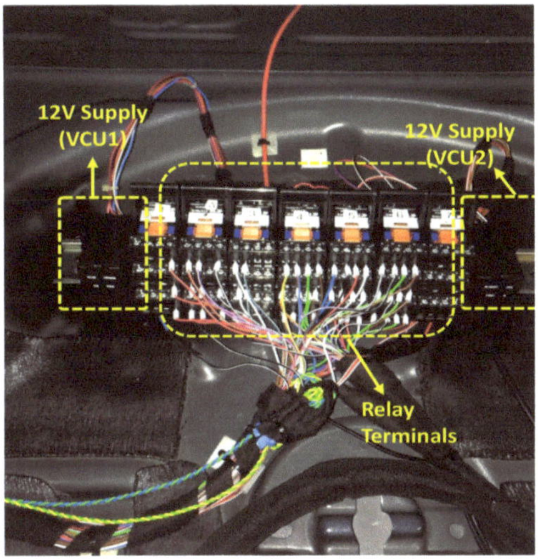

Figure 5. Required relay terminals to activate control of selected VCU.

3.5. *Software Modification for V2G in EV*

The bidirectional charging functionality has been implemented based on the ISO 15118-2 specification which is shown in Figure 6. The Power Line Communication (PLC) is employed at the physical and data-link layers. User Datagram Protocol/TCP (Transmission Control Protocol) is utilized for the transport layer on top of PLC, while IPv6 is used for the network layer. Although the testing system does not explicitly test these protocols since they are properly implemented by commercial operating systems such as Windows or Linux, it will indirectly discover the poor implementation of these levels.

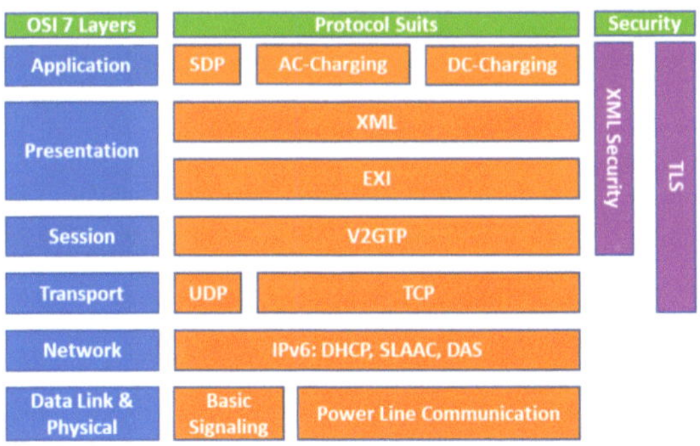

Figure 6. The OSI 7-layer structure of ISO 15118.20 standard.

The ISO/IEC 15118 standard defines the communication protocol during the V2G test, which is used for session management. All requests are sent via energy transfer protocol as a message group. The protocol just confirms the ISO/IEC 15118 version compatibility of the message request data packet. After the EV charging or discharging setup initialization is the request packet initiation; the EVCC broadcasts a request packet using this protocol, and an SECC responds by sending back a response. The response detects the charging controller's IP address and communication port identification. The later packet type is the application (charging/discharging) message, although the actual message in the transmission is multiplexed by the EXI method to reduce packet density. Every application message is internally represented in XML for higher versatility and adaptability. For each type of application message, the user sends a request message, and the ISO 15118 server responds with a response message. This is known as the client-server model of communication ISO standard [36]. The EV and charging supply unit exchange the essential messages prior to the whole charging process through a sequence of request and response messages, as seen in Figure 7. For instance, the EVSE receives the SSR messages from the EV for the set initialization message type and responds with the acknowledgment message. However, the sequence of the application message types in the table approximately corresponds to the actual message sequence. The message exchange sequence is shown in Figure 8.

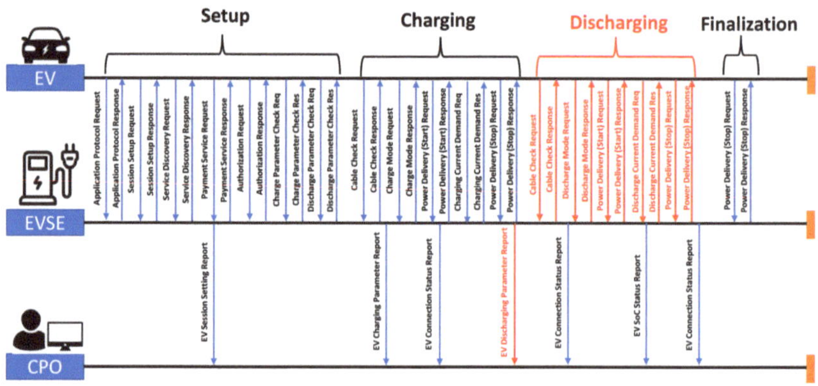

Figure 7. The communication request and response stream during charging/discharging test.

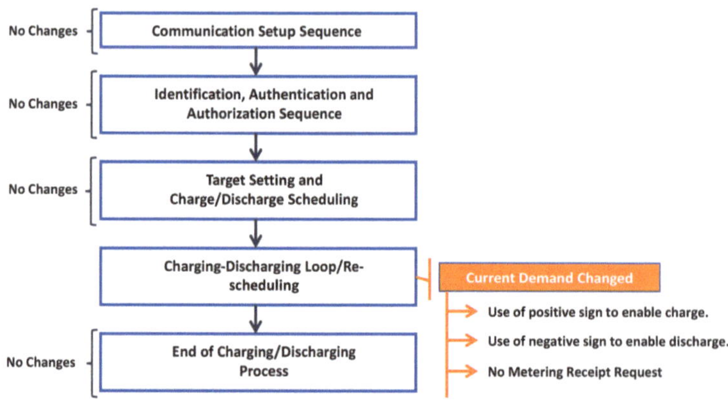

Figure 8. The software modification for V2G test.

Thus, during the "charging loop/re-scheduling", the ISO 15118-2 is extended to be able to handle negative current requests to enable the DC discharge functionality.

4. Testing Results and Discussions

The V2G test setup with EV and charging supply unit is shown in Figure 9. Tests are especially conducted to focus on V2G power transfer through the Combo CCS type 2 connector standard. The overall system is comprised of a battery emulator, EVCA, EVSE, and BOSCH passenger car with a Combo CCS2 connector and checking the ISO 15118-2 communication. The communications among the EV, the charging supply unit, and the operator were manually examined, and the timings of message deliveries were checked to ensure that the target EV charger is properly tested.

Figure 9. Experimental setup for V2G test with BOSCH passenger car.

The battery voltage and current are shown in Figure 10. The operator specifies the discharge current setpoints. Initially, the discharging current is set to −10 A, and then, it is increased up to −30 A. The battery voltage starts from around 323 V to decrease during discharging. The current waveforms from the AC side, which feeds the grid, are depicted in Figure 11. It can be observed that the AC voltage is around 220 V and is in phase opposition with the grid current. The THD level of the discharging current in the AC side is lower than 5%. The discharging power to feed the grid is around 10 kW. However, the discharging current is a bit distorted and has some noise.

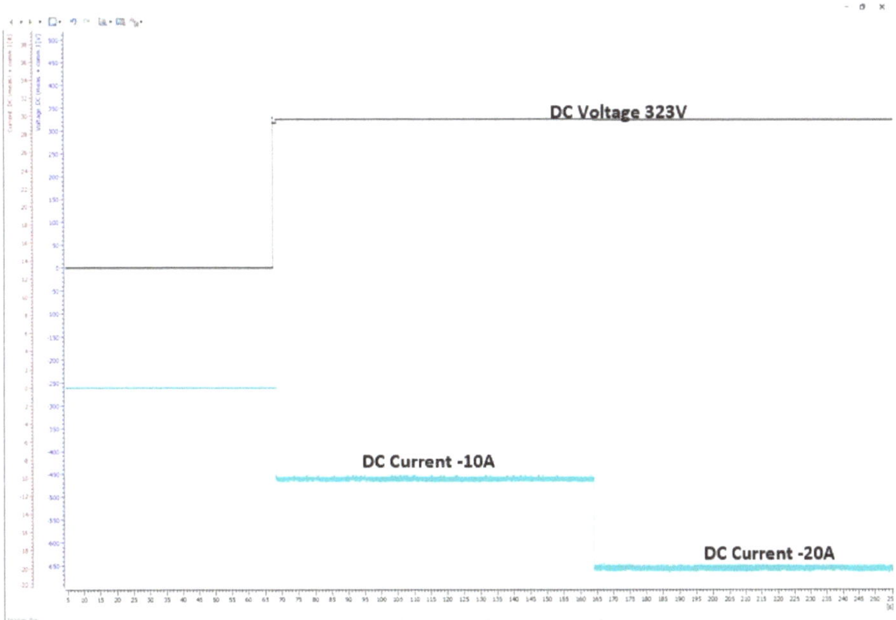

Figure 10. Measurement of DC voltage and current via CAN communication.

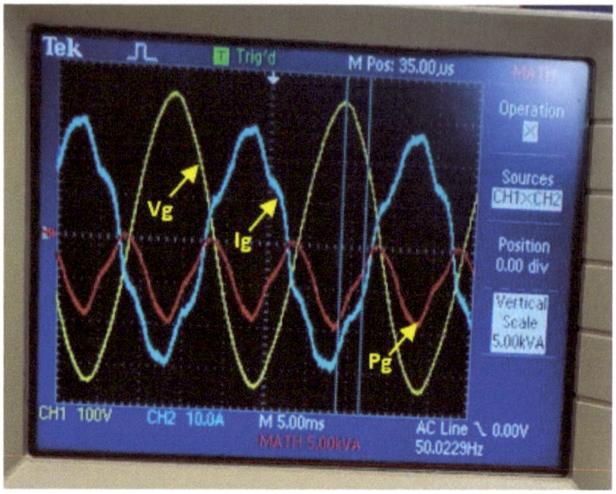

Figure 11. Measurement of AC voltage, current and power on the grid side. Here, Ig is grid current (Cyan), Vg is grid voltage (light green) and Pg is the instantaneous power (red).

The overall V2G test is performed for approximately 593 sec. The operator controls the discharge current setpoint over the entire period. The DC current is following the current setpoint with a delay of 2 sec, which is shown in Figure 12. The DC voltage from the battery side is decreased over the discharging period. The discharging power is the product of DC voltage and current, as shown in Figure 13. During the test, the limit of the discharged power is between 4.8 and 9.6 kW. Moreover, the discharged power on the grid side is estimated by the grid side current. The grid voltage is 230 Vrms and the power factor is approximately 0.975 leading. The grid current profile at two different vehicle discharge power levels is shown in Figures 14 and 15. Additionally, this current profile is recorded at 45% battery SoC level. So, the discharging efficiency is estimated at three different battery SoC levels, which are shown in Figure 16. The efficiency slightly varies with the initial SoC level. According to Figure 16, it is possible to obtain a higher discharging efficiency at a high initial battery SoC level.

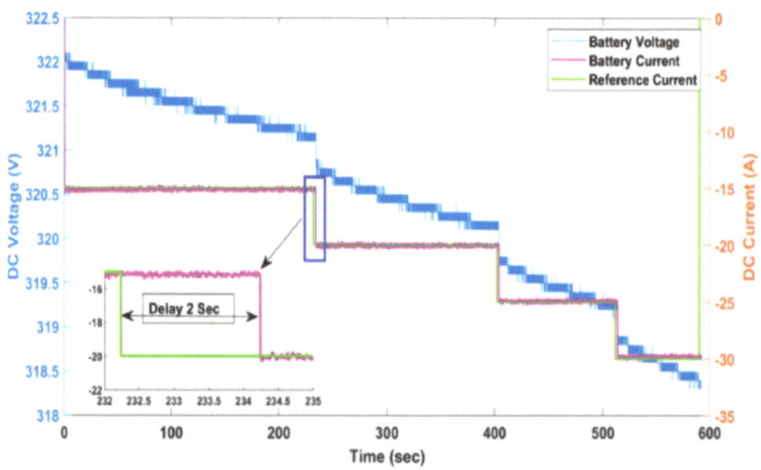

Figure 12. Measurement of DC voltage and current with reference value via CAN communication over the full testing period.

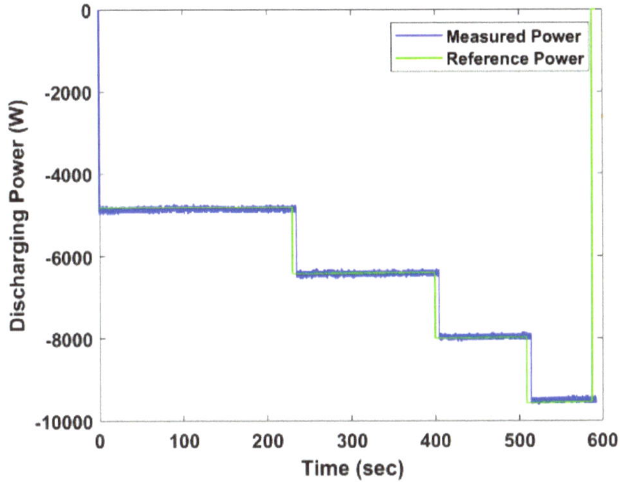

Figure 13. Measurement of DC power discharging from the battery.

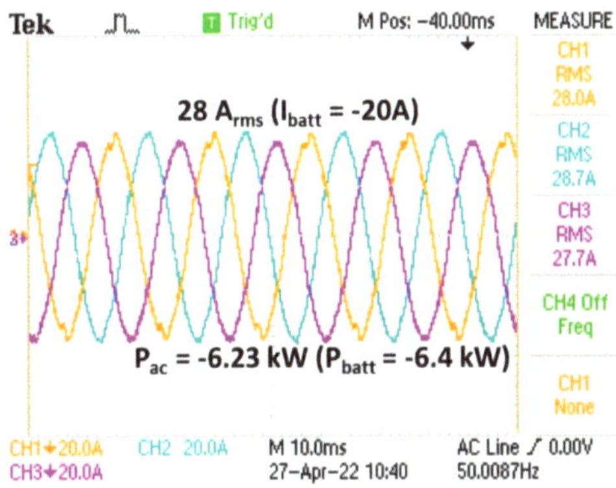

Figure 14. Grid current profile (20A/div) at 6.23 kW discharging power.

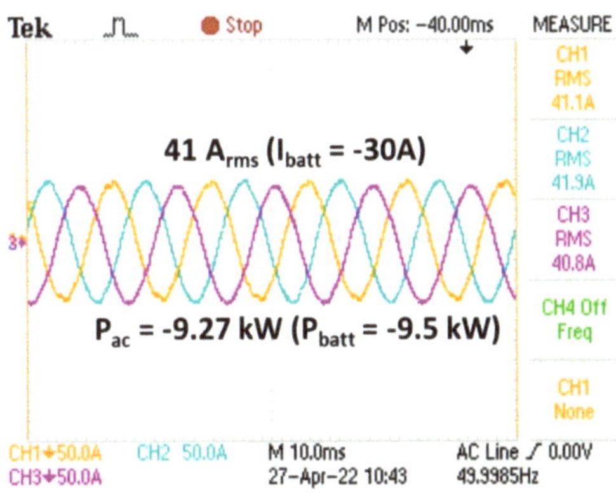

Figure 15. Grid current profile (50A/div) at 9.23 kW discharging power.

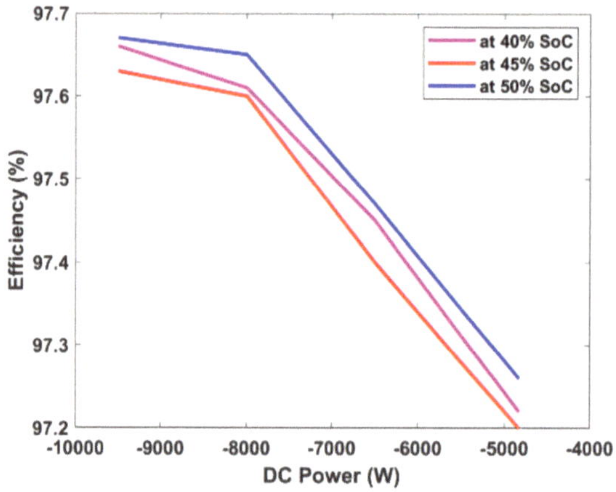

Figure 16. Estimated V2G efficiency at different initial battery SoC.

The signal delay between the reference and the actual signal indicates the controller response delay. It just provides the hardware response as it occurs, without any further delays brought on by miscellaneous setup communications between the reference and the measurement. The signal delay is determined using the cross-correlation method. Figure 17 depicted the correlation the reference and actual signal which describes how two signals correlate when one of them is subjected to various temporal shifts during the test. The peak is detected for a signal delay of around 2 sec, which is then taken into account as the activation time of the tested V2G apparatus. The charger communication testing architecture is implemented with the full stack of ISO/IEC 15118, including IEC 61851 standard signaling. In this way, the DC charger can exchange energy between an EV connected by a Combo CCSType 2 connector and the grid via standard communication. However, while the reference controller exchanges from the 15118 server and 61851 mobile applications, the main message for energy exchange is transmitted from the ISO 15118 server through different communication layers such as the physical layer, data link layer, etc. The communication architecture of the charger includes the 15118 servers, the 61851 module and the reference controller, as shown in Figure 18.

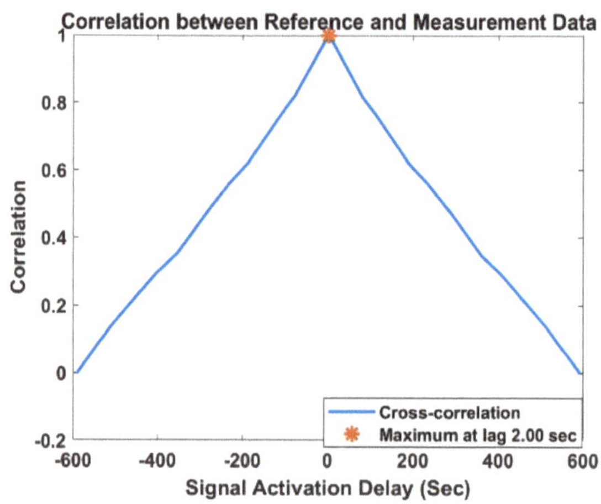

Figure 17. The correlation between reference and measurement current value.

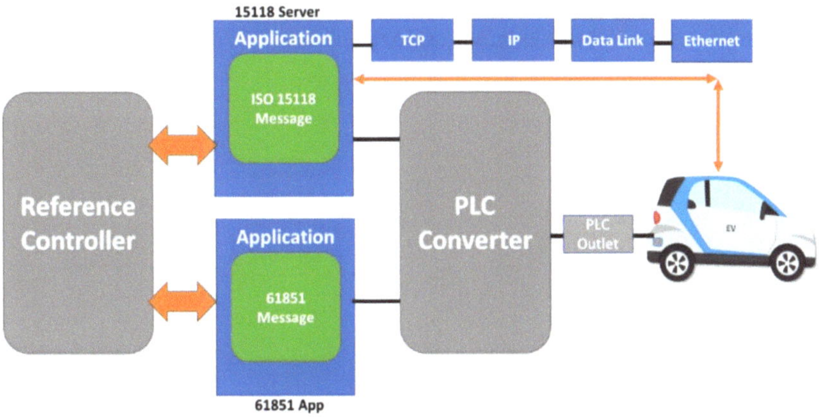

Figure 18. Communication architecture implemented during V2G testing.

The communication testing results for the specific DC charging cases and the reference design with different data transfer protocols are shown in Table 8. It is noted that the EV charger passed every request and response test, including the unfavorable ones. The test system was capable of concurrently acting in the capacities of the operator and the EV.

Table 8. The communication testing result with full stack ISO 15118 during V2G test.

Test No	Contents	Target	Protocol	Result
1	Slack Parameter Exchange	EVCC	SLAC	Pass
2	Signal Strength Measurement	EVCC	SLAC	Pass
3	Logical Network Parameter Exchange	EVCC	SLAC	Pass
4	Control Pilot Voltage Range	EVCC	IEC 61851	Pass
5	Control Pilot Frequency Range	EVCC	IEC 61851	Pass
6	Control Pilot Duty-Cycle Range	EVCC	IEC 61851	Pass
7	SECC Discovery Protocol (SDP)	EVCC	ISO 15118	Pass
8	Supported Application Protocol	EVCC	ISO 15118	Pass
9	Session Setup Message	EVCC	ISO 15118	Pass
10	Service Discovery	EVCC	ISO 15118	Pass
11	Payment Service Selection	EVCC	ISO 15118	Pass
12	Payments Details	EVCC	ISO 15118	Pass
13	Authorization	EVCC	ISO 15118	Pass
14	Charge Parameter Discovery	EVCC	ISO 15118	Pass
15	Cable Check	EVCC	ISO 15118	Pass
16	Pre-Charge	EVCC	ISO 15118	Pass
17	Power Delivery (Start)	EVCC	ISO 15118	Pass
18	Current Demand	EVCC	ISO 15118	Pass
9	Power Delivery (Stop)	EVCC	ISO 15118	Pass
20	Welding Detection	EVCC	ISO 15118	Pass
21	Session Stop	EVCC	ISO 15118	Pass
22	Power Delivery (Negative)	EVCC	ISO 15118	Pass

The reference controller interacts with the 15,118 servers and 61,851 modules to regulate the overall operation of the EVSE. The PLC converter, which changes the data to the control pilot and vice versa, receives communication from the 15,118 server and 61,851 modules to connect with the EV. The combo-type coupler is supported by the reference charger for connecting to the EV. Figure 19 displays the outcomes of messaging time to initiate and begin DC discharging with the improved EV.

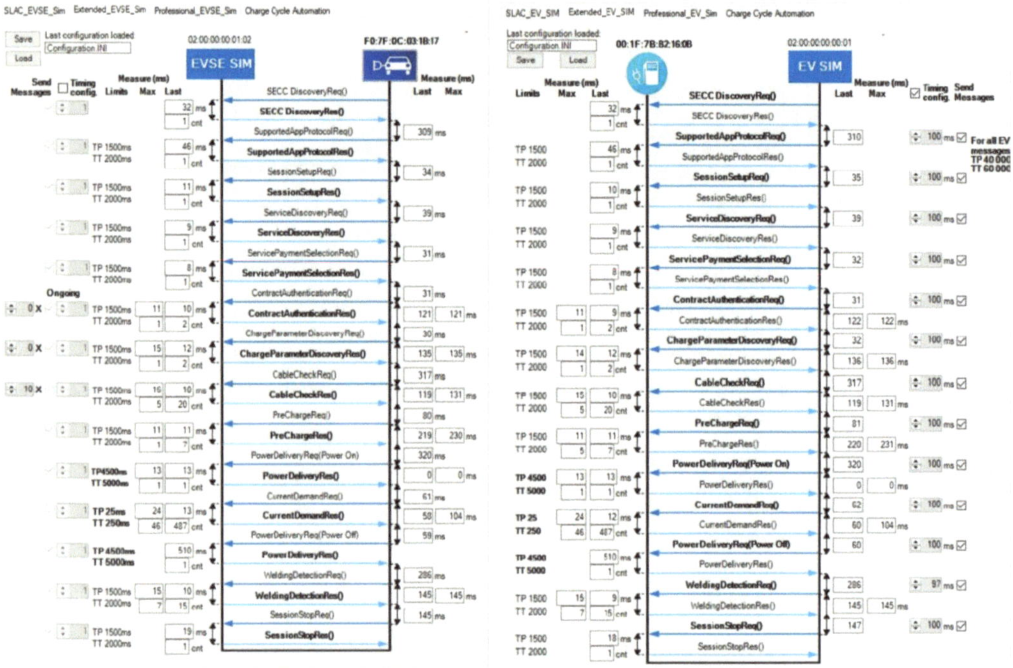

Figure 19. Panels of communication between real EV and EVSE.

5. Data-Driven V2G Charger Model Development

Investigation of the impact assessment of large-scale DC charging stations with V2G facility on the existing power system or microgrid is a popular research topic of recent years. This research reveals the positive and negative outcomes of the V2G facility in peak shaving, frequency regulation, the total cost of ownership reduction, and social barriers. Thus, a DC charging station simulation with a practical V2G charger model is very important. In this regard, a V2G charger modeling approach is described in [37]. A similar approach is followed to develop the real-time V2G charger model based on the tested data. This model is a low-fidelity model with less required simulation time requirement. Therefore, this model can be used for detailed simulation studies when investigating the impacts of DC-EV chargers with V2G features on the grid. The EV charger model is designed using the Simulink platform within the MathWorks 2019b software environment. The schematic representation is shown in Figure 20.

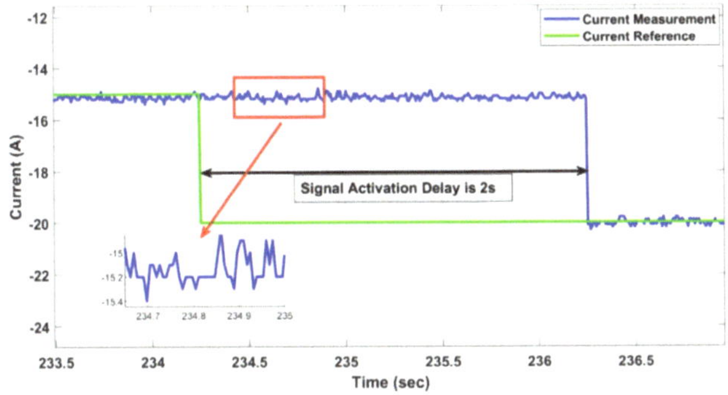

Figure 20. Actual hardware response delay compared to the reference signal.

The input of the model is the reference signal which is sent to the real-time controller from the user side. Using this reference input, the model processes this reference signal data based on the signal amplitude accuracy, signal relay, and noise precision of the real-time test output signal. Consequently, the output signal is estimated by appropriate actions that represent the charger's real operation.

The setpoint current signal and the corresponding current measurement are shown in Figure 20 where the activation delay information can be found. It can be noted that this signal delay is including the time delay of the battery emulator and EVCA. Therefore, this information has a significant role in the data driven V2G charger modeling.

The signal amplitude accuracy is calculated as the difference between the highest and the lowest values of the supplied current over the simulation time with stable setpoints as shown in Figure 21. The maximum and minimum magnitude are important data for exact noise modeling. The noise baseline can give an idea about the offset variation. It is found that the maximum signal baseline shift is about 0.2 A for the extreme operation, which is depicted in Figure 22. This value validates the choice of 0.2 A as the linearization constant that has been used to set the model linearity, which is shown Figure 22. This real-time response behavior is used to model the V2G charger which responds as an actual hardware setup.

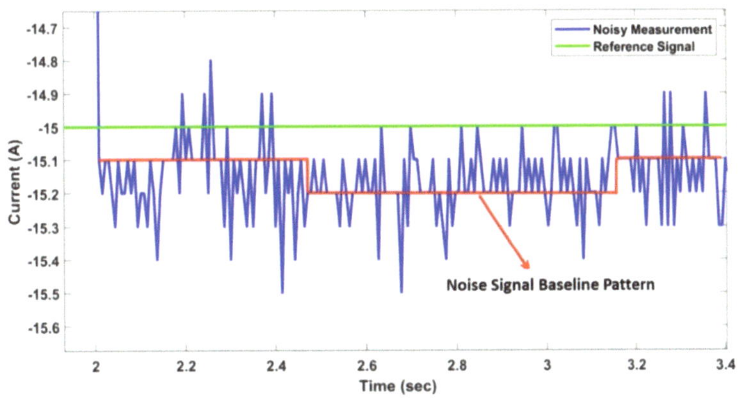

Figure 21. Measurement of noisy signal with the variable baseline.

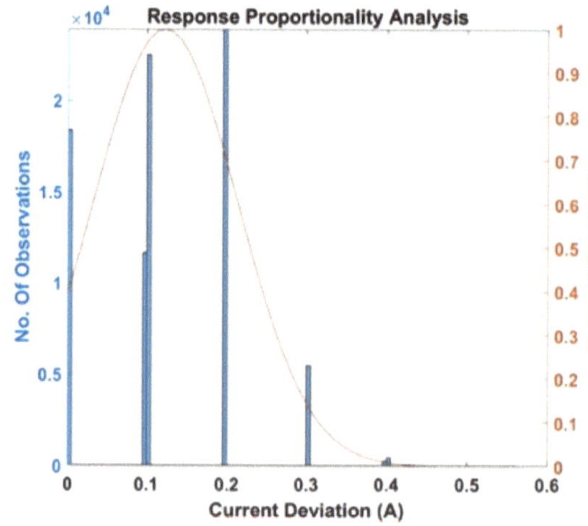

Figure 22. Number of observations of measurement signal deviation from the reference.

To develop the data-driven V2G charger model, the following relations are considered in MATLAB/Simulink:

- The response time is modeled as a transport delay, which is equal to 2 sec for the local control, which is shown in Figure 19.
- The setpoint proportionality is obtained by implementing Equation (1)

$$Y_{sp} = K_{sp} * X_{sp} * round\left(\frac{X_{sp}}{K_{sp}}\right) \quad (1)$$

where X_{sp} and Y_{sp} are the non-truncated and truncated current signals, respectively. Moreover, the Ksp is a proportionality gain equal to around 0.5% of maximum current setpoint.

- The relevant mean accuracy value, i.e., 0.2 A for a negative setpoint, is added to the current data points to perform the amplitude approximation. The implementation is obtained according to Equation (2):

$$Y_{ma} = X_{ma} + 0.67\% \; of \; rated \; Current \quad (2)$$

- By including a uniformly distributed noise into the computed setpoint, the noise precision is accomplished. For a set-point $\neq 0$, the noise typically has an amplitude of 0.3 A and 0.4 A. Equation (3) is used to determine the implementation.

$$Y_{np} = X_{np} + \alpha * \left(\frac{3}{10}\right) \quad (3)$$

where during the duration of the simulation, it is a uniformly distributed random integer between −1 and +1.

The developed V2G charger model has been validated via simulations in MATLAB/Simulink using the requested current signals from real hardware controller, i.e., the green line in Figure 23. Results show that the simulated results (red line) match rather adequately the actual current measured (blue line) via following the input signal, which was the same green test pattern. Furthermore, Figure 24 shows the error signal between the actual and simulated current data. It can be noticed that the error density is higher within the a very small band, i.e., 0–0.5A (1.66% of the charger rated power). Ultimately, Figure 25 reports the error distribution function for the entire simulation time relying on MATLAB/Simulink. It can be noticed that the mismatches that occur between actual and simulated current data are within the band 0.2 A. In particular, the error distribution figure shows that more than 80% of the error lies within the 0–0.2 A band. For the information, it is possible to achieve a good approximation of the normalized errors' distribution curve by using a normal distribution function with mean value $m = 0.0385$ A and standard deviation $s = 0.235$ A. The V2G facility test result of CHAdeMO and the Combo CCS Type 2 for EV charger application are compared in the Table 9 below.

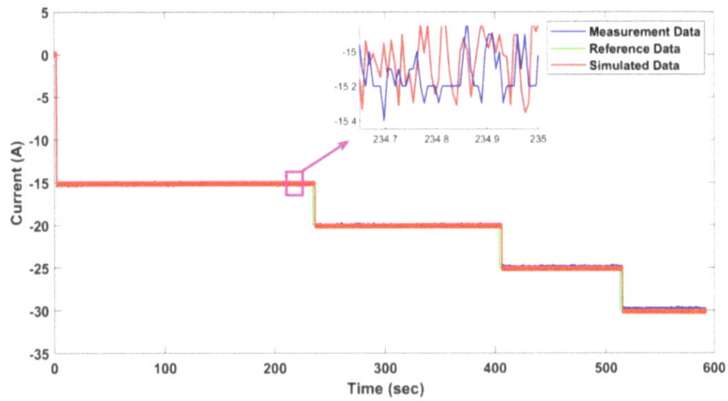

Figure 23. Simulation model results compared to the reference and actual hardware measurement data.

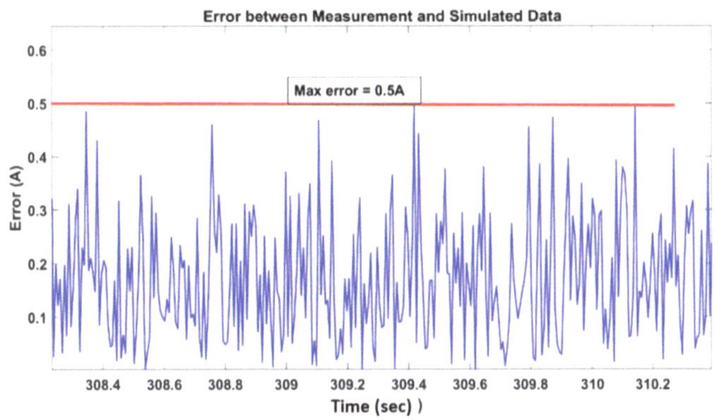

Figure 24. Error between simulated and hardware measurement.

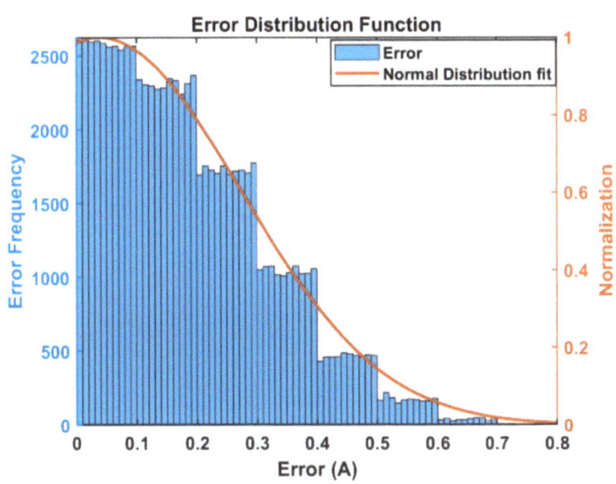

Figure 25. Measurement and simulated data error distribution function over the full testing period.

Table 9. The V2G testing result between CHAdeMO and Combo CCS Type 2 charger connector standard.

Performance Indices	V2G Test with CHAdeMO [37]	V2G Test with Combo CCS 2
Signal Activation Delay	4 s	2 s
Setpoint Linearity	<4% (Power Setpoint 400 W)	<1.5% (Current Setpoint 0.2 A)
Measurement Accuracy	4.4% (Power measurement)	2.56% (Current Measurement)
V2G Efficiency (@50% SoC and 8 kW Power Delivery)	Around 93%	Around 97%

6. Conclusions

In this paper, the technical feasibility of a commercial EV charger with V2G functionality with a Combo CCS Type 2 connector has been identified. The significance of initial SoC on the efficiency has been emphasized. It is found that the V2G efficiency with the Combo CCS Type 2 standard is around 97.7% at a 50% SoC level. The THD of the current level at the AC side is lower than 5%, which is IEC-1000-3-6 standard compliant. However, some noise is captured, which is the reason the current ripples around 0.3–0.5 A amplitude. The 0.1 A noise baseline offset is found during V2G power transfer. Furthermore, a data-driven model of the V2G charger has been developed by analyzing the testing data in a MATLAB/Simulink software environment. A thorough simulation research demonstrated the model's correctness when implementing the flexibility product's validated qualities. The suggested model is an illustration of how the results of hardware characterization tests may be used directly to construct realistic models that can be used for numerous further power system simulation activities.

Author Contributions: Conceptualization, S.J. and M.A.-M.; methodology, S.J.; software, S.J., O.H.G. and B.O.; validation, S.J., O.H.G., B.V. and B.O.; formal analysis, S.J.; investigation, S.J., O.H.G. and B.V.; resources, S.J. and M.A.-M.; data curation, S.J.; writing—original draft preparation, S.J.; writing—review and editing, B.V., T.G., M.A.-M. and O.H.; visualization, S.J.; supervision, O.H.; project administration, O.H.; funding acquisition, O.H. All authors have read and agreed to the published version of the manuscript.

Funding: This research was funded by European Commission under grant number 875131 (eCharge4Driver project).

Institutional Review Board Statement: Not applicable.

Informed Consent Statement: Not applicable.

Acknowledgments: This work was supported by the Horizon 2020 Programme of the European Commission under grant agreement no. 875131 (eCharge4Driver project). We, as authors, also acknowledge Flanders Make for the support to our research group.

Conflicts of Interest: The authors declare no conflict of interest.

Abbreviations

ACES	Adaptive Control of Energy Storage
ADR	Automatic Demand Response
CAN	Controller Area Network
CCS	Combined Charging Service
CHAdeMO	CHARGE de MOVE
CPO	Charge Point Operator
DAS	Direct-Attached Storage
DHCP	Dynamic Host Configuration Protocol
DUT	Device Under Test
EVCA	Electric Vehicle Charging Analyzer

EVCC	Electric Vehicle Communication Controller
EVSE	Electric Vehicle Supply Equipment
EXI	Efficient XML Interchange
FCR	Frequency Containment Reserves
G2V	Grid-to-Vehicle
GB/T	Guojia Biaozhun/Tuijian (China)
IEC	International Electromechanical Commission
IEEE	Institute of Electrical and Electronic Engineers
ISO	International Organization of Standardization
JARI	Japan Automotive Research Institute
JEVS	Japan Electric Vehicle Standard
JEVS	Japan Electric Vehicle Standard
OCPP	Open Charge Point Protocol
OEM	Original Equipment Manufacturer
PLC	Power Line Communication
SAE	Society of Automotive Engineers
SDP	SECC Discovery Protocol
SECC	Supply Equipment Communication Controller
SLAAC	Stateless Address Auto-Configuration
TCP	Transmission Control Protocol
THD	Total Harmonic Distortion
TLS	Transport Layer Security
UDP	User Datagram Protocol
V2G	Vehicle-to-Grid
VCU	Vehicle Control Unit
XML	Extensible Markup Language

References

1. EPA-United States Environmental Protection Agency. Sources of Greenhouse Gas Emissions. Available online: https://www.epa.gov/ghgemissions/sources-greenhouse-gas-emissions (accessed on 9 July 2022).
2. Xiamen Joint Tech Co. Ltd., Ag. *Advantages & Disadvantages of DC Charging Compared to AC Charging*; Xiamen Joint Tech Co. Ltd., Ag: Xiamen City, China, 2022; Available online: https://www.linkedin.com/pulse/advantages-disadvantages-dc-charging-compared-ac-jointevse (accessed on 13 September 2022).
3. Lauinger, D.; Vuille, F.; Kuhn, D. A review of the state of research on vehicle-to-grid (V2G): Progress and barriers to deployment. In Proceedings of the European Battery, Hybrid and Fuel Cell Electric Vehicle Congress, Geneva, Switzerland, 14–16 March 2017; pp. 1–8.
4. Kempton, W.; Letendre, S.E. Electric vehicles as a new power source for electric utilities. *Transp. Res. Part D Transp. Environ.* **1997**, *2*, 157–175. [CrossRef]
5. Bibak, B.; Tekiner-Moğulkoç, H. A comprehensive analysis of Vehicle to Grid (V2G) systems and scholarly literature on the application of such systems. *Renew. Energy Focus* **2021**, *36*, 1–20, ISSN 1755-0084. [CrossRef]
6. Schmutzler, J.; Andersen, C.A.; Wietfeld, C. Evaluation of OCPP and IEC 61850 for Smart Charging Electric Vehicles. *World Electr. Veh. J.* **2013**, *6*, 863–874. [CrossRef]
7. Bahrami, A. EV Charging Definitions, Modes, Levels, Communication Protocols and Applied Standards, Technical Report. Available online: https://www.researchgate.net/publication/338586995_EV_Charging_Definitions_Modes_Levels_Communication_Protocols_and_Applied_Standards/citations (accessed on 26 August 2022).
8. Neaimeh, M.; Andersen, P.B. Mind the gap- open communication protocols for vehicle grid integration. *Energy Inform.* **2020**, *3*, 1–17. [CrossRef]
9. Noel, L.; de Gerardo, Z.R.; Kester, J.; Sovacool, B.K. *Vehicle-to-Grid: A Sociotechnical Transition beyond Electric Mobility*; Springer: Cham, Switzerland, 2019.
10. Virta Global. The Two Sides of EV Charging Network Operators. 2019. Available online: https://www.virta.global/blog/emsp-cpo-ev-charging-roles-responsibilities (accessed on 14 July 2022).
11. Electric Vehicles, Volkswagen. Electric Vehicles | Volkswagen Newsroom. 2020. Available online: volkswagen-newsroom.com (accessed on 16 July 2022).
12. Energy Informatics. Table 2 Protocols for Vehicle Grid Integration. 2020. Available online: https://energyinformaticsspringeropen.com/articles/10.1186/s42162-020-0103-1/tables/2 (accessed on 16 July 2022).
13. V2G Global Roadtrip: Around the World in 50 Projects, Innovate UK. 2018. Available online: https://innovation.ukpowernetworks.co.uk/wp-content/uploads/2018/12/V2G-Global-Roadtrip-Around-the-World-in-50-Projects.pdf (accessed on 23 July 2022).

14. Vehicle-to-Grid (V2G): Everything You Have to Know. Available online: https://www.virta.global/vehicle-to-grid-v2g (accessed on 23 July 2022).
15. Schmidt, B. Volkswagen Wants to Stabilise the Grid by Adding V2G in All Its Electric Cars. *The Driven*, 7 April 2021. Available online: https://thedriven.io/2021/04/07/volkswagen-wants-to-stabilise-grid-by-adding-v2g-in-all-its-electric-cars (accessed on 23 July 2022).
16. Hanley, S. Volkswagen Plans to Offer V2G and Plug&Charge Technology in 2022. *Clean Technica*, 23 December 2021. Available online: https://cleantechnica.com/2021/12/23/volkswagen-plans-to-offer-v2g-and-plug-charge-technology-in-2022 (accessed on 23 July 2022).
17. Vehicle-to-Grid: Energy Storage on Wheels. *EIT Innoenergy*, 12 October 2021. Available online: https://www.eba250.com/vehicle-to-grid-energy-storage-on-wheels (accessed on 23 July 2022).
18. ACES Project—Across Continents Electric Vehicle Services. 2017. Available online: https://www.aces-bornholm.eu (accessed on 16 July 2022).
19. The Parker Project. 2016. Available online: https://parker-project.com (accessed on 16 July 2022).
20. INVENT Project—Intelligent Electric Vehicle Integration. 2017. Available online: https://backend.orbit.dtu.dk/ws/portalfiles/portal/77554391/thesis_Peter_Bach_Andersen.PDF (accessed on 16 July 2022).
21. Grid Motion Project—Reducing Electric Vehicle Usage Cost Thanks to Smart Charging Process. 2017. Available online: https://parker-project.com/wp-content/uploads/2018/12/VGI-Summit-Day-2-S2-Projects-Gridmotion.pdf (accessed on 16 July 2022).
22. Industry Learning Reports. Available online: https://www.v2g-hub.com/reports (accessed on 24 July 2022).
23. Wikipedia Contributors. CHAdeMO. Wikipedia, The Free Encyclopedia, 20 August 2022. Available online: https://en.wikipedia.org/w/index.php?title=CHAdeMO&oldid=1105514885 (accessed on 24 July 2022).
24. Leemput, N.; Geth, F.; Juan Van, R.; Jeroen, B.; Johan, D.; Marinelli, M. Reactive power support in residential LV distribution grids through electric vehicle charging. *Sustain. Energy Grids Netw.* **2015**, *3*, 24–35. [CrossRef]
25. Zecchino, A.; Marinelli, M. Analytical assessment of voltage support via reactive power from new electric vehicles supply equipment in radial distribution grids with voltage-dependent loads. *Int. J. Electr. Power Energy Syst.* **2018**, *97*, 17–27. [CrossRef]
26. Arias, M.N.B.; Hashemi, S.; Andersen, P.B.; Træholt, C.; Romero, R. V2G Enabled EVs Providing Frequency Containment Reserves: Field Results. In Proceedings of the 2018 IEEE International Conference on Industrial Technology (ICIT 2018), Lyon, France, 20–22 February 2018; pp. 1–6.
27. Yoshida, M.D.; Blech, T. CHAdeMO DC Charging: Successful Transition Towards Nation- and Region-Wide Zero-Emisson. CHAdeMO Association, April 2021. Available online: https://movelatam.org/portfolio-item/chademo-charging-protocol-for-electric-vehicles/ (accessed on 24 July 2022).
28. Myles, B.; Robert, E. Making the right connections General procurement guidance for electric vehicle charge points. 2015. Available online: http://ukevse.org.uk/resources/procurement-guidance/ (accessed on 13 July 2022).
29. Luyten, S.; Leemput, N.; Geth, F.; Van Roy, J.; Büscher, J.; Driesen, J. Standardization of Conductive AC Charging Infrastructure for Electric Vehicles. In Proceedings of the 22nd International Conference and Exhibition on Electricity Distribution (CIRED 2013), Stockholm, Sweden, 10–13 June 2013; Volume 2013. [CrossRef]
30. Schwarzer, V.; Ghorbani, R. *Current State-of-the-Art of EV Chargers*; EVTC Electric Vehicle Transportation Centre: Cocoa, FL, USA, 2015; pp. 1–13.
31. Quílez, M.G.; Abdel-Monem, M.; El Baghdadi, M.; Yang, Y.; Van Mierlo, J.; Hegazy, O. Modelling, analysis and performance evaluation of power conversion unit in G2V/V2G application-a review. *Energies* **2018**, *11*, 1082. [CrossRef]
32. Cuma, M.U.; Tümay, M.; Yirik, E.; Ünal, E.; Dericioğlu, Ç.; Onur, B. A review of charging technologies for commercial electric vehicles. *Int. J. Adv. Automot. Technol.* **2018**, *2*, 61–70.
33. China is Developing New GB/T Fast Charging Standard at 900kW. *Insideevs Press Release*, 19 Jun 2018. Available online: https://insideevs.com/news/338620/china-is-developing-new-gb-t-fast-charging-standard-at-900-kw (accessed on 13 September 2022).
34. SAE International. *SAE Electric Vehicle and Plug in Hybrid Electric Vehicle Conductive Charge Coupler*; SAE International: Warrendale, PA, USA; Troy, MI, USA, 2013.
35. International Electrotechnical Commission. *Communication Networks and Systems for Power Utility Automation-Part 90-8: IEC 61850 Object Models for Electric Mobility*; International Electrotechnical Commission: Geneva, Switzerland, 2014.
36. Shin, M.; Kim, H.; Kim, H.; Jang, H. Building an Interoperability Test System for Electric Vehicle Chargers Based on ISO/IEC 15118 and IEC 61850 Standards. *Appl. Sci.* **2016**, *6*, 165. [CrossRef]
37. Zecchino, A.; Thingvad, A.; Andersen, P.B.; Marinelli, M. Test and Modelling of Commercial V2G CHAdeMO Chargers to Assess the Suitability for Grid Services. *World Electr. Veh. J.* **2019**, *10*, 21. [CrossRef]

energies

Article

Influence of Battery Aging on the Operation of a Charging Infrastructure

Natascia Andrenacci [1], Mauro Di Monaco [2,*] and Giuseppe Tomasso [2]

1 ENEA C.R. Casaccia, Via Anguillarese, 301, 00193 Roma, Italy; natascia.andrenacci@enea.it
2 Department of Electrical and Information Engineering, University of Cassino and Southern Lazio, Via G. Di Biasio 43, 03043 Cassino, Italy; giuseppe.tomasso@unicas.it
* Correspondence: m.dimonaco@unicas.it; Tel.: +39-07762994372

Abstract: The increasingly widespread use of electric vehicles requires proper planning of the charging infrastructure. In addition to the correct identification of the optimal positions, this concerns the accurate sizing of the charging station with respect to energy needs and the management of power flows. In particular, if we consider the presence of a renewable energy source and a storage system, we can identify strategies to maximize the use of renewable energy, minimizing the purchase costs from the grid. This study uses real charging data for some public stations, which include "normal" chargers (3 kW and 7 kW) and "quick" ones (43 kW and 55 kW), for the optimal sizing of a photovoltaic system with stationary storage. Battery degradation due to use is included in the evaluation of the overall running costs of the station. In this study, two different cost models for battery degradation and their influence on energy flow management are compared, along with their impact on battery life.

Keywords: charging infrastructures; PV and battery integration; EV charging; battery aging

Citation: Andrenacci, N.; Di Monaco, M.; Tomasso, G. Influence of Battery Aging on the Operation of a Charging Infrastructure. *Energies* **2022**, *15*, 9588. https://doi.org/10.3390/en15249588

Academic Editor: Javier Contreras

Received: 5 November 2022
Accepted: 13 December 2022
Published: 17 December 2022

1. Introduction

The transition from fossil fuel-powered transport to electric mobility is considered a necessary step to tackle greenhouse gas emissions. However, the widespread use of electric vehicles may cause some concern, mainly related to the impact of battery charging and the need to carefully plan the charging infrastructure. The development of the charging infrastructure has an impact on the electricity grid, particularly on the ability to respond to requests for power. One way to limit these impacts is to integrate the charging structure with stationary storage systems and photovoltaic sources (PV), in order to contain power peaks on the grid, through on-site energy production. The factors influencing the choice of sizes and possible implementation are many, including the space available, the additional cost, the extent of the reduction in the withdrawal from the network, the diversity of the storage, and PV sizes.

Several studies have approached the problem of planning and operation schedules of charging infrastructures. There are several issues to consider in the correct design of charging infrastructures for electric road vehicles. For a public charging infrastructure, parameters such as the number of users, the number of charging points and the influx time distribution influence the quality of the service, as investigated in [1]. The complex relationship between charging services and electricity network requires careful planning and operation of the charging infrastructure, which takes into account traffic congestion, loads on the electricity grid and charging costs [2]. An approach for optimizing the siting and sizing of charging stations considering the interactions between the energy and transport industries is presented in [3]. The work takes into account the load capacity constraints to evaluate the feasibility of the implementation, also taking into account the economic convenience. In [4], electric physical constraints are included in a mathematical profit maximization model, which is designed to distribute appropriate capacities at the charging stations. The charging infrastructure could be integrated with smart grids, as well as with different types of renewable energy sources and stationary electric storage systems [5].

The use of renewable energy in power recharging would lead to an increase in acceptance by potential users of EVs [6]. In this context, Energy Storage Systems (ESS) can be used to balance the generation of electricity from renewable energy sources. This solution is particularly interesting for fast charging stations [7] or in stand-alone solutions [8]. In [9], an energy management scheme is introduced which provides uninterrupted daytime charging at a constant price, for a Charging Infrastructure (CI) equipped with a photovoltaic (PV) system and Battery Energy Storage System (BESS). In addition, Vehicle-to-Grid (V2G) technology is incorporated into the charging system to improve the payback schedule of the existing PV grid system. For a similar configuration of the charging station, an intelligent optimization and control algorithm is proposed in [10], which aims to reduce operating costs. The algorithm foresees: the optimization of the energy management programs of the day before; the determination of the charge price based on participation in the V2G; the optimization of energy management programs of the hour before; and real-time control. In [11], an optimal management of power flows in real time for a CI-PV-BEES system is presented, formulated as integer mixed linear programming, for an intelligent infrastructure based on microgrids for charging Electric Vehicles (EV), with the aim of minimizing the total cost of energy in different weather conditions. Particular attention is also paid to a rapid charge of electric vehicles and related issues [12]. In [13], the authors consider a charging station for electric vehicles, with PV and BESS installed. The approach involves examining first the problem of the operation of the charging station for electric vehicles and then the optimal planning of the stations. The load is synthetized using a stochastic model based on an M/M/c/K birth–death Markov model. In [14], the configuration optimization process aims to maximize the expected profit of the station considering the investment and operational costs, maintenance, and penalties linked to customer dissatisfaction due to waiting times or lack of service. The determination of the optimal size and charging station management was performed for various application environments. The EV flow was simulated using M/G/N/K queues.

Despite the large body of studies, there appears to be a lack of studies based on real data to improve the sizing and control strategies of charging stations [15]. In the present work, the study is based on data on real charges recorded for one year at eleven public stations installed in the metropolitan area of Barcelona. Although the optimization of CI-PV-BESS operation has been extensively studied, the influence of stationary battery duration has not been fully explored. In [16], a proposal was made to take battery duration into account in optimizing the power flow in a CI-PV-BESS system with V2G. The optimization problem is faced using a multilayer procedure, according to a set of priorities aiming at maximizing the use of photovoltaics and minimizing the amount of energy purchased from the grid. The procedure involves a stochastic approach to addressing uncertainty in the load and production of renewable energy. However, it only enters into the determination of the BESS operating energy cost by means of a function dependent on the State of Charge (SOC), while the impact of the reduction in battery capacity over time on the functionality of the infrastructure was not taken into account. In [17], the aging of the lead–acid batteries that compose the storage was taken into account to avoid overloads and overdischarges in a stand-alone configuration for the battery charger, although no issues related to the profitability of the investment were investigated. In [18], a hypothetical network of fast charging stations on highways is considered for a simulated charging request. The charging station consists of three 120 kW chargers and a BESS of two possible sizes: 250 and 650 kW. The comparison between the investments for these two BEES sizes is made with the net present value. The aging of the BESS is taken into account, considering the current, the operating temperature and the time of use, using models calibrated on cells for specific types of batteries for electric cars. However, the parameters of the aging models are specific to chemistry and battery type. Unlike the approaches proposed so far in the literature, this paper focuses on the degradation costs for the BESS and their influence on the CI power fluxes and impact on investment costs. In particular, we compare a linear cost function, in which the degradation cost is simply proportional to the energy exchanged, with one dependent on SOC. Subsequently, we explicitly include battery aging in the system operations to correctly evaluate the time horizon of the investment. The aim of the authors is to show that different, albeit simple, degradation models can influence the determination of the optimal energy fluxes, which in turn has consequences for battery usage and thus for battery life. A graphical representation of the proposed approach is reported in Figure 1. Furthermore, the proposed analysis and results are based on real data derived from public CI in the metropolitan area of Barcelona (AMB). As far as the authors know, this is the first time that the impact of different battery aging models can have on the operation of the charging station and on the actual investment time horizon has been studied. In Section 2, we present an approach for the optimal sizing of a Renewable Energy Source (RES) and BESS for a CI an urban area. In Section 3, we evaluate the power flow at the station on typical days and highlight the influence of different battery models on power exchanges. Section 4 presents the conclusions.

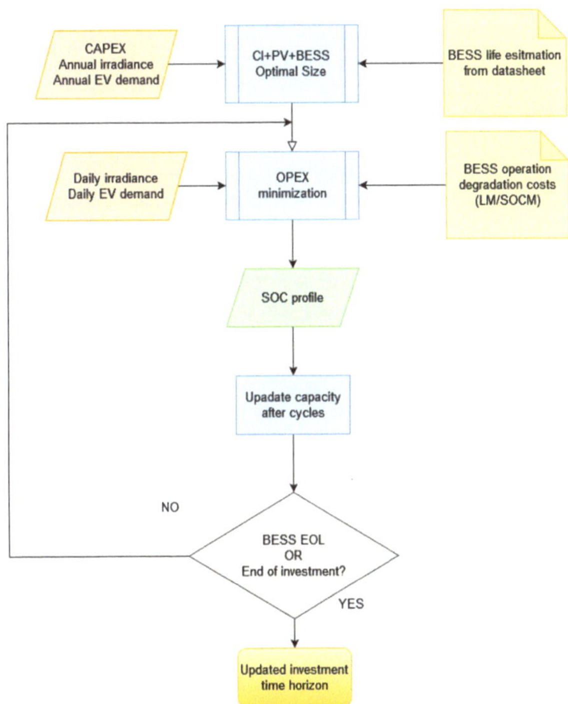

Figure 1. Flowchart of the proposed approach.

2. Charging Infrastructure Sizing

In the present section, we determine the optimal size of a RES with a BESS system for a CI in an urban area. In particular, the considered RES is a photovoltaic (PV) system, since it is the most suitable for an urban environment. The analysis uses real data on charging events at public CI in AMB. The PV system consists of a finite number of identical solar panels, with inverters and Balance-of-System (BoS). The BESS consists of a finite number of identical battery units of common technology integrated with a BoS. In our model, we will disregard BoS in the analysis, as it usually does not affect too much the project costs [19]. The influences of soft costs on the charging station construction can also have a big impact on the deployment of the infrastructure, both on the final costs and the installation time [20]. However, these costs can greatly vary from one country to another, and in time, making an exact estimation difficult. For this reason, soft costs are not included in the following cost analysis.

2.1. BESS and PV Cost

In this study, we are interested in Behind-The-Meter (BTM) applications for BESS, i.e.: commercial and residential use, for which capacity ranges from 0.01 to 0.25 MWh [21]. Usually, the battery technologies used for BTM applications are Li-ion and lead–acid. To compare different technologies, the Levelized Cost of Storage (LCOS) metric was introduced by Belderbos, et al. [22], which is analogous to levelized cost of energy for energy sources. This index takes into account many parameters, such as Capital Expenditure (CAPEX), Operational Expense (OPEX), life duration, and discount rate. In Table 1, we summarize LCOS for Li-ion and lead–acid batteries for BTM applications, according to different sources. Although the reported costs show some discrepancies, the LCOS for Li-ion and lead–acid batteries are overall comparable. A study of the International Renewable Energy Agency (IRENA) [23] reports that the CAPEX will decrease for all technologies by the year 2030, estimating a drop of about 60% for Li-ion system capital costs, and of 50% for lead–acid systems. Instead, OPEX is assumed to remain unchanged, as the maintenance costs are already very low compared to CAPEX. With this consideration, we will focus on Li-ion technology for the BESS. The CAPEX is calculated as an average of the 2018 and 2025 costs (0.35 EUR/Wh) [24]. We did not consider OPEX costs, since they are negligible compared to the CAPEX costs. We did, however, consider degradation costs, as illustrated in the next session. As for the PV system, the European Technology and Innovation Platform for Photovoltaics estimates that the cost for a 50 MWp utility system is about 65% of the residential one; for a 1 MWp system, it is 80% and for a 50 kWp (commercial) system, it is 90% of the residential cost [25]. Since the cost in 2014 of a turnkey photovoltaic utility scale was 955 EUR/kWp with a reduction to 823 and 724 in 2020 and 2025, respectively [26], we can deduce that the PV price in 2020 for residential is around 1.2 EUR/Wp, while for commercial, it is around 1.0 EUR/Wp. We did not include any analysis of COVID-19 influence on PV system price. Thereafter, CAPEX for the PV system is estimated at 1.2 EUR/Wp.

Table 1. LCOS for different battery systems.

Source	Battery Type	CAPEX [USD/kW]	OPEX [USD/kW]	LCOS [USD/kWh]
Apricum [27]	Li-ion	500	10	0.53
Lazard [28]	Li-ion	804–1289	0	0.89–1.27
Lazard	Lead-acid	556-835	0	1.06–1.24
SolarPro [29]	Li-ion	1406	0	0.56
World Energy Council [30]	Li-ion	300–3700	7–74	0.15–0.7
World Energy Council	Lead-acid	500–1700	10–34	0.1–0.4

2.2. Charging Events Dataset and Analysis

The optimization problem has been applied to the case study of Barcelona, which is a demo city in the H2020 UserChi Project [31]. The dataset used in the present work contains the registration of the charging events AMB in 2019 [32]. It consists of two files, the first including charging events information such as: Charging Point Identification (CP id), Charging Point (CP) name, type of connector, charging start time and stop time, duration, energy consumption, vehicle, and model. The last two fields are optional. The other file contains static information about the charging points, such as address, longitude, latitude, and type of connectors at the CP, CP makers. Each CP id corresponds to one of the following configurations:

(a) Two Schuko 3 kW 16A mode 1 sockets;
(b) One Mennekes 43 kW 63 A mode 3, one CHAdeMO 55 kW 125 A mode 4, one COMBO CCS 55 kW 125 A mode 4;
(c) Two Schuko 3 kW 16 A mode 1 and two Mennekes 7 kW 16 A mode 3.

In ten CP locations, the two configurations (a) and (b) are present, so we can refer to this combination as a charging infrastructure (CI). The configuration (c) is present only in one location, so it will be considered in the statistical analysis, but not included in the CI analysis and size determination. In the following, the term "normal" chargers includes 3 kW and 7 kW, while "quick" refers to 43 kW and 55 kW. The statistical analysis of the data can be found in [33]. After the filtering procedure, we obtained a data set with 36,194 records. Table 2 reports some figures on the usage of the CPs, while the relevant statistical parameters for the charges (mean value and standard deviation of charge duration and energy delivered) are reported in Table 3. As an input to the procedure of system sizing, for each month of the year, a synthetic daily load profile representative of all load profiles in the month was created. From the database, we extract the number of charges, the energy required, and the duration of the charges as a function of the hour of the day, for each month. For quick charges, we excluded charges that delivered less than 1 kWh and/or with power less than 1 kW. For normal charge, we excluded those with less than 0.1 kWh delivered. We reconstruct the time distribution of the number of users at the stations, according to the arrival time and taking into account the duration (expressed in hours, which represents the time scan unit) and the energy delivered during the charges, to obtain the a posteriori frequency distribution for the arrivals at the CIs and the average power required. Using this distribution, we can construct a "most-probable" load profile for each month and a given number of users per day. To emphasize the role of PV and BESS, we selected the most crowded station: 14 users per day for quick charges and 2 for normal ones. The average energy of the charge profile obtained is set equal to the average energy of the entire set. Some examples of load profiles are reported in Figure 2.

Table 2. Usage rate of charging points for different typologies.

Charge Typology	No. Charge/Year	No. Charge/Day	Max Charge/Day	Min Charge/Day
Normal	286.7	0.79	1.69	0.15
Quick	3302.9	9.05	14.03	3.78

Table 3. Statistical parameters for charges.

Dataset	Mean Duration (min) (min)	Stdev Duration (min)	Mean Energy (kWh)	Stdev Energy (kWh)
Entire	41.99	90.37	10.09	7.76
Normal	181.56	262.96	4.11	4.91
Quick	28.67	17.57	10.66	7.74

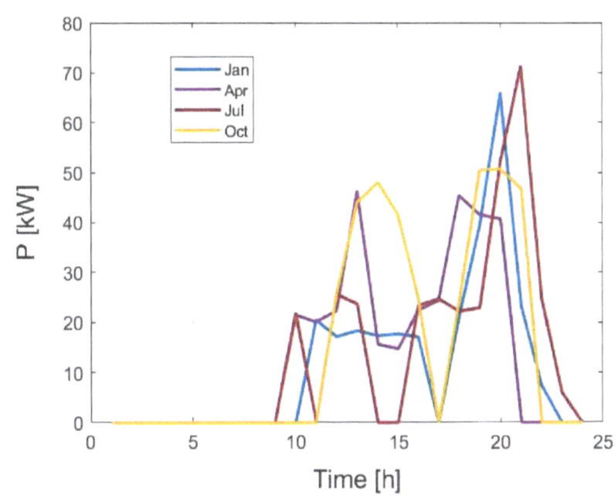

Figure 2. Examples of synthetic load profiles for Barcelona for some months of the year.

The procedure illustrated above is one of the possible options for creating a synthetic profile, and different approaches can lead to different results for the optimal dimension of the system [33].

2.3. Size Optimization Procedure

A diagram for the proposed system is shown in Figure 3. A DC bus connects the different energy sources with the CI, while the inverters guarantee the voltage stability. The BESS stores the energy from the PV system, and/or from the grid when the electricity cost is low. The energy stored in the BESS is then used to charge electric vehicles when the PV source is absent and/or the cost of energy from the grid is high. Based on technical-economic constraints, an algorithm determines the optimal configuration of the system according to the operating conditions and plant costs [33], and the following input data:

- The market price of electricity;
- Productivity and cost of the photovoltaic system;
- Costs of the storage system;
- Charging profiles.

Through the economic criterion of the Net Present Value (NPV) [34,35] by projecting the analysis over the entire depreciation period of the infrastructure, it is possible to determine the optimal system solution and identify the basic information for the preliminary design of the CI. An investment is advantageous when the NPV is positive, i.e., when the project incomes are sufficient to repay the initial outlay by providing a net gain. The NPV is assessed over a time horizon (N in years) and is a function of the cash flows relating to the investment made and the interest rate considered (k); it can be expressed as follows:

$$NPV = \sum_{t=1}^{N} \frac{F_t}{(1+k)^t} - I \tag{1}$$

where:

N: time horizon of the investment in years, fixed in 15 years;

F_t: cash flows in the t-th year, calculated as the difference among cash flow without and with the PV + BESS system;

I: initial investment, calculated as the sum of the PV and BESS CAPEX;

k: interest rate, fixed in 3%.

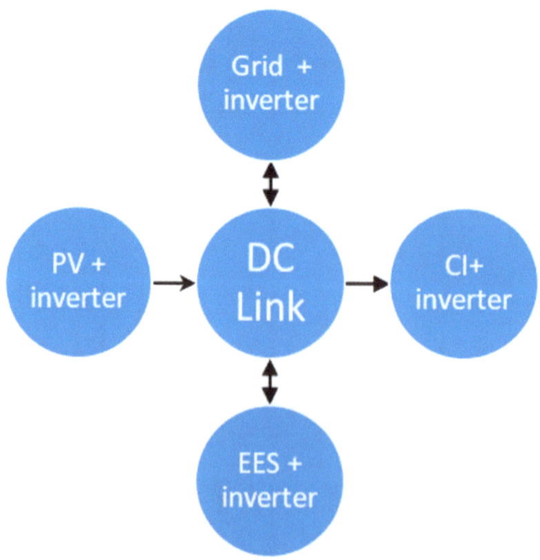

Figure 3. Proposed system for charging infrastructure with a renewable energy source.

In the monthly analysis, a representative day of the average monthly production of photovoltaics and the average monthly cost of electricity purchased from the grid is considered. The 15-year NPV is used as a basis for comparison of the different investment projects, which is a trade-off between the duration of a PV system (20–30 years) and a CI and BESS (10–15 years).

The power flows coming from the sources through the conversion systems are summed up in the common DC bus. In particular, it is possible to write the balance of the input powers coming to the DC bus:

$$P_{grid}\eta_{grid} + P_{PV}\eta_{PV} + P_{B,D}\eta_{B,D} - \frac{P_{B,C}}{\eta_{B,C}} = \frac{P_L}{\eta_L} \quad (2)$$

where:

P_{grid}: power withdrawn from the grid;
P_{PV}: power supplied by the photovoltaic system;
$P_{B,D}$: discharge power of the battery pack;
$P_{B,C}$: charging power of the battery pack;
P_L: power required by the load;
η_{grid}: efficiency of the network converter;
η_{PV}: efficiency of the photovoltaic system converter;
$\eta_{B,D}$: efficiency of the bidirectional converter of the battery pack, discharge mode;
$\eta_{B,C}$: efficiency of the bidirectional converter of the battery pack, charge mode;
η_L: efficiency of the output converter.

The optimization problem aims at minimizing the objective function represented by the daily operating cost of the charging infrastructure (C_e) consisting of two components: the cost of the energy supplied by the grid and the cost of the degradation of the storage system. In particular, the optimization problem was expressed as follows:

$$minC_e = min\sum_{h=1}^{24}(C_r(h)P_{grid}(h) + C_{deg}(h))\Delta t \quad (3)$$

where:

$C_r(h)$: is the price of the energy at time h [EUR/kWh];
$P_{grid}(h)$: the power withdrawn from the grid at time h [W];
Δt: the sample time, in the present 1 h;
$C_{deg}(h)$: the degradation cost for the EES [EUR/h].

The system model must respect a series of constraints, listed below:

$$P_{grid}^{min} \leq P_{grid}(h) \leq P_{grid}^{max} \quad (4)$$

$$P_{PV}^{min} \leq P_{PV}(h) \leq P_{PV}^{max} \quad (5)$$

$$P_{B,D}^{min} \leq x_{B,D}(h)P_{B,D}(h) \leq P_{B,D}^{max} \quad (6)$$

$$P_{B,C}^{min} \leq x_{B,C}(h)P_{B,C}(h) \leq P_{B,C}^{max} \quad (7)$$

$$x_{B,D}(h) + x_{B,C}(h) \leq 1 \quad (8)$$

$$P_{PV}(h) \leq P_{mppt}(h) \quad (9)$$

$$SOC(h) = SOC(h-1) + \left(\frac{\eta_{B,C}P_{B,C}(h)\Delta t}{E_B} + \frac{P_{B,D}(h)\Delta t}{\eta_{B,D}E_B}\right)100\% \quad (10)$$

$$SOC^{min} \leq SOC(h) \leq SOC^{max} \quad (11)$$

$$SOC(0) = SOC(24) \quad (12)$$

Equations (4)–(7) define the operating limits of the system based on the minimum and maximum power of the sources. $x_{B,D}$ and $x_{B,C}$, Equation (8), are binary variables necessary to prevent the battery pack from being simultaneously charged and discharged. Equation (9) limits the power that can be drawn from the photovoltaic system to the maximum extractable power P_{mppt} in the h-th hour. The constraint (10) links the state of charge at the $h - th$ hour ($SOC(h)$) to that one of the previous hour ($SOC(h-1)$) as a function of the charge/discharge power used, the nominal energy stored by the battery pack E_B and the charge and discharge efficiency of the storage system. According to the constraint (11), the SOC must be within a minimum (SOC^{min}) and a maximum (SOC^{max}) to limit the depth of discharge and, therefore, the cost of degradation. The stability of the state of charge between one day and another is imposed by Equation (12). The following inputs have been used in the optimal size procedure:

- Price of electricity: the data for 2019 were downloaded from the Comision Nacional de los Mercados y la Competencia (CNMC) [36];
- Productivity of the PV system: data were taken from the "Performance of grid-connected PV" tool of the Photovoltaic Geographical Information System (PVGIS) [37]. The data contain the monthly production for installed peak PV power of 1 kWp and system loss of 14% (mounting configuration: tilt 35°, azimuth 0°) from the PVGIS-SARAH data base for the selected locations;
- Efficiencies of the conversion systems: efficiency values of 0.9 were used for all systems, except BESS, for which the value was set to 0.8;
- Required charging power profiles, as illustrated in the charging events dataset and analysis.

The battery size depends on many characteristics: in particular, the power is one key parameter in the size determination. However, battery performance is generally expressed in terms of different parameters, such as the maximum charge and discharge currents (that usually vary with temperature and aging), maximum, minimum, and operating voltage, and capacity. Since voltage V is not a linear function of SOC, power is also not a linear function of SOC, which makes it difficult to implement a simple algorithm for size optimization. This problem is partially overcome by the fact that the curve V vs. SOC for Li-ion batteries can be approximated by a straight line within the SOC range of 10–90%.

2.4. System Degradation

The degradation rate of the photovoltaic system is set at an average value of 0.5% per year of loss of productivity [38].

The battery duration is taken into account in two ways: for the calendar life, the duration is fixed to 15 years, while for life cycle it is assumed that the battery can last for fixed to N_{life} = 10,000 charges and discharges cycles at 70% of Depth of Discharge (DOD), which leads to a $N_{life} = 5000$ full cycles at 70% DOD. This assumption considers current storage solutions performance, with up to 20 years and 5000 cycles at 100% DOD [39–41].

To model the degradation cost in the optimization procedure, we use a very simple model based on cycles counting (similar to the rain-flow counting). This choice allows the optimization model to be mixed linear integral, while a more realistic model would make it non-linear. However, in the size determination phase, this simple model is sufficient to consider the costs that derive from the battery usage.

Taking as cost per watt-hour of the battery degradation the price for the battery pack: CB = 0.35 EUR/Wh [24], we can obtain the cost for a full cycle with DOD = 70% as: $CC = CB * E_B / N_{life}$.

The degradation cost for a generic cycle "i" is given by: $C_i = CC * DOD_i / 70\%$.

2.5. Growth in Demand

In the analysis, we will also consider the projections in the growth of the electric car market. The share of electric vehicles is growing fast. According to the European Automobile Manufacturers' Association (ACEA), the market share of electric cars in 2019 was 3.0%, which is one percentage point more than in 2018 [42]. According to forecasts by the International Energy Agency, electric cars, which accounted for 2.6% of global car sales and about 1% of global car stock in 2019, registered a 40% year-on-year increase. As a result of the continuous increase in sales, EVs are expected to account for about 7% of the global vehicle fleet by 2030 [43]. We will include a growing rate in the range 0–10% in the analysis.

2.6. Results

In the following, the results of the optimization problem for AMB case study are presented. The optimization problem was solved in MatLab using the "solve" command, which is an equations and system solver, in a linear programming problem defined by an optimization problem. Even though Equation (3) with constraints (4)–(12) represent a mixed-integer linear programming problem, it can be reduced to a linear programming problem, since the constraint (8) is implicitly satisfied when minimizing battery degradation costs.

Costs for the installation of CI are not considered as, for AMB, the local administrations promote the installation of CI in the streets. In this case, the municipalities act as a Charge Point Operator (CPO). Moreover, we consider the possibility of introducing subsidies programs oriented to the installation of PV panels for CI, similar to those already existing for residential PV systems [44].

As a first step, a PV system without BESS is considered. Without incentives for investment costs, the renewable source is not convenient. However, applying a 20% incentive on the PV investment cost, the NPV becomes positive and the optimal size for the PV system is around 23 kWp for the ideal case of converters' efficiency equal to 1. In the more realistic scenario of efficiencies lower than 1, the optimal size is around 25.3 kWp. If the degradation of the PV system is included, the optimal size grows to 27.25 kWp. In the following, only a non-ideal system will be considered, unless explicitly stated.

If we allow increasing the demand with time, we obtain an increase of the PV size, as well as an increase in the corresponding NPV. In Figure 4, we show the results for different constant yearly growth rates. The NPV increases when the growth rate is higher than a threshold, around 5%. Correspondingly, the growth of PV size over this threshold slows down.

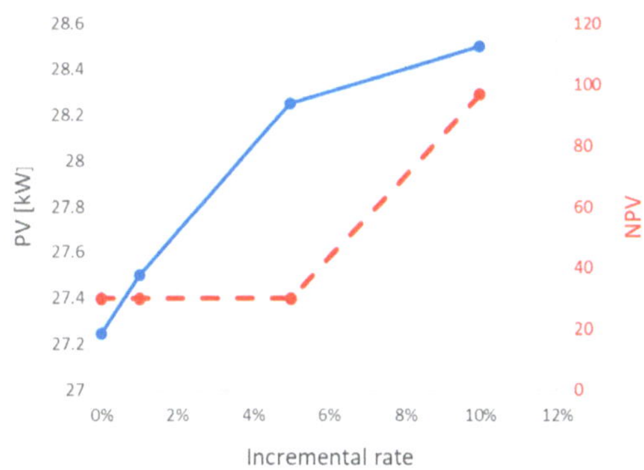

Figure 4. Optimal PV sizing with no BESS (principal axis) and corresponding NPV (secondary axis) as a function of demand growth rate.

The increment of the NPV can be correlated to the increase in the self-consumption of the PV energy when the load demand increases.

To evaluate the size of the system components when a BESS is present, we initially consider a zero CAPEX for the BEES plant. We obtain that the investment is feasible if the incentives on PV costs are preserved: the optimal value for the PV increases to 39.5 kW with a BEES as large as 98 kWh. This BEES size guarantees that all the energy produced by the PV is entirely consumed by the CI. However, when the investment cost for the BESS is considered, there is no convenience in designing it to obtain the total self-consumption of the PV energy. In particular, if the BESS CAPEX is equal or greater than 0.35 EUR/kWh, the incentives of 20% should be extended to the total system investment cost to obtain a positive NPV. If a load demand increase is included, the optimal BESS size increases along with the PV size. If the load increases with a 5%, 1%, and 0% growth rate per year, and the BESS CAPEX is equal to 0.35 EUR/kWh, the optimal size for the PV in the is around 37 kW, 35 kW, and 33.5 kW, respectively. The relative values for the BEES are: 12.5 kWh, 13 kWh, and 13.5 kWh. The BESS optimal size increases as the load increases. This trend holds for all the scenarios investigated, independently from the values of CAPEX BESS and the incentives put in place. It is interesting noting that the ratio between the optimal BESS and PV sizes is constant with respect to the incremental rate, at a given BESS CAPEX. The dependence of the ratio from the CAPEX is reported in Figure 5. We can conclude that for the Barcelona scenario, the use of BEES with PV can lead to positive NPV values, if some incentive policies are put in place, and increase the self-consumption of renewable energy.

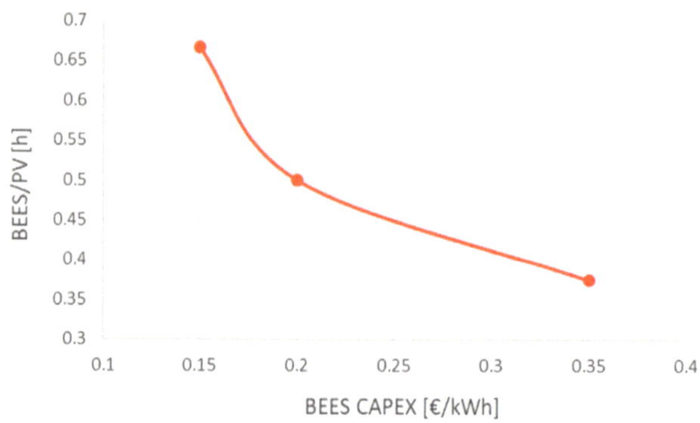

Figure 5. BEES to PV ratio as a function of the BEES CAPEX.

3. Charging Infrastructure Operations

When the PV system is coupled with a BEES, there is the need for a control strategy to optimize the resources usage. In the present approach, the aim is to minimize the operating cost, which includes the purchase of energy from the grid and the degradation of the battery due to cycling, using Equation (2). Moreover, the BESS can participate to power-shifting from the grid, i.e., charge the BESS when the energy price of the grid is low and discharge it when the price is high. To solve the optimization problem for the CI operation, we use a linear programming solver that optimizes the use of the BESS according to the minimization of the costs, Equation (3), and maximization of the PV usage. To determine the optimal functioning of the CI, we used a probabilistic approach. In fact, some of the system inputs, such as PV production and the power required by vehicles in charging processes, are intrinsically random. The approach adopted is numerical and is based on the Monte Carlo (MC) method: for the charge method, the input data are obtained in a pseudo-random way starting from the Non-Parametric Distribution Function (NPDF) of the arrivals at the CI, and using the mean value and the variance of the duration of the charges and the energy required for each time interval. The workflow of the procedure is illustrated in Figure 6. Given the duration and average number of daily arrival, for each MC run, we construct a load profile from the a posteriori distribution of arrivals, assuming a Gaussian distribution for the charge duration and energy. In a similar way, from the monthly irradiation, we extract a most-probable profile and add Gaussian noise. These randomized synthetic charging and irradiation profiles are used in the algorithm to obtain the corresponding optimal BESS SOC distributions that minimize the operational costs. The ensemble of SOC distributions and power flows are then used in the statistical analysis to identify the statistically most favorable usage of the BESS to determine the optimal activity schedule for the following day.

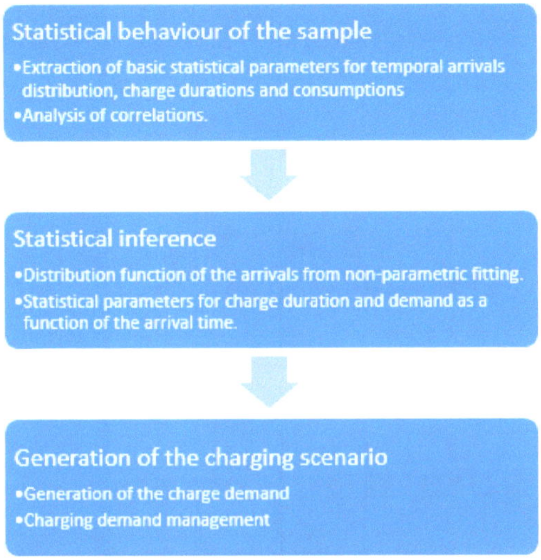

Figure 6. Workflow of the scenarios creation procedure.

Influence of Battery Modeling

The goal of day-to-day management is to minimize operating costs—the sum of the cost of energy purchased from the grid—and the cost of battery degradation due to cycling. The quantification of BEES degradation costs during the operations influences the outcome of the procedure. In the following, we compare the results for different BEES degradation models. In particular, we analyze a linear cost model (LM), where the energy price is constant in all the SOC range: this means that the degradation is assumed to be unrelated to the depth of charge/discharge. The cost was determined as follows:

$$C^L_{deg} = \frac{C_{Wh}}{N_{life}} \tag{13}$$

where C_{Wh} is the price for battery cell, and N_{life} is the life cycle expressed in number of full cycle. We allow the SOC to vary between 20% and 90%, to minimize battery degradation.

Applying Equation (3) to each run of the MC procedure, and averaging over the output, we can determine the most probable trend for the power fluxes in a given day. As an example, we illustrate the energy fluxes for two different months. For both sets of runs, we use the set PV size equal to 33.5 kW, with a BESS of 12.5 kWh. The energy price for the battery cell is C_{Wh} = 0.15 EUR/Wh, which is an average price for 2019, although price is decreasing rapidly [45]. The results for the power distribution among the sources are reported in Figure 7. The top stacked histogram chart represents the contribution of the different energy sources to the power demand of the load (black line). In particular, the red bar is the grid contribution, the blue bar is the PV contribution, and the cyan is the BEES contribution. The yellow bar represents the PV power exceeding the load demand that can be sold to the grid, while the magenta is the power withdrawn from the grid or from the exceeding PV energy, to charge the battery. The blue line is the PV production. The green line represents the grid energy price. The bottom chart represents the SOC of the battery during the day. With the chosen parameters, the battery takes part in the power flux exchange helping the self-consumption. In fact, the surplus of PV power (the part of the blue bar exceeding the black line) is used to charge the battery, as it can be seen during the day of January (Figure 7a). From Figure 7b, it can be seen that a part of the PV energy produced in July is not absorbed by the BEES (yellow bar) since the battery SOC has already reached its maximum value. Although the details of the system behavior depend on the approaches used to model the battery degradation, it can be concluded that, for the operation of a CI with PV, the presence of the BESS is beneficial, since it decreases the energy withdrawn from the grid by increasing the exploitation of the PV production.

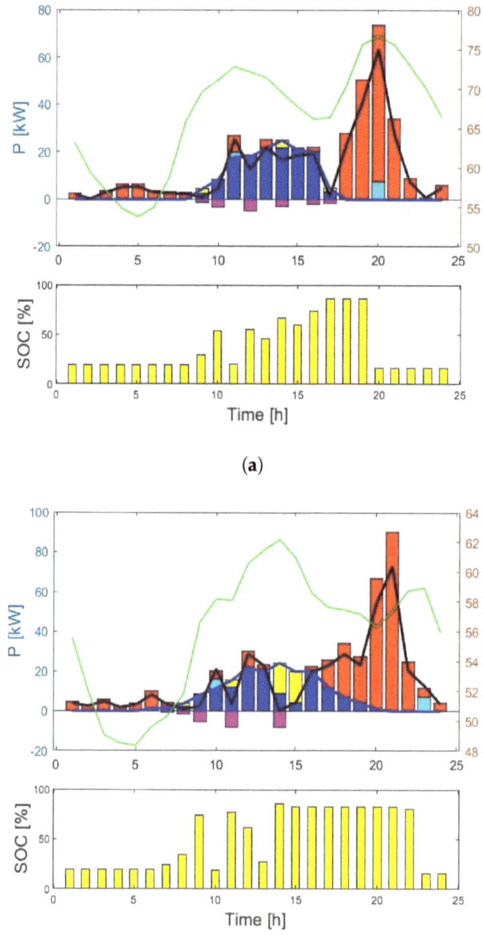

Figure 7. Power fluxes and optimal SOC distribution for LM, of a CI with a 33.5 kW PV system and a 12.5 kWh BEES for AMB: (**a**) January, (**b**) July.

We compare the results thus obtained with the ones for a cost model where the battery energy price depends on the state of charge, as in [33]. We will refer to this model as the State of Charge Cost Model (SOCM). As in the LM, we made the hypothesis that charge and discharge have the same energy cost. In particular, we adopt the following cost function:

$$C_{deg}^{SOC} = \frac{C}{(SOC + \epsilon) * (100 + \epsilon - SOC)} \tag{14}$$

where C is a coefficient that is chosen so that the cost of a full cycle (0–100% SOC) is the same for LM and SOCM:

$$\frac{C}{100 + 2\epsilon}\left[ln\left(\frac{100 + 2\epsilon}{\epsilon} - 1\right) - ln\left(\frac{100 + 2\epsilon}{100 + \epsilon} - 1\right)\right] = \frac{100 * C_{Wh}}{N_{life}} \tag{15}$$

and ϵ is a constant introduced to avoid divergences. Its value influences the cost function shape. We use here the value 1, so that the SOCM function is as in Figure 8. We apply the MC procedure using either the LM or the SOCM model, over the same sets of random load and insulation profiles, and compare the average results over the runs. First of all, we calculate the CI OPEX for a typical day when BESS degradation cost is not considered, $C_{Wh} = 0$. The results for four different months are reported in Table 4, along with the partition of PV and grid energy utilized to satisfy the load demand, and the number of complete cycles of the BESS in a day, obtained with the rain flow counting method. As it can be seen, the CI daily operating costs are reduced when a PV + BESS system is present, although the saving is different for each month. The percentage of PV energy used directly to satisfy the load demand or to charge the BESS depends more on the temporal distribution of the load rather than the seasonal production. Since we considered converters' efficiencies smaller than one, the sum of the PV and grid contribution is greater than 1. In general, the BESS usage is more intense during summer and spring with respect to winter and autumn. The different models used for the battery degradation cost influence directly the average daily CI OPEX. A sensibility analysis is conducted for different life cycle span: 1250, 2500, and 5000 full cycles, both for LM and SOCM. We stress here that when aging is not explicitly included in the model, the life span enters only in the OPEX cost (13) and (14). We analyzed the yearly average values of the following variables: CI OPEX, partition of PV energy used, partition of grid energy used, and number of complete BESS cycles. If we limit the BESS depth of discharge to 20% and the upper SOC to 90% in the SOCM, the energy price is always lower than in LM (see Figure 8). We then relax any constraints for the SOC available range when using SOCM, since the highest price for high and low SOC values should limit the usage in these regions. As an example, Figure 9 reports the optimal SOC distribution for SOCM with $N_{life} = 5000$ for a typical day of January (a) and July (b). Comparing with the SOC distribution obtained for LM with SOC boundaries, Figure 7, we see that for SOCM, the range used in a typical day is extended well below the 20% lower limit, although the upper SOC limit is very close to 90%. The influence on the CI OPEX of BESS life cycle is illustrated in Table 5, which reports some results obtained for LM and SOCM, for different life cycle durations. It is evident from the results in Table 5 that SOCM is not as sensitive to battery life duration as LM. Furthermore, the application of the SOCM model results in the lowest daily costs. By observing the energy flows, we can verify that the largest share of energy purchased from the grid belongs to the LM model when the battery life is 1250 cycles, and the smallest to SOCM, when the battery life is 5000 cycles. Indeed, the application of the SOCM model leads to an average CI OPEX lower than that obtained by LM with a null BESS OPEX and SOC in the range 20–90%, and equal to that of LM with a null BESS OPEX and no limitation on SOC, although with a slightly different partition of energy among the sources.

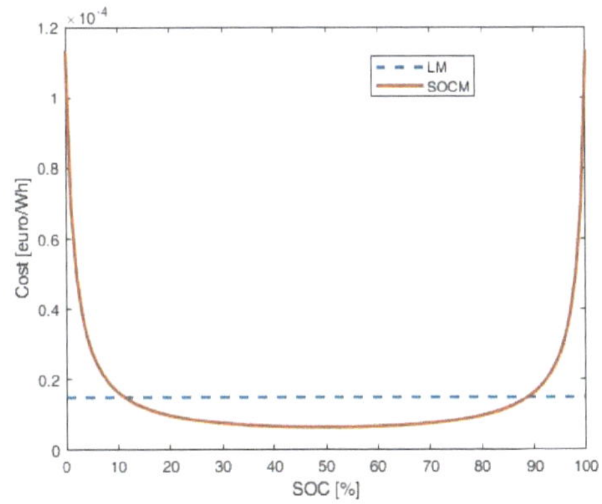

Figure 8. Cost function for LM (dashed blue line) and SOCM (solid red line).

Table 4. CI OPEX and source power partition when BESS OPEX is null.

Month	January	April	July	October
CI OPEX w/o PV + BESS [€/kWh]	0.054	0.047	0.045	0.051
CI OPEX [EUR/kWh]	0.049	0.043	0.041	0.048
EPV/EL	0.531	0.482	0.523	0.358
EG/EL	0.806	0.873	0.852	0.961
COMPLETE BESS CYCLES in a day	0.99	1.24	1.39	0.86

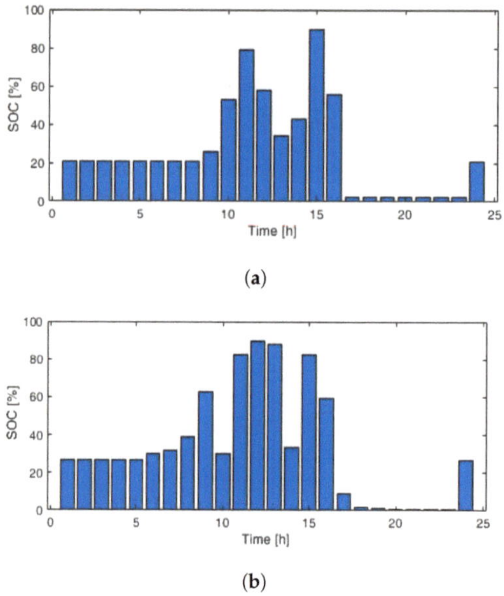

Figure 9. Optimal SOC distribution for SOCM model of a CI with 33.5 kW PV and 12.5 kWh BEES for: (**a**) January, (**b**) July.

Table 5. Average yearly figures for LM OPEX with different BESS life cycles, and for SOCM OPEX.

OPEX BEES = 0.15 EUR/Wh	CI OPEX [EUR/kWh]	EPV/EL	EG/EL	Complete BESS Cycles in a Day
LM-1250	0.049	0.422	0.791	0.805
LM-2500	0.047	0.461	0.752	1.021
LM-5000	0.046	0.463	0.751	1.039
SOCM-1250	0.044	0.492	0.716	1.452
SOCM-2500	0.044	0.492	0.716	1.452
SOCM-5000	0.044	0.492	0.715	1.452
BESS OPEX = 0; 20% < SOC < 90%	0.045	0.472	0.751	1.120
BESS OPEX = 0; SOC < 100%	0.044	0.493	0.733	1.345

This shows that using SOCM the battery is exploited for greater depths of discharge, allowing to use a larger fraction of the energy produced by the PV, compared to the LM case, when SOC is limited to 20–90%. Furthermore, since the battery works mostly in the range around SOC = 50%, where on average the SOCM price is lower than LM (see Figure 7), even removing the SOC limitation, the CI OPEX continues to be lower when using SOCM than LM. The different use of the BESS in the two model influences the duration of the BESS itself. We thus explicitly include the capacity degradation in the models in order to evaluate the impact of the different usages on the BESS life duration. Let us consider a simple degradation model, in which the capacity decreases in proportion to the number of equivalent complete cycles performed. This model does not take into account the influence of depth of discharge or intensity of current on battery degradation. However, since a very broad time horizon is being analyzed, even a simplified model like this one can give some indications on the life expectancy of the BESS. If the declared BESS cycle life stated that after N_{life} full equivalent cycles, the capacity has decreased by a $x\%$, such that the final capacity is $(1 - x/100) * C_0$, where C_0 is the initial (nominal) capacity and x is the capacity reduction, the proposed aging model can be mathematically represented as follows:

$$C_{ag}(k) = C_{ag}(k-1) * \left(1 - \frac{N_k * x/100}{N_{life}}\right) = C_0 \prod_{j=1}^{k} \left(1 - \frac{N_j * x/100}{N_{life}}\right) \quad (16)$$

where $C_{ag}(k)$ is the capacity at step k and N_k is the number of full equivalent cycles at step k. Using the model (16), the trend of the relative capacity over the 15 years of the investment period is as reported in Figure 10, with $x\%$ = 40%.

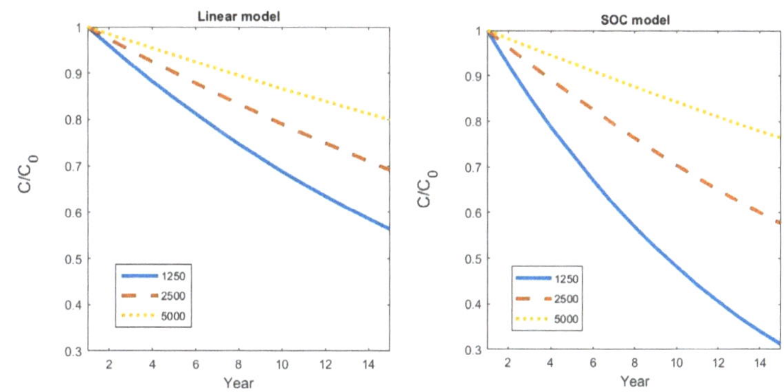

(**a**) (**b**)

Figure 10. Capacity degradation as a function of years, for different battery life durations and for (**a**) LM and (**b**) SOCM. Solid blue line: 1250 cycles; red dashed line: 2500 cycles; yellow dotted line: 5000. The trend is polynomial of degree 2 in all the cases.

As expected, the capacity diminishes at a higher rate when the expected life cycle is shorter. The SOCM leads to a faster degradation since it exploits a larger SOC range than LM. As an example, for a 1250 life cycle, the BESS reaches 60% of the initial capacity after 7 years with SOCM, while it takes 13 years with LM; for 2500 life cycle, the limit is reached after 14 years using SOCM, while it is not reached in the investment horizon of 15 years using LM. This means that the profit obtained in the OPEX CI with SOCM could be canceled by the need to purchase a second BESS, due to the achievement of the end-of-life condition of the first one.

4. Conclusions

In the present work, we consider the optimal dimensioning of a charging station equipped with a renewable energy source, and a storage system, based on the real charging requests of some public stations in the metropolitan area of Barcelona. Next, the procedure analyzes daily operations, identifying the power flows that maximize the use of renewable energy and minimize operating costs, including battery degradation. We analyzed the influence of different BESS degradation models that are used as OPEX functions. We use some simple approaches for battery modeling that could, however, be suitable in the evaluation of CI investment over a time horizon. We use some simple approaches for battery modeling that could, however, be suitable in the evaluation of CI investment over a time horizon. The main contributions of the paper are:

1. To highlight the influence that different battery aging models can have on the operations and functioning of a charging station;
2. To demonstrate that different BESS usage policies affect the investment time horizon. In this sense, a careful analysis of BESS performance is recommended during the design phase, in order to correctly calibrate the range of use and maximize the return on investment.

When we consider a degradation price model that depends on the SOC (SOCM), we obtain lower values for the operating CI costs than using a constant cost (LM). The SOCM is implemented without range restriction for BESS since the increment of the degradation costs toward low and high SOC values prevents the system usage in those regions. On the other hand, LM explicitly includes SOC limitation in the range [20–90%]. The results show that using SOCM lead to a greater number of cycles per day, and allows exploiting better the PV production with respect to LM. However, this translates into a higher number of full equivalent cycles per year, which brings to a shorter BESS life. This result directly influences the time horizon of the investment and should be considered in the economic evaluation of the investment. Furthermore, it could influence the operating policies at the CI by tuning the relative importance of the different components with respect to the CI operating costs. A more detailed and complex aging model can be included to take into account specific battery characteristics.

Author Contributions: Conceptualization, N.A. and M.D.M.; methodology, N.A., M.D.M. and G.T.; software, N.A. and M.D.M.; validation, N.A. and M.D.M.; formal analysis, N.A.; investigation, N.A., M.D.M. and G.T.; resources, N.A.; data curation, N.A.; writing-original draft preparation, N.A.; writing-review and editing, M.D.M. and G.T.; visualization, N.A. and M.D.M.; supervision, G.T.; project administration, N.A.; funding acquisition, N.A. All authors have read and agreed to the published version of the manuscript.

Funding: This work has been supported in part by the European Union's Horizon 2020 research and innovation programme under grant agreement No. 875187. The results reflect only the authors' view and the Agency is not responsible for any use that may be made of the information it contains.

Data Availability Statement: Data on charging events are available in Zenodo. https://doi.org/10.5281/zenodo.5721233, accessed on 4 November 2022.

Conflicts of Interest: The authors declare no conflict of interest.

Nomenclature

$\eta_{B,C}$	Efficiency of the bidirectional converter of the battery pack, charge mode
$\eta_{B,D}$	Efficiency of the bidirectional converter of the battery pack, discharge mode
η_{grid}	Efficiency of the network converter
η_L	Efficiency of the output converter
η_{PV}	Efficiency of the photovoltaic system converter
AMB	Metropolitan Area of Barcelona
BESS	Battery Energy Storage System
BoS	Balance of System
BTM	Behind the Meter
C_r	Price of the energy
C_{deg}	Degradation cost for the BEES
CAPEX	Capital Expenditure
CI	Charging Infrastructure
CP	Charging Point
DOD	Depth of Discharge
ESS	Energy Storage System
EV	Electric Vehicle

LCOS	Levelized Cost of Storage
LM	Linear Cost Model
MC	Monte Carlo
NPDF	Non-Parametric Distribution Function
NPV	Net Present Value
OPEX	Operational Expenditure
$P_{B,C}$	Charging power of the battery pack
$P_{B,D}$	Discharge power of the battery pack
P_{grid}	Power withdrawn from the grid
P_L	Power required by the load
PPV	Power supplied by the photovoltaic system
PV	Photovoltaic System
RES	Renewable Energy Source
SOCM	SOC Dependent Cost Model
SOC	State of Charge
V2G	Vehicle to Grid

References

1. Gunther, M.; Fallahnejad, M. Strategic Planning of Public Charging Infrastructure. In *IUBH Discussion Papers—IT & Engineering, No. 1/2021*; IUBH Internationale Hochschule: Erfurt, Germany, 2021. Available online: http://hdl.handle.net/10419/229137 (accessed on 24 October 2022).
2. Jawad, S.; Liu, J. Electrical Vehicle Charging Services Planning and Operation with Interdependent Power Networks and Transportation Networks: A Review of the Current Scenario and Future Trends. *Energies* **2020**, *13*, 3371. [CrossRef]
3. Xiang, Y.; Liu, J.; Li, R.; Li, F.; Gu, C.; Tang, S. Economic planning of electric vehicle charging stations considering traffic constraints and load profile templates. *Appl. Energy* **2016**, *178*, 647–659. [CrossRef]
4. Chen, Z.; Li, C.; Chen, X.; Yang, Q. Towards Optimal Planning of EV Charging Stations under Grid Constraints. *IFAC-PapersOnLine* **2020**, *53*, 14103–14108. [CrossRef]
5. Veneri, O.; Ferraro, L.; Capasso, C.; Iannuzzi, D. Charging infrastructures for EV: Overview of technologies and issues. In Proceedings of the 2012 Electrical Systems for Aircraft, Railway and Ship Propulsion, Bologna, Italy, 16–18 October 2012; pp. 1–6. [CrossRef]
6. Li, X.; Chen, P.; Wang, X. Impacts of renewables and socioeconomic factors on electric vehicle demands—Panel data studies across 14 countries. *Energy Policy* **2017**, *109*, 473–478. [CrossRef]
7. Elma, O. A dynamic charging strategy with hybrid fast charging station for electric vehicles. *Energy* **2020**, *202*, 117680. [CrossRef]
8. Ali, A.; Shakoor, R.; Raheem, A.; Muqeet, H.A.U.; Awais, Q.; Khan, A.A.; Jamil, M. Latest Energy Storage Trends in Multi-Energy Standalone Electric Vehicle Charging Stations: A Comprehensive Study. *Energies* **2022**, *15*, 4727. [CrossRef]
9. Bhatti, A.R.; Salam, Z. A rule-based energy management scheme for uninterrupted electric vehicles charging at constant price using photovoltaic-grid system. *Renew. Energy* **2018**, *125*, 384–400. [CrossRef]
10. Yan, Q.; Zhang, B.; Kezunovic, M. Optimized Operational Cost Reduction for an EV Charging Station Integrated With Battery Energy Storage and PV Generation. *IEEE Trans. Smart Grid* **2019**, *10*, 2096–2106. [CrossRef]
11. Cheikh-Mohamad, S.; Sechilariu, M.; Locment, F. Real-Time Power Management Including an Optimization Problem for PV-Powered Electric Vehicle Charging Stations. *Appl. Sci.* **2022**, *12*, 4323. [CrossRef]
12. Khan, W.; Ahmad, A.; Ahmad, F.; Alam, M.S. A Comprehensive Review of Fast Charging Infrastructure for Electric Vehicles. *Smart Sci.* **2018**, *6*, 256–270. [CrossRef]
13. Park, S.; Probstl, A.; Chang, W.; Annaswamy, A.; Chakraborty, S. Exploring planning and operations design space for EV charging stations. In Proceedings of the 36th Annual ACM Symposium on Applied Computing (SAC '21), Virtual Event, 22–26 March 2021; Association for Computing Machinery: New York, NY, USA, 2021; pp. 155–163. [CrossRef]
14. Nishimwe H.; Fidele, L.; Yoon, S.G. Combined Optimal Planning and Operation of a Fast EV-Charging Station Integrated with Solar PV and ESS. *Energies* **2021**, *14*, 3152. [CrossRef]
15. Alkawsi, G.; Baashar, Y.; Abbas, U.D.; Alkahtani, A.A.; Tiong, S.K. Review of Renewable Energy-Based Charging Infrastructure for Electric Vehicles. *Appl. Sci.* **2021**, *11*, 3847. [CrossRef]
16. Abronzini, U.; Attaianese, C.; D'Arpino, M.; Di Monaco, M.; Tomasso, G. Cost Minimization Energy Control Including Battery Aging for Multi-Source EV Charging Station. *Electronics* **2019**, *8*, 31. [CrossRef]
17. Cherif, A.; Jraidi, M.; Dhouib, A. A battery ageing model used in stand alone PV systems. *J. Power Sources* **2002**, *112*, 49–53. [CrossRef]
18. Richard, L.; Petit, M. Fast charging station with battery storage system for EV: Grid services and battery degradation. In Proceedings of the 2018 IEEE International Energy Conference (ENERGYCON), Limassol, Cyprus, 3–7 June 2018; pp. 1–6. [CrossRef]
19. Cervantes, J.; Choobineh, F. Optimal sizing of a nonutility-scale solar power system and its battery storage. *Appl. Energy* **2018**, *216*, 105–115. [CrossRef]

20. Bernard, M.R.; Hall, D. *Efficient Planning and Implementation of Public Chargers: Lessons Learned from European Cities*; International Council on Clean Transportation (ICCT): San Francisco, CA, USA, 2021. Available online: https://theicct.org/sites/default/files/publications/European-cities-charging-infra-feb2021.pdf (accessed on 24 October 2022).
21. Khalilpour, R.; Vassallo, A. Planning and operation scheduling of PV-battery systems: A novel methodology. *Renew. Sustain. Energy Rev.* **2016**, *53*, 194–208. [CrossRef]
22. Belderbos, A.; Delarue, E.; Kessels, K.; D'haeseleer, W. Levelized cost of storage—Introducing novel metrics. *Energy Econ.* **2017**, *67*, 287–299. [CrossRef]
23. IRENA. *Electricity Storage and Renewables: Costs and Markets to 2030*; International Renewable Energy Agency: Abu Dhabi, United Arab Emirates, 2017. Available online: https://www.irena.org/publications/2017/oct/electricity-storage-and-renewables-costs-and-markets (accessed on 24 October 2022).
24. Mongird, K.; Viswanathan, V.V.; Balducci, P.J.; Alam, M.J.E.; Fotedar, V.; Koritarov, V.S.; Hadjerioua, B. *Energy Storage Technology and Cost Characterization Report*; Pacific Northwest National Lab. (PNNL): Richland, WA, USA, 2019. Available online: https://energystorage.pnnl.gov/pdf/PNNL-28866.pdf (accessed on 24 October 2022).
25. Vartiainen, E.; Masson, G.; Breyer, C.; Moser, D.; Roman, Medina, E. Impact of weighted average cost of capital, capital expenditure, and other parameters on future utility scale PV levelised cost of electricity. *Prog. Photovolt. Res. Appl.* **2020**, *28*, 439–453. [CrossRef]
26. Fraunhofer ISE. Current and Future Cost of Photovoltaics. Long-Term Scenarios for Market Development, System Prices and LCOE of Utility-Scale PV Systems. Study on Behalf of Agora Energiewende. 2015. Available online: https://www.agora-energiewende.de/en/publications/current-and-future-cost-of-photovoltaics/ (accessed on 24 October 2022).
27. Mayr, F.; Beushausen, H. Navigating the maze of energy storage costs. *PV Tech. Power* **2016**, *7*, 84–88.
28. Lazard. Lazard's Levelized Cost of Storage Analysis—Version 3.0, u.o.: Lazard. 2017. Available online: https://www.lazard.com/perspective/levelized-cost-of-storage-2017/ (accessed on 24 October 2022).
29. Matthias, B.; Brearley, D. *Residential Energy Storage Economics*; SolarPro: Sofia, Bulgaria, 2016; pp. 22–36.
30. World Energy Council. *E-Storage: Shifting from Cost to Value, Wind and Solar Applications*; World Energy Council: London, UK, 2016. Available online: https://www.worldenergy.org/assets/downloads/Resources-E-storage-report-2016.02.04.pdf (accessed on 24 October 2022).
31. Available online: https://www.userchi.eu/ (accessed on 24 October 2022).
32. Andrenacci, N.; Bosch, R.; Kulla, A. Supporting data to the paper "Modelling charge profiles of electric vehicles based on charges data" [Data set]. *Zenodo* **2021**. [CrossRef]
33. Andrenacci, N.; Karagulian, F.; Genovese, A. Modelling charge profiles of electric vehicles based on charges data. *Open Res. Eur.* **2021**, *1*, 156. [CrossRef]
34. Abronzini, U.; Attaianese, C.; D'Arpino, M.; Di Monaco, M.; Genovese, A.; Pede, G.; Tomasso, G. Multi-source power converter system for EV charging station with integrated ESS. In Proceedings of the 2015 IEEE 1st International Forum on Research and Technologies for Society and Industry Leveraging a better tomorrow (RTSI), Turin, Italy, 16–18 September 2015; pp. 427–432. [CrossRef]
35. Lin, G.C.I.; Nagalingam, S.V. *CIM Justification and Optimisation*; Taylor & Francis: London, UK, 2000; p. 36, ISBN 0-7484-0858-4.
36. Comision Nacional de los Mercados y la Competencia. Available online: https://www.cnmc.es (accessed on 24 October 2022).
37. Available online: https://ec.europa.eu/jrc/en/pvgis (accessed on 24 October 2022).
38. Jordan, D.C.; Kurtz, S.R. Photovoltaic Degradation Rates-an Analytical Review. *Prog. Photovolt. Res. Appl.* **2013**, *21*, 12–29. [CrossRef]
39. Melin, H.E. The lithium-ion battery end-of-life market–A baseline study. In Proceedings of the World Economic Forum, Cologny, Switzerland, 23–26 January 2018; Volume 2018.
40. Available online: https://energy.ec.europa.eu/system/files/2022-01/vol-6-009.pdf (accessed on 24 October 2022).
41. Beltran, H.; Ayuso, P.; Pérez, E. Lifetime Expectancy of Li-Ion Batteries used for Residential Solar Storage. *Energies* **2020**, *13*, 568. [CrossRef]
42. Available online: https://www.acea.be/statistics/article/Share-of-diesel-in-new-passenger-cars (accessed on 18 January 2021).
43. Available online: https://www.iea.org/reports/global-ev-outlook-2020#prospects-for-electrification-in-transport-in-the-coming-decade (accessed on 24 October 2022).
44. Available online: https://www.cmh.cat/web/cmh/ajuts/programes/programa-projectes-2020 (accessed on 24 October 2022).
45. Ziegler, M.S.; Trancik, J.E. Re-examining rates of lithium-ion battery technology improvement and cost decline. *Energy Environ. Sci.* **2021**, *14*, 1635–1651. [CrossRef]

energies

Article

How Can EVs Support High RES Penetration in Islands

Ioannis Karakitsios *, Dimitrios Lagos, Aris Dimeas and Nikos Hatziargyriou

School of Electrical and Computer Engineering, National Technical University of Athens, 15773 Athens, Greece
* Correspondence: jkarak@power.ece.ntua.gr

Abstract: The electrification of the transportation sector contributes to a cleaner environment in non-interconnected island (NII) systems or standalone islands. Moreover, e-mobility can significantly contribute to achieving very high renewable energy source (RES) penetration levels in islands, allowing a reduction both in the emissions due to the conventional generation and the system's cost. Increased RES penetration, however, can pose technical challenges for an island's system. In order to overcome these challenges, new technologies like grid-forming converters are important. Moreover, the provision of new ancillary services in relation to battery storage systems might be considered, while novel control and protection schemes are needed to ensure secure operation. E-mobility can also contribute to solving technical problems that arise from very high RES penetration by providing frequency containment reserves or reactive power compensation. Since EV charging demand introduces modifications in the system's load curve, e-mobility may affect the power grid for long-term planning and short-term operation, i.e., line loading and voltages. The application of specifically developed smart charging methodologies can mitigate the relevant grid impact, while effective exploitation of EV–RES synergies can achieve higher RES penetration levels. This paper examines how e-mobility can contribute to increasing RES penetration in islands while considering the technical issues caused. In particular, this paper takes into account the distinct characteristics of NIIs towards the identification of solutions that will achieve very high RES penetration while also addressing the relevant technical challenges (voltage control, frequency control, short circuit protection, etc.). The effect of e-mobility in the power grid of NII systems is evaluated, while smart charging methodologies to mitigate the relevant impact and further increase RES penetration are identified.

Keywords: non-interconnected islands; high RES penetration; electric vehicles; smart charging

Citation: Karakitsios, I.; Lagos, D.; Dimeas, A.; Hatziargyriou, N. How Can EVs Support High RES Penetration in Islands. *Energies* **2023**, *16*, 558. https://doi.org/10.3390/en16010558

Academic Editor: Tek Tjing Lie

Received: 29 November 2022
Revised: 22 December 2022
Accepted: 29 December 2022
Published: 3 January 2023

1. Introduction

The transportation sector has a considerably high reliance on fossil fuels and is responsible for around 20% of the global CO_2 emissions [1]. Moreover, emissions from the transportation sector have been growing with an annual average rate of around 1.7% from 1990 to 2021 [2]. Europe has set significant goals toward the decarbonization of the transportation sector with the 2019 EU Green deal that aims to achieve a clear path for a 90% reduction in transport-related GHG emissions by 2050 [3]. Moreover, the REPowerEU plan concentrates on adopting legislative packages on greening the transportation sector while encouraging Member States to consider measures like tax reductions and exemptions from vehicle taxation for the purchase and use of electric vehicles [4]. The increased requirements for the decarbonization of transport have led to high interest in electric vehicles (EVs) all around the world with electric car sales reaching 9% of all vehicles sales in the global car market in 2021 (four times their market share in 2019) [5].

The electrification of the transportation sector is a path towards achieving a cleaner environment in islands around the world. For instance, the ARISE project in Maldives explores the potential of new technologies like EV charging stations and Vehicle-to-Grid (V2G) [6]. E-mobility is also promoted in the island of Astypalea in Greece with plans to replace all types of vehicles in the island with electric ones [7,8]. Similar activities towards a significant promotion of E-mobility are noted in islands all around the world, like the Azores [9], the Ghoramara island in India [10], the Aran islands in Ireland [11], the Sanso island in Denmark [12], or the Porto Santo island in Portugal [13].

E-mobility can also contribute to achieving very high RES penetration levels at NII systems or standalone islands. In particular, NIIs are characterized by high generation costs related to the thermal power generation [13–15]. However, when considering the potential for exploitation of EV–RES synergies, e-mobility can lead to a reduction in the system's generation cost while also achieving a cleaner environment [16]. Pilot programs are significantly important towards indicating the appropriate solutions (e.g., hybrid stations) that will allow increased RES penetration in non-interconnected islands [14], while they can also serve as test beds for innovative technologies and management solutions [17].

Increased RES penetration could introduce technical challenges in an island's system operation, e.g., due to reduction of physical inertia and short circuit currents. In order to overcome these challenges, ancillary services by central battery storage system should be provided. Additionally, novel control and protection schemes can be utilized to strengthen system security [18,19]. E-mobility can contribute to the mitigation of these technical challenges, considering that EVs can provide services related to frequency containment reserves or reactive power compensation [16,20–22].

Although e-mobility can increase the RES penetration in an island, the additional EV charging demand will modify the system's load curve. The impact of EV charging in the grid has been studied in the literature [23–27], while several management strategies have been proposed to mitigate this impact. For instance, the application of a management scheme that indicates the availability of charging stations, taking into account grid-related issues, is described in [28]. Strategies to lower the peak demand are studied in [29], while management schemes to avoid voltage issues and mitigate grid overloading are studied in [30]. Moreover, EV charging strategies considering the transformers' overloading are described [31,32].

Studies have also been performed to address some of the issues that are expected when considering the introduction of e-mobility in islands. For instance, [33] proposes solutions to the problems of imbalance between power supply and demand, while [34] suggests demand response programs to lower the operating costs of generation and reserves in an island. Smart charging methodologies that have been considered in island case studies have also been developed aiming to lower the peak demand [35,36], increase RES penetration [13,37], and reduce CO_2 emissions [16] while also examining V2G solutions [13,35–37].

None of the aforementioned studies related to e-mobility in island systems [13,16,33–37] considers all the above aspects. This paper provides an overview of the ways electric vehicles can contribute to increasing the RES penetration in non-interconnected islands. More specifically, the paper:

- Evaluates the effect of EV charging in the power grid of non-interconnected islands. The paper focuses on smart charging methodologies that can mitigate the relevant impact and increase RES penetration.
- Provides an overview of the challenges expected in the case of very high RES penetration in non-interconnected islands. A review of the relevant solutions to these challenges is presented, focusing on the EVs' contributions. The ways EVs can provide services related to frequency containment reserves or reactive power compensation are discussed.

Section 2 presents the distinct characteristics of NIIs related to their operation and power supply while also identifying effective solutions to achieve very high RES penetration. Section 3 indicates the challenges that should be expected in high-level RES penetration while also identifying potential solutions. Section 4 evaluates the effect of e-mobility on the power grid, focusing on smart charging methodologies that can mitigate the relevant impact and further increase RES penetration. Conclusions are drawn in Section 5.

2. Achieving Very High RES Penetration in Non-Interconnected Islands

Non-interconnected islands (NIIs), or standalone islands, have distinct characteristics compared to those of an interconnected system when considering power generation, resulting in high operational costs.

High RES penetration can reduce the high generation costs in such systems while also significantly reducing CO_2 emissions [16]. RES operation, however, especially wind farms, are constrained by the technical minima of thermal units. In particular, a sufficient number of generators should be in operation to meet the reserve requirements in each island, while the cumulative technical minimum production levels cover a significant amount of load. A penetration level limit is therefore usually applied for stability purposes (e.g., 20–35% of the total demand). Different limits are imposed according to the island and the demand levels.

Several studies have been performed on the operation of islands with high RES penetration, both regarding economic and technical aspects. A study focusing on the Astypalea island [14] has examined the optimal combination of RES technologies that provide high internal rates of return (IRR) for the installation of hybrid station investments in order to achieve RES penetration levels beyond 60%. The system topology for Astypalea is depicted in Figure 1. The indicative results considering various sizes for wind farms and PV stations are depicted in Figure 2a. Considering that the capacity factor for wind generators, which is typically 30–35%, is considerably higher than the relevant capacity factor for PVs (15–18%) at similar investment costs per MW, wind appears to be the dominant RES-generation technology (as is evident in Figure 2a) [14].

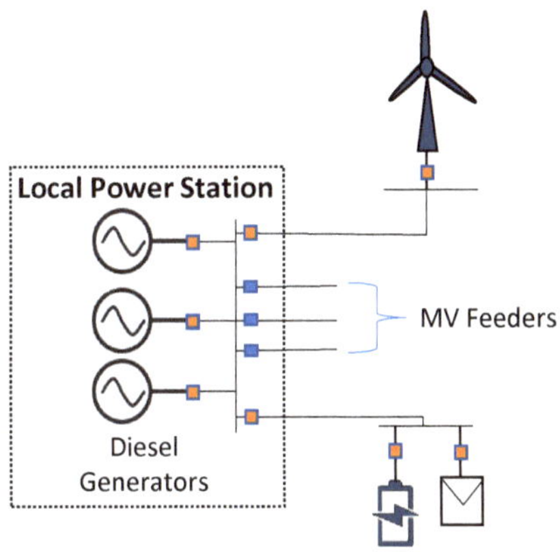

Figure 1. System topology for Astypalea.

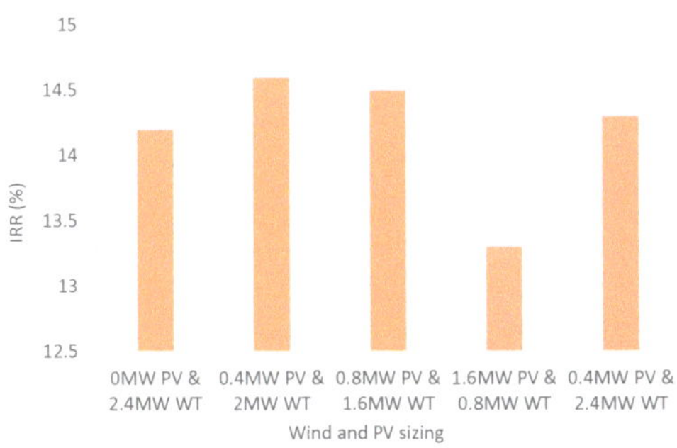

(a)

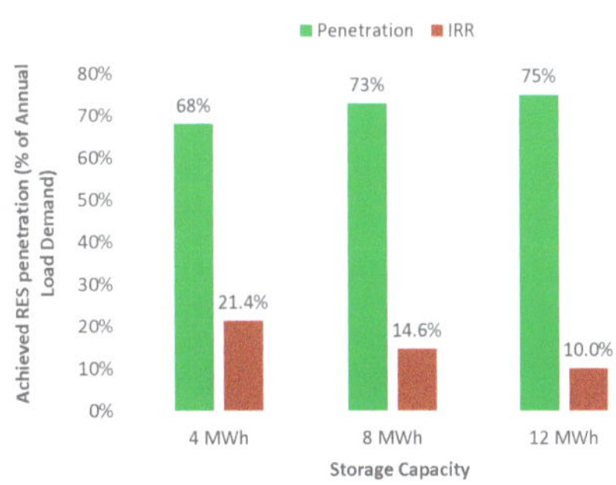

(b)

Figure 2. (**a**) Internal rate of return (IRR) for different RES capacity combinations in Astypalea; (**b**) IRR and achieved RES penetration for different storage capacities [14].

It should be also noted that the sizing of the storage capacity has a major impact on project profitability due to the high cost related to storage components [14]. According to Figure 2b, in case the storage capacity is increased beyond a certain level, slightly higher RES penetration is achieved, while the economic cost of the relevant investment is significantly higher. According to Figure 2b, a hybrid station that comprises a 0.4 MW PV station, a 2 MW wind farm, and a 2 MW/8 MWh battery storage system appears to be a very good option for the system of Astypalea. This selection results in RES penetration levels as high as 73% of the annual demand [14].

3. Challenges and Solutions Related to High RES Penetration Levels

Although e-mobility can assist in increasing the RES penetration in NII systems, technical challenges may arise. In particular, existing protection and control schemes on island systems have been designed according to the characteristics of the installed conventional oil-fired generators. However, the replacement of conventional oil-fired generators with inverter-based generators will introduce critical challenges (e.g., low inertia, low short circuits currents) that need to be addressed. To overcome them, new technologies like grid-forming converters are necessary providing new ancillary services, especially for a central battery storage system. Additionally, novel control and protection schemes should be expected from the smart grids and microgrids domain to ensure secure operation.

In frequency control, RES and battery energy storage technologies, which are inverter-based resources, do not provide physical inertia. Frequency containment reserves (FCR) can be provided and could be requested in an island's grid codes for inverter-based units; however, they are rarely applied in practice. The same holds true for frequency regulation reserves (FRR), which are usually not requested in island grid codes by inverter-based resources. Decarbonization of islands requires the replacement of synchronous generators (conventional units) with inverter-based technologies, which reduces system's inertia and increase the disturbance levels. Thus, critical frequency transients can occur, which can lead to disconnection of generators or loads [18,38]. Fast support of frequency containment reserves by the central battery storage system can mitigate the frequency transient. There is an ongoing trend by system operators to include a fast FCR service in their codes, focusing on battery energy storage (BES) technologies to address the reduced inertia concern [39,40] However, existing requirements for fast provision that have been introduced in interconnected system codes might be insufficient for the case of island systems. Additional services might be required for island systems, i.e., synthetic inertia to address the small amounts of physical inertia in them [39]. In addition, in several island projects, the installation of synchronous condensers with flywheels has been deployed or proposed to increase the physical inertia [18]. Finally, the absence of frequency restoration reserves (FRR) due to the reduced number of conventional generators operating in high RES penetration can be addressed by the provision of automatic FRR by the central battery storage unit.

To meet the annual targets for RES penetration greater than 60%, however, small islands may have to operate with 100% RES penetration at specific hours. Under those conditions, the central BES should form the grid for the RES grid-following inverters. In this respect, the converter of central battery storage units must be "grid-forming", controlling the voltage and frequency at their terminals. There are prominent pilot projects with grid-forming battery units providing black start and ancillary services [41]. To ensure seamless transition between island states, the battery storage unit should be able to operate in grid-forming control even in the presence of conventional units as presented in [42].

The small length of lines in the islands makes their voltage control less challenging compared to that of interconnected systems. Therefore, the most critical task in voltage control is to ensure the power balance in reactive power and the presence of adequate reactive power reserves. Currently, conventional thermal units provide/absorb reactive power by controlling the voltage at their terminals through their automatic voltage regulators. High RES penetration levels can result in less conventional units providing/absorbing reactive power that can lead to their overload. For instance, improper reactive power sharing has led to blackout events in island systems [43]. Inverter-based units can provide reactive power according to commands from the island energy management system and according to the relevant Q(V) characteristics.

The replacement of synchronous generators with inverter-based technologies also introduces a reduction in short-circuit current levels. At the same time, multiple short-circuit current sources are expected to be present on the island. The protection equipment in the island systems comprises digital overcurrent relays in the main feeders and fuses on the lateral lines. The characteristics of this equipment are dictated by the short-circuit current source and the local power station, where the installed synchronous generators can provide multiple times (e.g., 4–5) the nominal current during faults. On the other hand, inverter-based resources are characterized by reduced short-circuit current levels (e.g., 1–1.5 the nominal current) and unconventional behavior (e.g., suppressed negative sequence currents) dictated by the deployed manufacturer control. Thus, in high RES penetration, and specifically in the case where no synchronous generators are operating, the reduction of short-circuit currents increases the fault detection times compared to that in a scenario where diesel generators are connected. There are also additional concerns that arise due to the presence of various short-circuit sources in the island, due to the installation of inverter-based resources or the coordination of protection with FRT. Such issues are described in detail for the island of Astypalea in [44], where mitigation measures are proposed.

4. The Effect of E-Mobility on the Power Grid and Application of Smart Charging Methodologies to Mitigate Impact

Replacing conventional vehicles with electric ones improves the environmental conditions in islands if a reduction in conventional production is also taken into account. In this respect, the impact of the EV charging demand in the long-term planning of an island's network should be considered, evaluating how smart charging can increase the RES penetration levels as well as the economic feasibility of the relevant RES-related investments. Moreover, it is important to evaluate the potential of EVs to contribute to the mitigation of technical challenges in islands (providing frequency containment reserves, etc.). The potential impact of EV charging in the line loading and the voltage values in the islands' system should also be considered, and it is essential to evaluate how smart charging methodologies could mitigate them. E-mobility in island systems allows the exploitation of synergies among EVs and RESs. In particular, smart charging or V2G methodologies can be employed in order to increase both EV and RES hosting capacity in an island network.

4.1. *Impact of EV Charging in Long-Term Plannig*

EV charging demand is expected to introduce an additional load in the system that can modify the load curve, affecting the expected RES penetration levels and the economic feasibility of the relevant RES-related investments (for instance, those related to the installation and operation of hybrid stations). Additional EV charging demand may negatively affect the cost for the operation of an island's network, particularly when conventional generation is required to cover the additional demand. In this respect, it is important to study the impact of EV charging demand in the long-term planning of an island's network.

The long-term planning in the presence of EVs for the Greek island of Astypalea has been studied in [45]. As noted in Section 2, Astypalea is one of the three Pilot Projects in Greece that have been selected to promote increased RES penetration and sustainable solutions. During recent years, Astypalea has also been selected to significantly promote e-mobility [7]. In particular, 12 publicly available charging stations have already been installed in the island, while a number of public-authority conventional vehicles have been replaced with electric vehicles. E-mobility services improving mobility on the island compared to the previous public transport system are already available, and there are plans to replace all types of private vehicles (grants towards this purpose have already been set) [7,8].

The expected EV charging demand in Astypalea (Figure 3) has been calculated in [45], taking into account the simulation tools for the estimation of EV demand in [23,24,26,27]. Different types of vehicles have been considered (two-wheelers, passenger cars of various sizes, mini buses, etc.), while also taking into account charging at various locations (home, work, publicly available stations, etc.) and various charging levels.

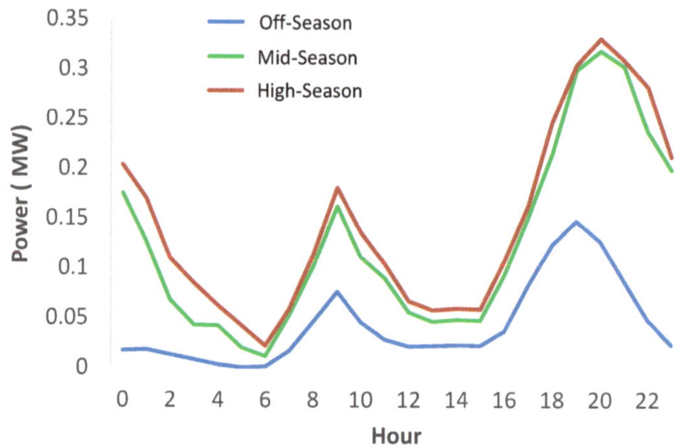

Figure 3. EV charging demand when considering the Greek island of Astypalea [45].

Due to high tourism activity during the summer months, different time-seasons are taken into account: (1) off-season (October–April), (2) high season (July–August), and (3) mid-season (May, June, and September). As depicted in Figure 3, considerable demand is introduced during the evening hours, while a significant difference is also noted when considering the different seasons. In particular, peak EV demand of 150 kW is noted during the off-season period. The demand is significantly increased during the summer months, particularly during the high-season period (peak EV demand during the high-season period is equal to around 0.33 MW).

According to the results in [45], hybrid stations in Astypalea should comprise at least 1 MW PVs, at least 2 MW from wind generation, and a battery capacity of at least 7.2 MWh in order to achieve a high IRR and a considerable RES penetration level (greater than 60%).

In this respect, two scenarios are considered for the hybrid stations in Astypalea:

- Scenario 1: PVs: 1.3 MW, wind: 2 MW, battery 7.2 MWh;
- Scenario 2: PVs: 2.3 MW, wind: 2 MW, battery 9.6 MWh.

The results in Figure 4 concern a time period of one year. As depicted in Figure 4a,b uncontrolled charging of EVs increases the RES energy injected in the grid while also increasing the IRR of the relevant investment. However, RES penetration, as a percentage of the system's demand, decreases (Figure 4c), since EV charging takes place without considering the hours with increased RES penetration. Moreover, according to Figure 4d, the average production cost of the system increases.

In order to further increase RES penetration in the island, a smart charging methodology has been developed in [45], shifting the charging demand to hours with increased RES penetration. According to Figure 4a,b, smart charging of EVs further increases the RES energy injected in the grid and the IRR of the relevant investment. As is depicted in Figure 4c, smart charging manages to increase RES penetration (as a percentage of the system's load), which is reduced in the case of uncontrolled charging.

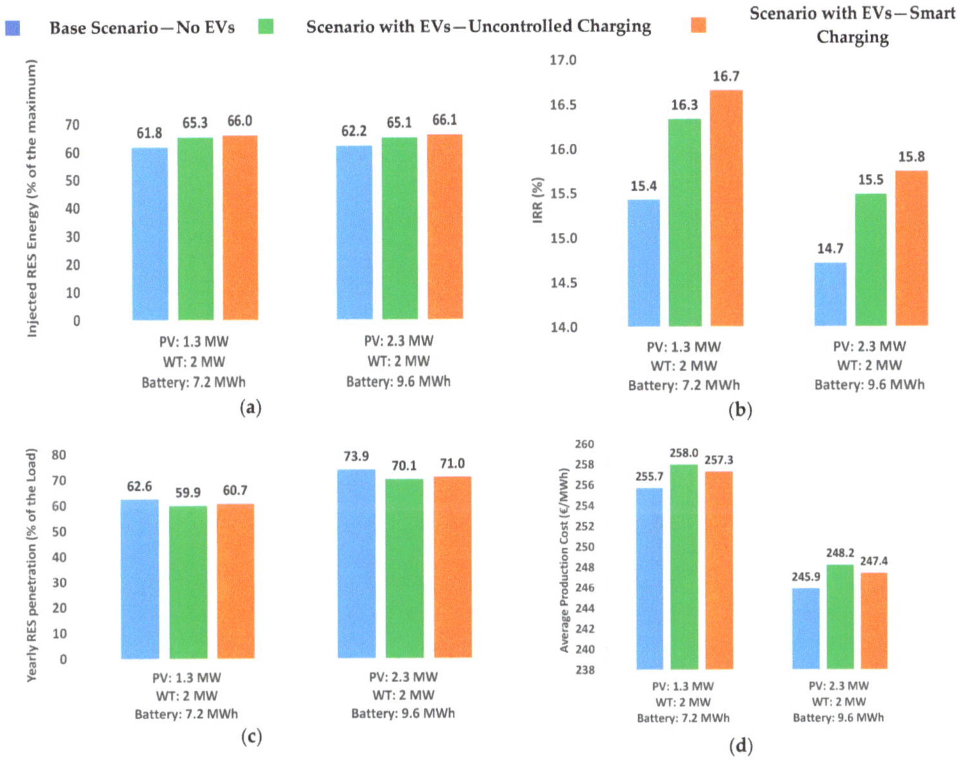

Figure 4. The impact of EV charging in Astypalea [45] regarding (**a**) injected RES energy (% of the maximum available), (**b**) IRR, (**c**) RES penetration (% of the system's demand), and (**d**) average production cost [45].

Additionally, smart charging reduces the system's cost (compared to that of the uncontrolled charging scenario), as depicted in Figure 4d. The effect of smart charging methodologies in reducing the system's cost is also evident in similar studies. For instance, the analysis of the EV load impact on the power system of Maldives has indicated that smart charging methodologies can reduce the additional system costs that should be expected in the case of uncontrolled charging by around 17% [46]. Similarly, the results in [47], indicate that the application of V2G methodologies can decrease the generation cost of the power system in the Greek islands by 10% in 2030 (compared to that with uncontrolled charging) when considering that the EV share (i.e., the percentage of EVs over the total number of vehicles) is equal to 4%. In case of an increased EV share by 2040 (equal to 20% of the total number of vehicles), the application of V2G methodologies could significantly decrease the system's generation cost by 20% (compared to the generation cost in the case of uncontrolled charging) [47].

4.2. Contribution of E-Mobility to the Mitigation of Technical Challenges in Islands

EVs can contribute to mitigating the technical challenges due to high RES penetration. For instance, interfaces that allow power quality services like reactive power compensation to be offered are described in [20], while EV chargers that allow reactive power control have been developed in [21,22].

EVs can alter their charging power to provide frequency containment reserves (FCR). More specifically, regarding underfrequency or overfrequency events, the charging power of EVs can be dynamically reduced or increased according to the measured frequency in order to provide FCR. Figure 5 depicts the provision of FCR services considering the

Astypalea island. In particular, an operating scenario is considered where the island demand is equal to 1.6 MW, and an additional demand of 0.6 MW is considered for EV charging needs. A hybrid station is considered to operate in the island comprising a wind turbine (WT) and a battery energy storage (BES) system. During the examined scenario, the wind turbine provides power equal to 2 MW, while the battery is charging at 0.4 MW. A diesel generator is also operating in the island with a power of 0.6 MW in order to cover the rest of the system's demand.

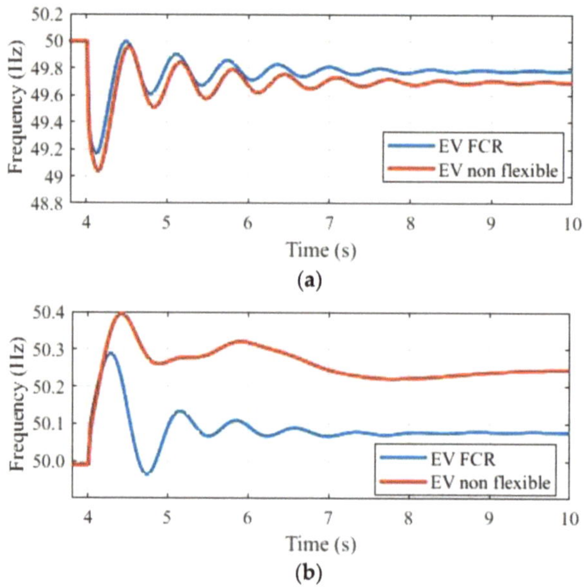

Figure 5. Contribution of EV charging in (**a**) underfrequency and (**b**) overfrequency transients.

In case the wind turbine is disconnected, an underfrequency event is evident in Figure 5a. In case EV charging demand is considered as a flexible load (blue line in the diagram), the charging power of the vehicles can be decreased, and the system's frequency can be restored to a value closer to the nominal one, compared to the case where EV charging is uncontrolled (red line in Figure 5a).

In case the central BES is disconnected, an overfrequency event should be expected, as is evident in Figure 5b. In this case, if the EVs are considered as a nonflexible load (i.e., in case of uncontrolled charging—red line in Figure 5b), the frequency is increased to values far from the nominal one. However, in case the vehicles are considered as flexible loads providing FCR, the charging power of the vehicles can be increased, allowing the system's frequency to be restored to a value closer to the nominal one.

Similar results are also evident in other studies. For instance, [16] has evaluated the potential of EVs to support the frequency in case of disturbances in the Greek non-interconnected island of Crete. Two different disturbances are evaluated in [16], as depicted in Figure 6:

- Simulation of a three-phase short circuit at a high-voltage bus;
- Simulation of a three-phase short circuit at a high-voltage bus with simultaneous loss of two generators.

In each case three, different scenarios are considered. In the base scenario, EV charging is considered as a nonflexible load. In the second scenario, EVs are considered as flexible loads with unidirectional power flow capabilities (i.e., energy can only be provided from the grid to the stations). In the third scenario, bidirectional power flow capabilities are considered, with the EVs being able to provide power to the grid when needed. The contribution of each EV to the primary frequency regulation depends on its charging energy needs (charge level—SoC) [16].

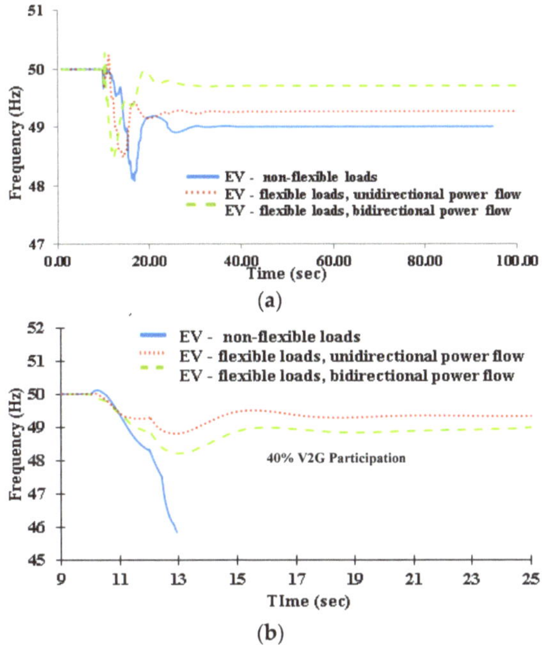

Figure 6. Simulation of (**a**) a three-phase short circuit at a high-voltage bus and (**b**) a three-phase short circuit at a high-voltage bus with simultaneous loss of two generators in the Greek non-interconnected island of Crete [16].

Regarding the base scenario in Figure 6a (simulation of a three-phase short circuit at a high-voltage bus), EVs are considered as nonflexible loads, and the frequency after the disturbance and the activation of the primary reserves is stabilized in the value of 49 Hz (the lowest value among the scenarios examined). The best results are achieved when adopting a management scheme that exploits the bidirectional exchange of energy with the EVs [16].

In case of a three-phase short circuit at a high-voltage bus with simultaneous loss of two generators (Figure 6b), when EVs are considered as nonflexible loads (blue line in Figure 6b), the frequency drops to unacceptable levels. In this case, relay protections of conventional units are tripped, and the system collapses. This blackout can be prevented when EVs offer frequency support, either by curtailing their charging demand or by injecting power to the grid [16] (red and green lines in Figure 6b).

4.3. Impact of EV Charging on Line Loading and Voltages Values

EV charging can affect the loading of lines as well as the voltage values in the island system's buses. The voltage values and the loading of the lines have been studied for the case of the Greek island of Astypalea. In particular, the distribution grid of Astypalea, which comprises three feeders, has been simulated in Dlgsilent-Powerfactory. Simulations for the time period of one year were performed, and two scenarios were examined:

- Scenario 1: Normal operation of the system;
- Scenario 2: Operation in case of a fault requiring the interconnection of two lines.

Considering a penetration level of EVs that will increase the island's peak demand by 5% and energy requirements by 15%, the maximum loading of the lines, as obtained from hourly simulations for one year, is depicted in Figure 7. It is evident that when the normal operation of the system is considered, the lines are loaded well below their thermal limits, even when EV charging is taken into account. A fault that requires the interconnection of two of the system's lines results in increased line loading, since the loads of two lines are only supplied by a single line. In this scenario, as depicted in Figure 7, a maximum line loading of 71% is noted when considering that no EVs are present. When EV charging is taken into account, the maximum line loading is increased to 74%.

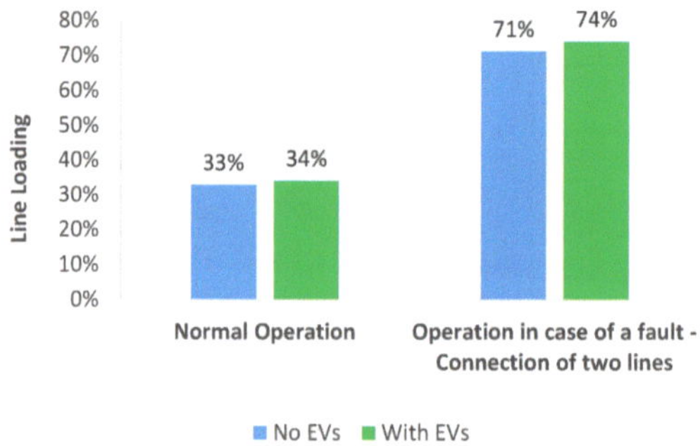

Figure 7. Maximum loading of the lines in Astypalea when considering the presence of EVs.

The minimum voltage values that were noted when applying simulations for the time period of one year are depicted in Figure 8. Due to the small length of the lines and the relatively small increase in peak demand (~5%), no issues should be expected with minimum voltages. In particular, EV charging reduces voltages; however, the relevant values remain within the required limits.

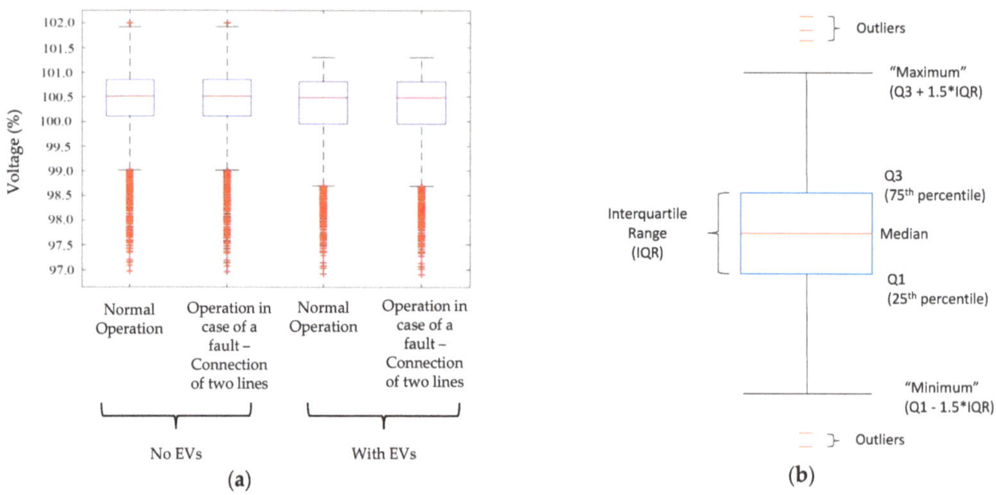

Figure 8. (**a**) Boxplot for the minimum voltage values in Astypalea when considering the presence of EVs, (**b**) Explanation of the boxplot.

However, higher EV penetration can introduce higher peak demand, particularly in the case of uncontrolled charging. For instance, according to Figure 9a, when considering the island of Ikaria [48], uncontrolled charging of EVs results in a violation of the minimum voltage limit in case more than 200 EVs enter the island. The number of EVs that can enter the island can be increased to 300 when considering the application of dual tariff schemes (red bars in Figure 9a). Similar results are observed when considering a voltage deviation (Figure 9b). In this case, uncontrolled charging of 150 EVs results in a violation of the voltage deviation limit. In this respect, penetration of EVs greater than the maximal number of EVs that the system can support will require additional grid investments.

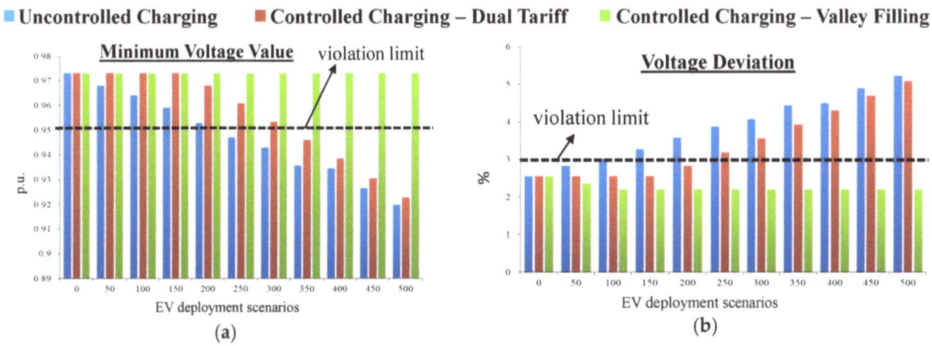

Figure 9. Voltage values for various EV penetration scenarios in the island of Ikaria (a 1 MW wind generator is considered to operate in the island for the scenarios examined): (**a**) minimum voltage value and (**b**) voltage deviation [48].

However, such investments can be avoided with smart charging strategies. For instance, when considering the application of a valley filling scheme (i.e., EV charging is distributed during the night hours in order to avoid spikes in the system's load) in the case of Ikaria [48] (green bars in Figure 9a,b), an increased number of EVs can enter the grid without causing any voltage violation issue.

4.4. Exploiting EV–RES Synergies towards Increasing the RES Penetration

Simple charging schemes can be applied to shift EV charging demand towards hours with increased RES penetration [25]. In this respect, EV charging exploits the available RES production that otherwise would not be utilized.

Advanced EV management strategies can be applied to achieve maximal RES penetration, taking into account the EV hosting capacity alongside grid-related parameters (grid losses, voltage fluctuations, etc.) [37]. For instance, the flexible EV demand management strategy in [37] results in reduced grid losses in the island of Ikaria, as is depicted in Figure 10a. Three hundred EVs are considered to enter the network of Ikaria in the case examined in Figure 10a. The maximum allowable RES penetration that can be achieved in each charging strategy examined in [37] is determined by the maximum voltage fluctuation and is represented by the continuous cycle in Figure 10a. The intermittent circles in Figure 10a indicate the more efficient RES penetration in the examined network (i.e., the maximum RES penetration that can be achieved with the minimum system losses). It is evident that the exploitation of EV–RES synergies alongside V2G capabilities (Strategy D in Figure 10a) allow for increased RES penetration alongside reduced system losses.

The maximal EV hosting capacity in the network can also be increased when exploiting EV–RES synergies, as is depicted in Figure 10b. More specifically, Figure 10b depicts the voltage deviation for various levels of EV penetration when considering that 2.1 MW RESs are installed in the island of Ikaria [37]. The maximal EV hosting capacity for the examined network of the island of Ikaria is determined by the maximum voltage deviation [37]. According to Figure 10b, the management strategy that exploits EV–RES synergies allows an increment in the grid's EV hosting capacity without violating technical network restrictions.

Similar results are shown in [47], where the application of V2G methodologies results in increasing the RES hosting capacity in the Greek islands, as indicated in Figure 11. More specifically, in case of slow growth in EV penetration in 2030, where EV shares as a percentage of the total number of vehicles are equal to 4%, 120 MWh of additional RES capacity will be available when considering the application of V2G methodologies. The additional RES capacity can be further increased in 2040 when considering an EV share equal to 20%. A considerable increase in RES capacity can be achieved if high growth in EV penetration (24% of vehicles are EVs in 2030) is considered (orange bars in Figure 11).

Strategy A: Uncontrolled charging
Strategy B: Dual tariff charging
Strategy C: Valley-filing
Strategy D: EV–RES coordination with V2G

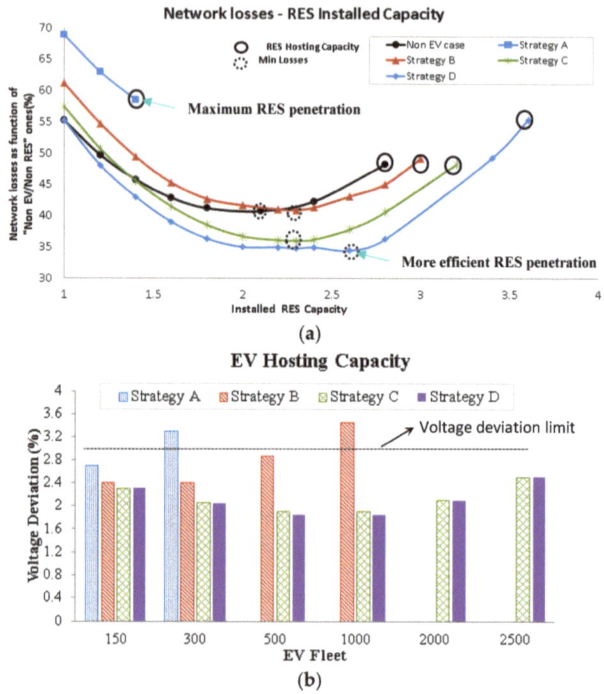

Figure 10. (**a**) Network losses for various cases of RES penetration when considering the introduction of 300 EVs in the island of Ikaria; (**b**) voltage deviation for various levels of EV penetration when considering that 2.1 MW RESs are installed in the island of Ikaria [37].

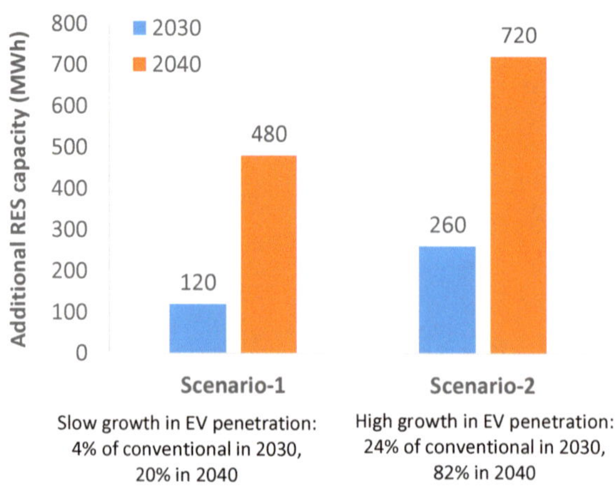

Figure 11. Increase in RES hosting capacity in the Greek islands with the application of V2G methodologies [47].

It should be noted that in the case of uncontrolled charging, increased RES penetration can limit the number of EVs that the grid can support, particularly due to violations in voltage deviation [49]. As depicted in Figure 12, when the RES installed capacity is increased, the number of EVs that can enter the island of Ikaria without causing any grid-related issue is reduced for uncontrolled (dumb charging) and valley filling charging (blue and red lines in Figure 12). However, more efficient smart charging schemes that also consider the maximum system load and RES penetration (green line in Figure 12), can increase both the grid's RES hosting capacity as well as the number of EVs that can enter the grid. More specifically, in this case, EV charging demand is shifted towards hours with low load or large wind power production, achieving increased EV hosting capacity in addition to increased RES penetration.

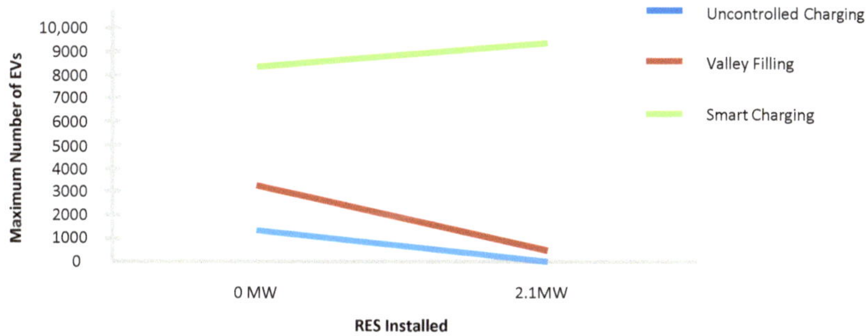

Figure 12. Relation between the maximum allowed number of EVs and the RESs installed in the network of Ikaria (based on the results of the Merge EU project) [49].

4.5. Reduction in CO_2 Emissions

Smart charging schemes can also reduce the CO_2 emissions related to the conventional energy production in an island. In particular, according to [16], an increase in the load of the system due to uncontrolled EV charging demand in the Greek island of Crete results in increased CO_2 emissions of conventional production (Figure 13). CO_2 emissions are slightly decreased when applying dual tariff schemes, as is depicted in Figure 13. However, valley filling schemes can significantly decrease the CO_2 emissions (green bars in Figure 13).

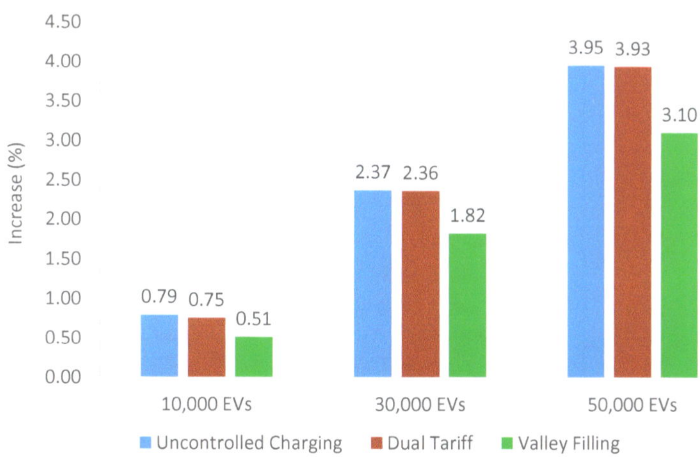

Figure 13. Generation CO_2 footprint in Crete, considering different charging schemes [16].

Similar results are evident in [46] for the case study of the Maldives islands. In particular, the optimized smart charging scheme that is applied in [46] reduces the evening spike, introduced in the system demand due to the uncontrolled charging of EVs, and distributes the EV charging load throughout the whole day. In this respect, the developed charging management scheme smoothens the load profile while also reducing the costs of the system. However, EV load is mainly covered by diesel generators in this case, resulting in increased CO_2 emissions [46]. In particular, as is depicted in Figure 14, CO_2 emissions are increased by 3.6% compared to those in the scenario where no EVs are considered in the examined network (baseline scenario). In this respect, a "carbon-neutral scenario" has been developed in [46] that successfully constrains the CO_2 emissions to be equal to the emissions in the baseline scenario (non-EV case), as depicted in Figure 14.

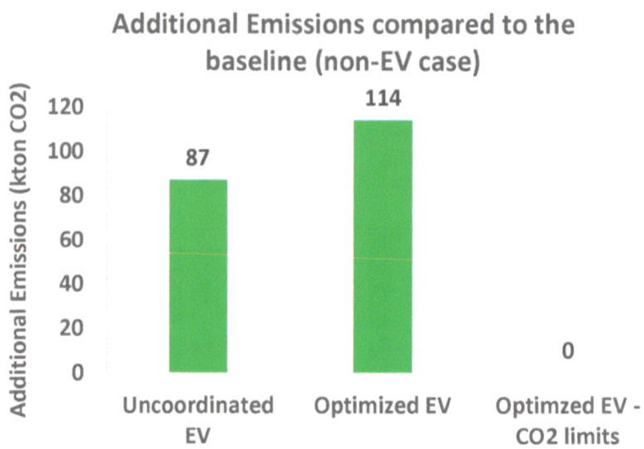

Figure 14. Additional emissions (kton CO_2) compared to the baseline scenario that considers no EVs for different charging schemes for the case study of the Maldives islands [46].

An overview of references indicating the services that can be offered by EVs alongside the smart charging strategies to mitigate the grid impact and increase RES penetration is provided in Table 1.

Table 1. References indicating the services that can be offered by EVs alongside the smart charging strategies to mitigate the grid impact and increase RES penetration.

Services Offered by EVs and Smart Charging Strategies to Mitigate the Grid Impact and Increase RES Penetration	References
Frequency support in case of grid disturbances	[16]
Power quality services like reactive power compensation and EV chargers that allow reactive power control	[20–22]
Mitigation of issues related to voltage values and line loading	[37,48]
Charging strategies to lower the peak demand	[29,35,36]
Charging strategies to avoid transformer overloading	[31,32]
EV–RES synergies to increase the RES penetration	[25,37,47,49]
Smart charging strategies to reduce CO_2 emissions	[16,46]

5. Conclusions

The effective introduction of e-mobility in NII systems can contribute to the reduction of costly and polluting operation of conventional generation units by achieving high RES penetration. Pilot studies have indicated that RES penetration greater than 70% can be achieved with the installation of hybrid stations based on centralized storage facilities. The relevant solutions also appear to be economically feasible with a high IRR related to the investments for the hybrid station.

Although e-mobility can assist in increasing the RES penetration in NII systems, technical challenges may arise. In order to overcome these challenges, new technologies like grid-forming converters are important, while novel control and protection schemes are needed to ensure secure operation. EVs can contribute to providing solutions in the technical challenges that are noted due to high RES penetration in an island. In particular,

EVs can alter their charging power to provide frequency containment reserves (FCR), while solutions to offer reactive power control have also been proposed.

The additional EV charging demand in an island with hybrid stations (comprising variable RES production and an energy storage system) can increase the RES-injected energy, while an increase is also noted in the IRR of the relevant investment for the hybrid station. The application of smart charging schemes further increases the RES injected energy and the IRR while also increasing the RES penetration as a percentage of the system's demand and reducing the system's cost (in comparison to that with uncontrolled charging).

Small EV penetration that does not significantly increase an island's peak demand is not expected to introduce issues related to voltages and line loading. However, higher EV penetration would introduce higher peak demand, particularly in the case of uncontrolled charging. In this case, smart charging schemes can lower peak demand. Methodologies that manage to exploit the synergies among EVs and RESs can effectively increase the RES hosting capacity in an island while also taking into account grid-related parameters (grid losses, voltage fluctuations, etc.).

Author Contributions: Formal analysis, I.K., D.L., A.D. and N.H.; methodology, I.K., D.L., A.D. and N.H.; supervision, N.H.; writing—original draft, I.K., D.L., A.D. and N.H. All authors have read and agreed to the published version of the manuscript.

Funding: This research received no external funding.

Data Availability Statement: The link to data that has been used in this paper has been provided by indicating the relevant references within the manuscript.

Conflicts of Interest: The authors declare no conflict of interest.

References

1. Ritchie, H. Cars, Planes, Trains: Where Do CO_2 Emissions from Transport Come From? Our World in Data. Available online: https://ourworldindata.org/co2-emissions-from-transport (accessed on 22 November 2022).
2. Teter, J. IEA Transportation Report, Transport, IEA, Paris. 2022. Available online: https://www.iea.org/reports/transport (accessed on 22 November 2022).
3. European Commission. Transport and the Green Deal. Available online: https://ec.europa.eu/info/strategy/priorities-2019-2024/european-green-deal/transport-and-green-deal_en (accessed on 22 November 2022).
4. European Commission. REPower EU Plan. 2022. Available online: https://eur-lex.europa.eu/resource.html?uri=cellar:fc930f14-d7ae-11ec-a95f-01aa75ed71a1.0001.02/DOC_1&format=PDF (accessed on 22 November 2022).
5. International Energy Agency (IEA). Global EV Outlook. 2022. Available online: https://www.iea.org/reports/global-ev-outlook-2022 (accessed on 22 November 2022).
6. World Bank. World Bank Supports Maldives to Accelerate Renewable Energy Transition. 2020. Available online: https://www.worldbank.org/en/news/press-release/2020/12/12/world-bank-supports-maldives-to-accelerate-renewable-energy-transition (accessed on 22 November 2022).
7. E-Astypalea Services. Available online: e-astypalea.gov.gr (accessed on 22 November 2022).
8. Astypalea: Smart and Sustainable Island. Available online: https://www.astypalea-sustainable-island.gr/en/ (accessed on 22 November 2022).
9. Azores Exceed Incentives for Electric Mobility Allocated in 2021 in June. Available online: https://sustainable.azores.gov.pt/en/2022/07/28/azores-exceed-incentives-for-electric-mobility-allocated-in-2021-in-june/ (accessed on 25 November 2022).
10. Renewable Energy Empowering European & Indian Communities, REEmpowered EU Project. Available online: https://reempowered-h2020.com/ (accessed on 22 November 2022).
11. Clean Energy for EU Islands Secretariat. *Clean Energy Transition Agenda Aran Islands*; Clean Energy for EU Islands Secretariat: Brussels, Belgium, 2019.
12. Mathiesen, B.V.; Hansen, K.; Ridjan, I.; Lund, H.; Nielsen, S. *Samsø Energy Vision 2030 Converting Samsø to 100% Renewable Energy*; Aalborg University: Aalborg, Denmark, 2015.
13. Torabi, R.; Gomes, Á.; Morgado-Dias, F. Energy Transition on Islands with the Presence of Electric Vehicles: A Case Study for Porto Santo. *Energies* **2021**, *14*, 3439. [CrossRef]
14. Hatziargyriou, N.; Margaris, I.; Stavropoulou, I.; Papathanassiou, S.; Dimeas, A. Noninterconnected Island Systems: The Greek Case. *IEEE Electrif. Mag.* **2017**, *5*, 17–27. [CrossRef]
15. Stavropoulou, E. The Electrical Systems of the Greek Non-Interconnected Islands. In Proceedings of the 4th International Hybrid Power Systems Workshop, Heraklion, Greece, 22–23 May 2019.

16. Karfopoulos, E.L.; Voumvoulakis, E.M.; Hatziargyriou, N. Steady-state and dynamic impact analysis of the large scale integration of plug-in EV to the operation of the autonomous power system of Crete Island. In Proceedings of the MedPower 2014, Athens, Greece, 2–5 November 2014.
17. Stavropoulou, E. Smart Islands—Smart Networks. In Proceedings of the European Forum on Clean Energy for Islands Naxos, Naxos, Greece, 9–11 July 2018.
18. Gatta, F.M.; Geri, A.; Lauria, S.; Maccioni, M.; Palone, F.; Portoghese, P.; Buono, L.; Necci, A. Replacing Diesel Generators with Hybrid Renewable Power Plants: Giglio Smart Island Project. *IEEE Trans. Ind. Appl.* **2019**, *55*, 1083–1092. [CrossRef]
19. Vukojevic, A.; Lukic, S. Microgrid Protection and Control Schemes for Seamless Transition to Island and Grid Synchronization. *IEEE Trans. Smart Grid* **2020**, *11*, 2845–2855. [CrossRef]
20. Monteiro, V.; Lopes, J.P.; Moreira, C.; Afonso, J.L. New Operation Opportunities for the Solid-State Transformer in Smart Homes: A Comprehensive Analysis. In Proceedings of the IECON 2021—47th Annual Conference of the IEEE Industrial Electronics Society, Toronto, ON, Canada, 13–16 October 2021; pp. 1–6.
21. Monteiro, V.; Soares, T.; Lopes, J.P.; Matos, M.; Afonso, J.L. Interfacing Power Electronics Systems for Smart Grids: Innovative Perspectives of Unified Systems and Operation Modes. In Proceedings of the IECON 2021—47th Annual Conference of the IEEE Industrial Electronics Society, Toronto, ON, Canada, 13–16 October 2021; pp. 1–6.
22. Carrión, M.; Zárate-Miñano, R.; Domínguez, R. Integration of Electric Vehicles in Low-Voltage Distribution Networks Considering Voltage Management. *Energies* **2020**, *13*, 4125. [CrossRef]
23. Karakitsios, I.; Karfopoulos, E.; Hatziargyriou, N. Impact of dynamic and static fast inductive charging of electric vehicles on the distribution network. *Electr. Power Syst. Res.* **2016**, *140*, 107–115. [CrossRef]
24. Karakitsios, I.; Karfopoulos, E.; Hatziargyriou, N. Grid impact of static and dynamic inductive charging and its mitigation through effective management. In *Cable Based and Wireless Charging Systems for Electric Vehicles: Technology and Control, Management and Grid Integration*; IET: London, UK, 2021; Chapter 2. [CrossRef]
25. ENTSO-E. *Position Paper, Electric Vehicle Integration into Power Grids*; ENTSO-E: Brussels, Belgium, 2021.
26. Hatziargyriou, N.; Lopes, J.P.; Bower, E.T.; Strunz, K.; Rivier, M.; Lioliou, V.; Wu, J.; Papathanassiou, S.; Karfopoulos, E.; Willums, J.O.; et al. Impact of electric and plug-in hybrid vehicles on grid infrastructure—Results from the merge project. In Proceedings of the 44th International Conference on Large High Voltage Electric Systems, Paris, France, 26–31 August 2012.
27. Hatziargyriou, N.; Karfopoulos, E.; Tsatsakis, K. The impact of EV charging onthe System Demand. In *Electric Vehicle Integrationinto Modern Power Networks*; Spinger: New York, NY, USA, 2012; Chapter 3; ISBN 978-1-4614-0134-6.
28. Karakitsios, I.; Karfopoulos, E.L.; Hatziargyriou, N. Static and dynamic fast inductive charging: The FastInCharge project concept. In Proceedings of the MedPower 2014, Athens, Greece, 2–5 November 2014; pp. 1–6. [CrossRef]
29. Said, D.; Mouftah, H.T. A Novel Electric Vehicles Charging/Discharging Management Protocol Based on Queuing Model. *IEEE Trans. Intell. Veh.* **2020**, *5*, 100–111. [CrossRef]
30. Nizami, M.S.H.; Hossain, M.J.; Mahmud, K. A Coordinated Electric Vehicle Management System for Grid-Support Services in Residential Networks. *IEEE Syst. J.* **2021**, *15*, 2066–2077. [CrossRef]
31. Karakitsios, I.; Karfopoulos, E.; Hatziargyriou, N. An EV Management System Exploiting the Charging Elasticity of EV Users. *Int. J. Distrib. Energy Resour. Smart Grids* **2015**, *11*, 3.
32. Muñoz, E.R.; Razeghi, G.; Zhang, L.; Jabbari, F. Electric vehicle charging algorithms for coordination of the grid and distribution transformer levels. *Energy* **2016**, *113*, 930–942. [CrossRef]
33. Kim, G.; Hur, J. Methodology for Security Analysis of Grid- Connected Electric Vehicle Charging Station With Wind Generating Resources. *IEEE Access* **2021**, *9*, 63905–63914. [CrossRef]
34. Jeon, W.; Cho, S.; Lee, S. Estimating the Impact of Electric Vehicle Demand Response Programs in a Grid with Varying Levels of Renewable Energy Sources: Time-of-Use Tariff versus Smart Charging. *Energies* **2020**, *13*, 4365. [CrossRef]
35. Han, H.-S.; Oh, E.; Son, S.-Y. Study on EV Charging Peak Reduction with V2G Utilizing Idle Charging Stations: The Jeju Island Case. *Energies* **2018**, *11*, 1651. [CrossRef]
36. Mele, E.; Natsis, A.; Ktena, A.; Manasis, C.; Assimakis, N. Electromobility and Flexibility Management on a Non-Interconnected Island. *Energies* **2021**, *14*, 1337. [CrossRef]
37. Karfopoulos, E.L.; Hatziargyriou, N.D. Distributed Coordination of Electric Vehicles Providing V2G Services. *IEEE Trans. Power Syst.* **2015**, *31*, 329–338. [CrossRef]
38. Riesz, J.; Palermo, J. International Review of Frequency Control Adaptation Australian Energy Market Operator. 2016. Available online: https://www.google.com/url?sa=t&rct=j&q=&esrc=s&source=web&cd=&cad=rja&uact=8&ved=2ahUKEwiu6tKTqKr8AhVDVmwGHejPAvUQFnoECAgQAQ&url=https%3A%2F%2Faemo.com.au%2F-%2Fmedia%2FFiles%2FElectricity%2FNEM%2FSecurity_and_Reliability%2FReports%2F2016%2FFPSS-International-Review-of-Frequency-Control.pdf&usg=AOvVaw3JAH197RQAdR2xGQryRsML (accessed on 22 November 2022).
39. Meng, L.; Zafar, J.; Khadem, S.K.; Collinson, A.; Murchie, K.C.; Coffele, F.; Burt, G.M. Fast Frequency Response From Energy Storage Systems—A Review of Grid Standards, Projects and Technical Issues. *IEEE Trans. Smart Grid* **2019**, *11*, 1566–1581. [CrossRef]
40. Rodrigues, E.M.G.; Osório, G.; Godina, R.; Bizuayehu, A.; Lujano-Rojas, J.; Catalão, J. Grid code reinforcements for deeper renewable generation in insular energy systems. *Renew. Sustain. Energy Rev.* **2016**, *53*, 163–177. [CrossRef]
41. Aurecon. *Large-Scale Battery Storage Knowledge Sharing Report*; Australian Renewable Energy Agency: Canberra, Australia, 2019.

42. Lagos, D.; Kontou, A.; Kotsampopoulos, P. Protection in Island Systems Operating with High RES Penetration: Case Study Astypalea. In Proceedings of the CIGRE Session 2022, Paris, France, 28 August–2 September 2022.
43. Vournas, C.D.; Nikolaidis, V.C.; Tsourakis, G.I. Coordinated Countermeasures Against Overvoltage Instability in Autonomous Power Systems. *IEEE Trans. Power Deliv.* **2020**, *36*, 3329–3338. [CrossRef]
44. Gkiokas, G.; Lagos, D.; Korres, G.; Hatziargyriou, N.D. A hardware in the loop testbed for adaptive protection of non-inteconnected island systems with high RES penetration. In Proceedings of the 2022 International Conference on Smart Grid Synchronized Measurements and Analytics (SGSMA), Split, Croatia, 24–26 May 2022; pp. 1–7. [CrossRef]
45. Eleftherakis, K. Study of the Astypalea System in Cases of High EV and RES Penetration. Diploma Thesis, National Technical University of Athens, Athens, Greece, 2022. Available online: http://artemis.cslab.ece.ntua.gr:8080/jspui/bitstream/123456789/18273/1Ελευθεράκης_διπλωματική.pdf. (In Greek).
46. Suski, A.; Remy, T.; Chattopadhyay, D.; Song, C.S.; Jaques, I.; Keskes, T.; Li, Y. Analyzing Electric Vehicle Load Impact on Power Systems: Modeling Analysis and a Case Study for Maldives. *IEEE Access* **2021**, *9*, 125640–125657. [CrossRef]
47. Zafeiratou, E.; Spataru, C. Modelling electric vehicles uptake on the Greek islands. *Renew. Sustain. Energy Transit.* **2022**, *2*, 100029. [CrossRef]
48. Hatziargyriou, N. Impact of the electrification of the road transportation in the electricity grids. In Proceedings of the Conference of the Energy Committee of the Athens Academy, Athens, Greece, 2–3 February 2016; Available online: https://deddie.gr/el/kentro-enhmerwsis/nea-anakoinwseis/omilia-proedrou-akadimia-ahtinon/ (accessed on 22 November 2022). (In Greek).
49. Hatziargyriou, N. E-mobility for a clean future: Interaction of EV charging with the grid and future trends. In Proceedings of the TRI's 7th Annual Electric Vehicle Event 'Electric Vehicle Revolution Across the Globe', Online, 13–14 October 2021.

energies

Article

Optimal Placement and Sizing of Electric Vehicle Charging Infrastructure in a Grid-Tied DC Microgrid Using Modified TLBO Method

Nandini K. Krishnamurthy [1], Jayalakshmi N. Sabhahit [1,*], Vinay Kumar Jadoun [1], Dattatraya Narayan Gaonkar [2], Ashish Shrivastava [3], Vidya S. Rao [4] and Ganesh Kudva [1]

1 Department of Electrical & Electronics Engineering, Manipal Institute of Technology, Manipal Academy of Higher Education, Manipal 576104, Karnataka, India
2 Department of Electrical & Electronics Engineering, National Institute of Technology Karnataka, Surathkal 575025, Mangalore, India
3 Skill Faculty of Engineering and Technology, Shri Vishwakarma Skill University, Gurugram 122003, Haryana, India
4 Department of Instrumentation & Control Engineering, Manipal Institute of Technology, Manipal Academy of Higher Education, Manipal 576104, Karnataka, India
* Correspondence: jayalakshmi.ns@manipal.edu

Abstract: In this work, a DC microgrid consists of a solar photovoltaic, wind power system and fuel cells as sources interlinked with the utility grid. The appropriate sizing and positioning of electric vehicle charging stations (EVCSs) and renewable energy sources (RESs) are concurrently determined to curtail the negative impact of their placement on the distribution network's operational parameters. The charging station location problem is presented in a multi-objective context comprising voltage stability, reliability, the power loss (VRP) index and cost as objective functions. RES and EVCS location and capacity are chosen as the objective variables. The objective functions are tested on modified IEEE 33 and 123-bus radial distribution systems. The minimum value of cost obtained is USD 2.0250×10^6 for the proposed case. The minimum value of the VRP index is obtained by innovative scheme 6, i.e., 9.6985 and 17.34 on 33-bus and 123-bus test systems, respectively. The EVCSs on medium- and large-scale networks are optimally placed at bus numbers 2, 19, 20; 16, 43, and 107. There is a substantial rise in the voltage profile and a decline in the VRP index with RESs' optimal placement at bus numbers 2, 18, 30; 60, 72, and 102. The location and size of an EVCS and RESs are optimized by the modified teaching-learning-based optimization (TLBO) technique, and the results show the effectiveness of RESs in reducing the VRP index using the proposed algorithm.

Keywords: DC microgrid; electric vehicle charging station; fuel cell; renewable energy sources; teaching-learning-based optimization

Citation: Krishnamurthy, N.K.; Sabhahit, J.N.; Jadoun, V.K.; Gaonkar, D.N.; Shrivastava, A.; Rao, V.S.; Kudva, G. Optimal Placement and Sizing of Electric Vehicle Charging Infrastructure in a Grid-Tied DC Microgrid Using Modified TLBO Method. *Energies* **2023**, *16*, 1781. https://doi.org/10.3390/en16041781

Academic Editors: Michele Roccotelli, Evangelos Karfopoulos and Ioannis Karakitsios

Received: 9 December 2022
Revised: 30 December 2022
Accepted: 11 January 2023
Published: 10 February 2023

1. Introduction

The world transportation system is transforming from traditional gasoline to electric vehicles (EVs). A long-term charging infrastructure must be developed to deploy electric vehicles on a large scale. To satisfy long-term energy demands, countries are intentionally lowering their reliance on conventional energy and moving their emphasis to renewable sources. Choosing renewable sources of energy depends on a country's climate and environmental factors. A DC microgrid (MG) is an integrated energy infrastructure composed of sources of energy generation, storage options, and users of energy.

A two-stage joint chance-constrained program design is modeled for minimizing the operation cost of a multi-energy microgrid (MEMG) consisting of various energy carriers such as thermal power, electricity, and gas, which work independently [1]. However, until all vehicles are fully electrified, gas vehicles and electric vehicles (EVs) will exist in an MEMG. The MEMG's electric and gas sides should be run in tandem to meet their energy demands. It is proposed to use a mixed-integer linear procedure (MILP) [2] to decide the best operation planning for a DC-based EV charging station (EVCS). The approach reduces daily operational costs by predicting photovoltaic production and EV usage. The suggested MILP solution has been determined to be economical and immediately meet the system's time needs.

A multi-objective context is used to analyze the problem with charging station employment that considers cost, distribution network functioning constraints such as voltage stability, dependability, power loss, charging station availability, and waiting period in the charging station [3]. Teaching-learning-based optimization (TLBO) and chicken swarm optimization (CSO) were employed to address the placement of EVCSs. The optimal size and position of renewable energy sources (RESs) and EVCSs are instantaneously performed, and the handling vehicle charging method is delivered by [4]. A problem of multi-objective optimization is created to decrease power losses, voltage variations, demand and charging-providing outlays, and EV storage price. The performance on an IEEE 33-bus unit was studied to validate the efficacy of the new genetic algorithm-particle swarm optimization (GA-PSO) technique to optimize placement and capacity of RESs and EVCSs concurrently. The hybrid grey wolf optimization (GWO) and PSO technique are applied to determine the appropriate position of EVCSs, and distributed generations (DGs) in a stable IEEE 33- and 69-bus distribution unit analyze the consistency of the electrical distribution network [5]. The assistance for the MG's rate of recurrence through a control method has been provided in order to address the sudden change in EV load at the CS. A smart CS has been planned where EVs contain a variety of battery voltage ratings to offer frequency assistance to the MG. The MG load profile is maintained by controlled discharging and charging of EVs. EVs parked long term at the CS take part in the process of charging and discharging their batteries [6]. References [1–6] highlight the multi-energy microgrid, DC-based EVCSs, and smart charging station distribution network operating parameter optimization by considering RESs and EVCSs together. The placement and sizing of EVCSs and DGs are performed parallelly on test systems.

In [7–10], the effect of EVCSs on the distribution network is emphasized along with CS operational planning without any traffic network. The charging station's location and size are determined by factors such as solar energy variation, construction cost, and operator requirement. The EVCS's overall profit is maximized by considering both construction and operation costs. In [7], the effect of EVCS load on a distribution unit is described. The findings revealed that locating quick-charging stations near inefficient buses had an impact on the power distribution network's smooth operation. If quick-charging stations were put at the network's fragile buses, a significant amount of money would be lost. A strategy for the placement of EVCSs has been offered based on the voltage stability, reliability, and power loss index. The best placement and size of the CS in the town of Allahabad, India is performed in [8]. For optimal charging infrastructure placement in the Allahabad distribution unit, a hybrid approach centered on a GA and an upgraded form of predictable PSO is used. The significance of including protective device upgrades in charging station planning in urban systems is described in [9] to reduce the overall price of charging stations, extend the distribution network and voltage control, and upgrade costs for the safety system after placement of EVCS. A proposal is made for an EV supply infrastructure's best operation planning by [10]. The operation cost is optimized based on photovoltaic (PV) production and EV exploitation. The suggested method is tested on a DC MG on various functional days, and their efficacy is related using a genetic algorithm. An MG-based EV charging scheme and typical issues that arise during the charging of an EV, such as suspension operation and standby mode, are discussed in [11]. The proposed strategy is verified using MATLAB software.

The microgrid-based EVCS location problem is formulated in [11–14], considering both the distribution network and transport unit. The possibility of expanding EVs as a provisional energy storage system (ESS) contributing to structure requirements is investigated by [12]. To discover the amount of EVs that would minimize the total electricity obtained from the grid, an algorithm is suggested for charging and discharging. To demonstrate the proposed algorithm, a case study at Jordan University is used for the MG. The charging station sizing problem [13] is solved utilizing an optimization outline that reduces the asset price of CS operators while maintaining a given level of quality of service for EV owners. A two-stage planning model for charging infrastructure employment has been planned in view of the distance, grid stability, and traffic [14]. The places with the least amount of energy loss are determined. The smart DG is also allotted for additional energy loss decline and voltage enhancement. To cope with the uncertainties connected with EVs and SDGs, the 2 m point estimation method has been used [15]. The expenses of conveyance loss, accumulation, and substation energy, as well as power loss, are factored into optimal rapid CS design using a unique approach. To overcome the optimization challenge, the binary lightning search algorithm is developed [16]. The four different reliability indices are assessed on an IEEE 33-bus test system by increasing the load in equal and unequal mode on every node. An unequal increase in load is more suitable than an equal increase. As all the indices remain at an acceptable range of values with the increase in load, the system operation is considered as healthy and reliable [17]. References [15–17] describe EVCS operation strategy and placement considering uncertainties associated with EVs and DGs.

A reduction in the overall cost of microgrid operations is achieved. To reduce the cost of energy for consumers, the idea of dynamic pricing has also been implemented. A load variance index has been taken into account to ensure the stability of the microgrids, and the fuzzy-based technique has been employed for cost and load variance minimization to lower operating costs without sacrificing the stability of the MG. The grid-to-vehicle (G2V) and vehicle-to-grid (V2G) operations of EVs are incorporated, which will aid in shifting loads during off-peak and peak hours [18]. A methodology for sizing a DC-MG for optimal techno-economic coverage of EV mobility needs is developed. In each case, the best operation is achieved with a unique approach to EV commitment at various stations. The sizing technique is capable of handling MG device's modular structure [19]. Different constraints are used to mathematically express the economic and environmental concerns. The multiple charging scenarios of EVs are implemented by multiagent reinforcement learning (RL). The efficacy of RL has been proved by integrating five and fifteen thermal producing units, which make up the two conventional test functions [20].

The optimal placement of the EVCS with synchronized charging within the MG at Wroclaw University of Science and Technology is performed using ant colony optimization (ACO). The chosen place was determined by grouping and positioning the EVCS on various MG nodes made up of distributed energy sources and analyzing the energy reserves calculated on typical days of the year [21]. The placement and sizing of the EVCS is performed on a distribution system in Allahabad, India. The voltage profile enhancement index and active and reactive power loss minimization index are optimized, as well as the the initial development costs, to obtain the lowest possible installation cost and to deliver better power grid parameters. The proposed problem has been solved by a balanced mayfly algorithm [22]. The suitable placement and sizing of PV and wind energy sources along with EVs are determined using a multi-objective grey wolf optimizer. The uncertainty of PV, wind generation, state of charge, and arrival and exit hours of EVs are taken into account. The total voltage deviances and grid power are optimized under four different cases of operation [23]. The EVCS placement problem is performed on an IEEE 33-bus system based on the reliability index. The present approach is validated on an actual system of Kerala, India [24]. A linear program model is developed to reduce the costs associated with network reinforcement and energy loss and determine which distribution network branches need to be strengthened. The created models are put into practice on the IEEE 69-bus network, and the outcomes are looked at under various scenarios [25]. To assure the autonomous and reliable operation of the DC microgrid, an improved decentralized droop control with load power dispatch and SOC balance of the energy storage unit is developed. The developed control strategy is validated in the Simulink platform [26]. References [18–26] highlight

EVCSs integrated with the grid, MG-based EVCSs concerned with environmental factors, the distribution system loss minimization index, the uncertainty of PV and wind sources, the reliability index, and network reinforcement.

Table 1 gives the comparison of current work with the literature.

Table 1. Comparison of literature review.

Reference Number	Technique Applied	Remarks
[3]	CSO-TLBO	Considering voltage stability, reliability, power loss, the availability of charging stations, and the waiting time within the charging station as functional constraints, EVCS is positioned optimally.
[4]	GA-PSO	Optimal placing and sizing of RESs and EVCSs are performed on IEEE 33 test system to decrease power losses, voltage variations, and EV storage price.
[5]	GWO-PSO	The placement of EVCSs and DGs on IEEE 33- and 69-bus systems is performed to verify the reliability of the distribution network.
[7]	GA	The placement of EVCSs has been implemented on IEEE 33-bus system to optimize the VRP index of the distribution system.
[8]	GA-PSO	The installation of EVCSs in the Allahabad distribution unit is performed considering protective device upgrade cost.
[13]	Monte Carlo Simulations (MCS)	Charging station (CS) sizing is performed with smart charging competencies during operation to lessen the investment cost of charging station operators.
[14]	CSO-TLBO	The CS placement problem is solved by considering the VRP index, cost, waiting time, and accessibility index as objective functions. Two-stage planning exemplar is authenticated on IEEE 33-bus system and practical network in Tianjin, China.
[15]	Harris Hawk Optimization (HHO)	The CS and solar-based DG placement is performed on 33-bus systems to reduce losses with energy and improvement of the voltage profile.
[18]	GWO	The effect of EV placement on the 69-bus test system has been analyzed to enhance the voltage profile and minimize power losses.
[19]	Scenario based operation	Techno-economic sizing of a DC MG is performed, and EV commitment is developed.
Present paper	Modified TLBO	Optimal placement and sizing of EVCSs and RESs are performed simultaneously to optimize the cost and VRP index on IEEE 33- and 123-bus test units in a DC MG environment.

The history of the present work on placement and sizing of CSs is given in Table 2. It should be noted that the majority of the research has focused on economic constraints and distribution network operating parameters separately. This paper works on placement and sizing of CSs along with RESs, considering both cost and VRP index simultaneously on an IEEE 33- and 123-bus distribution system.

RESs are eco-friendly and assist in reducing the use of fossil fuels. Wind and PV are unreliable sources of electricity, with output varying depending on the amount of wind and sunshine available. Wind and PV fluctuate; hence, this necessitates a large amount of storage, which provides many possibilities for the fuel cell. It is very important to meet the required power demand of EVCSs and to reduce the burden on the utility grid. Hence, three different sources, PV, wind, and fuel cell, are used in the proposed DC MG. Large-scale EV incorporation and the location of EV charging and discharging stations present additional challenges to the power system unit, such as the VRP index. Evaluating the effect of EVCSs on the VRP index of grids and MGs is critical.

Table 2. History of the present topic.

Author	Approach	Objective Function
Hu et al. (2013)	Single-objective	Cost is optimized focusing solely on the distribution unit.
Zheng et al. (2014)	Single-objective	Cost is optimized focusing solely on the distribution unit.
Pazouki et al. (2015)	Multi-objective	Cost, power loss, and voltage profile are optimized on distribution unit.
Marcelo et al. (2015)	Multi-objective	Active power loss optimization considering distribution network.
Pan et al. (2016)	Multi-objective	Cost and momentary power supply index are optimized on distribution network.
Aljanad et al. (2018)	Multi-objective	Voltage deviation and power loss considering distribution network.
Zhang et al. (2018)	Multi-objective	Charging station construction cost, discontented charging demand penalty, and cost of energy considering both distribution and transportation network.
Rahmani Andebili et al. (2018)	Multi-objective	Cost, power loss, and unmet energy expectations considering both distribution and transportation network.
S. Deb et al. (2019)	Multi-objective	Cost, accessibility index, VRP index, and wait time considering both distribution and transportation network.
Sachan, Sulabh et al. (2021)	Multi-objective	Cost and VRP index with coordinated and uncoordinated charging considering both distribution and transportation network.

According to the existing literature, the location and sizing of the charging station are organized using various optimization strategies. Because of the widespread focus on RESs, it is vital to assess the effect of DC MG in terms of wind, PV, and fuel-cell-based EVCSs integrated with the grid. In the extant literature, there is little emphasis on optimizing the location and sizing of EVCSs integrated with the grid in a DC MG environment. The principal objective of this work is to decrease the VRP index of the 33- and 123-bus test systems after installing EVCSs with and without RESs. The cost of installation and operation of EVCSs are also optimized in this work. The placement and size of RESs and EVCSs are taken as the objective variables. The present problem is made more difficult by the non-linear properties of EVs when combined with DC MG. An efficient nature-inspired meta-heuristic modified TLBO technique is used to handle this challenging problem. The novelty of the work lies in the optimum placement and sizing of EVCSs with RESs in a DC MG environment. The present approach is tested on an extensive network of IEEE 123- and medium-scale 33-bus system with the objective of VRP index and cost optimization.

The key contributions of this work are as follows:

- A new problem formulation is proposed for a grid-connected DC microgrid to optimize the cost and VRP index on modified IEEE 33- and 123-bus radial distribution units after installing EVCSs.
- The proposed case of operation is performed on a DC MG integrated with the utility grid for optimal position and sizing of EVCSs with RESs under six diverse schemes to minimize the VRP index and cost with the related constraints. The minimization of the VRP index and cost is compared with the latest published results.
- The VRP index is optimized to account for 24 h variations in PV, wind, and EV loads.
- The operation strategy is proposed to minimize the cost and VRP index by performing distribution load flow on a DC microgrid. The proposed operation strategy helps to improve the voltage profile by the optimal location of RESs and EVCSs on modified IEEE test systems.
- Modified meta-heuristic optimization technique TLBO given numerous relevant constraints with the proposed operation strategy is employed to resolve this composite problem.
- The investigation of modified TLBO for the proposed operation strategy produces solutions of improved quality, which is evident in the numerical outcomes.

The remainder of the paper is systematized as follows. DC MG components are modeled in Section 2. The operational strategy, problem formulation, and objective function are detailed in Section 3. Section 4 illustrates the proposed optimization strategy. Section 5 outlines the results of the proposed case, and the complete work is concluded in Section 6.

2. Modeling DC MG Components

Figure 1 depicts the structure of a DC MG that includes wind, solar PV, fuel cells, EVCS, and the utility grid. Wind and PV are the prime energy sources in the DC MG, with the fuel cell as the backup source. Solar PV and the fuel cell produce DC electrical power, whereas the wind turbine produces electricity in the form of AC. The total amount of generated electricity by the PV system and wind turbine depends on many aspects, such as sun irradiance, ambient temperature, and wind speed, which are variable and unpredictable. All three sources are linked to the DC bus via a DC-DC type converter. The DC bus is interlinked to the conventional grid and EVCS through a bidirectional DC-AC inverter and DC-DC type converter, respectively.

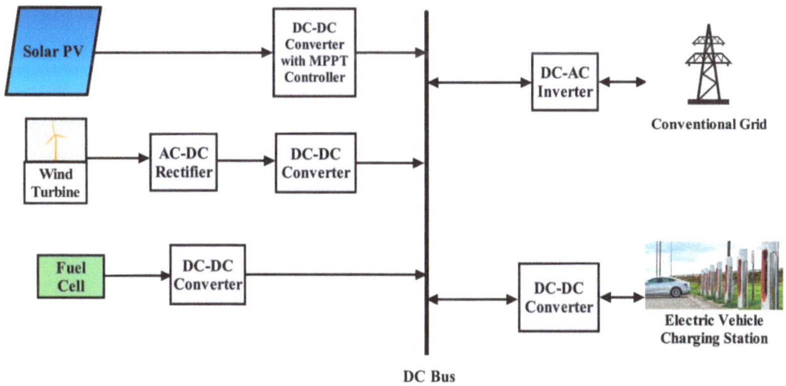

Figure 1. Structure of DC MG integrated with EVCS.

2.1. Solar PV

Solar power is essential and influential in addressing the energy issues of any nation. Before installing PV modules in any area, the modeling of the same helps diagnose the performance of PV in real-world climate conditions [27]. A photovoltaic array collection contains several solar cells to achieve the necessary output current and voltage [28]. In the present work, solar and wind energy are used as primary sources of supply. The fuel cell is used as a backup source of supply. With drastic variations in energy production, on some days when solar energy production is very low the required load demand is met through wind energy and the fuel cell. If all three sources fail to meet the required EVCS load demand, then the utility grid is used to supply the deficit power. Hence, the proposed method works properly irrespective of solar radiation and wind speed.

The power output of the PV system for each hour (t) can be expressed as:

$$PV_{out}(t) = FF(t) \times I_{sc}(t) \times V_{oc}(t) \tag{1}$$

$$I_{sc}(t) = \frac{s}{1000}\left[I_{sc,\ STC} + K_i(T_c(t) - 25)\right] \tag{2}$$

$$V_{oc}(t) = V_{oc,\ STC} + K_v(T_c(t) - 25) \tag{3}$$

$$FF(t) = FF_o(t) \times [1 - r_s(t)] \tag{4}$$

$$FF_o(t)\ \frac{V_{oc,0}(t) - ln\left[V_{oc,0}(t) + 0.72\right]}{V_{oc,0}(t) + 1} \tag{5}$$

$$V_{oc,0}(t) = V_{oc,0}(t) \times \frac{q}{nk[T_c\,(t) + 273.5]} \quad (6)$$

$$r_s(t) = R_s\,\frac{I_{sc\,(t)}}{V_{oc\,(t)}} \quad (7)$$

$$T_c(t) = T_a(t) + s\frac{NOCT - 20}{0.8} \quad (8)$$

where, $I_{sc}\,(t)$ and $V_{oc}(t)$ are the short circuit current and open circuit voltage under working circumstances and $PV_{out}(t)$ is the highest output power at any given time t. The sun irradiance at time t is represented as s. The module's actual and ideal fill factors are FF and FF_0. $V_{oc,0}(t)$ represents the regulated open circuit voltage at any time t, q represents an electron's charge. The ideality factor n is assumed to be equal to 1. Boltzmann's constant is K, R_s is the series resistance of the module, temperature of the modules at any time t is T_c. (t). $T_a(t)$ is the module ambient temperature, $r_s(t)$ is the standardized module series resistance at any time t, and $NOCT$ is the manufacturer's specified rated operating cell temperature [29,30]. In the present work, the parameters for the PV module are adapted from reference [31].

2.2. Wind Energy System

Wind power is considered one of the utmost rapidly growing RESs. Based on various types of wind zones, the wind turbines are installed, generating revenue in rural areas and consolidating the region's economic status. The hybridized scheme in which both PV and wind are combined is indeed a potential alternative, because when either system is not working at its peak, the other can compensate. Apart from that, it keeps on supplying power without any disruptions with reduced storage system size [32]. By using PMSG, variable speed operation is achieved. One of the benefits of WT functioning at multiple speeds with PMSG is that it considerably cuts the necessary design parameters, as it comes with fixed losses from the generator. In the power transformation process, iron losses (P_{iron}), mechanical losses (P_{mec}), and copper losses (P_{cu}) occur in the wind rotor. Inclusive of all the losses, PMSG (η_{con}) has a conversion efficiency of about 90% [33]. The mechanical power generated from the wind rotor at each hour (t) is expressed as in (9) adapted from [34].

$$P_m(t) = 0.5\rho A[v(t)]^3 C_p\,(\lambda, \beta) \quad (9)$$

The power coefficient $C_p\,(\lambda, \beta)$ is expressed as:

$$\mathbf{C_p}\,(\lambda, \boldsymbol{\beta}) = 0.22\left(\frac{\mathbf{116}}{\gamma} - \mathbf{0.4} * \boldsymbol{\beta} - \mathbf{5}\right).\exp\left(-\frac{\mathbf{12.5}}{\gamma}\right) \quad (10)$$

$$\frac{\mathbf{1}}{\gamma}\quad \frac{\mathbf{1}}{\lambda + \mathbf{0.089}}\quad \frac{\mathbf{0.035}}{\boldsymbol{\beta}^3 + \mathbf{1}} \quad (11)$$

$$\lambda = \frac{\mathbf{R}\,.\;\omega}{\boldsymbol{\vartheta}} \quad (12)$$

The torque acting on the shaft is calculated as:

$$\mathbf{T_m} = \frac{\mathbf{p_m}}{\omega} \quad (13)$$

The power output extracted from the PMSG is:

$$P_{wind}(t) = \mathbf{P_m(t)} * \eta_{con} \quad (14)$$

where ρ is the density of air, wind speed is referred to as v in m/sec, A is the rotor area, and Cp is the coefficient of performance. λ is the tip speed fraction, γ is the specific weight, and ß is the pitch angle. ω is the blade's angular speed and R is the rotor radius. The selected WT specification in the particular DC MG is mentioned in [35].

2.3. Fuel Cell

A fuel cell is a type of energy conversion method that directly changes chemical energy to electrical energy. Fuel cells are an excellent choice for applications requiring dispersed power production. A fuel cell cannot be utilized directly in most applications since it produces only unregulated DC power that varies with load. The Nernst expression used for determining cell potential is given as [36,37]:

$$E = \boldsymbol{E^0} - \frac{\boldsymbol{RT}}{\boldsymbol{nF}} log\, Q \tag{15}$$

Equation (15) represents a cell's electrical potential, which is determined by the reaction quotient Q, where E^0 denotes open circuit voltage, R is the collective gas constant, F is Faraday's number, and T refers to the absolute temperature of the cell in ^oK. When a load is functional, the external current (I_{ext}) flows cause the voltage drips with a current loss (I_{loss}). The overall current of the fuel cell is given as

$$I = I_{ext} + I_{loss} \tag{16}$$

$$\begin{aligned} V_{stk}(I) = \eta_{cell} V_{cell}(I) = \eta_{cell}(E - \Delta V_{act} - \Delta V_{ohm} - \Delta V_{con}) \\ = \eta_{cell}(E - [\xi_1 + \xi_2 T_s + \xi_3 T_s \ln(C_{O2}) + \xi_4 T_s \ln(I)] - R_{ion} I - (1 + \tfrac{1}{\alpha}) \tfrac{RT}{4F} \ln \tfrac{I_L}{I_L - I}) \end{aligned} \tag{17}$$

where η_{cell} is the number of cells linked in series, taken as 30, ΔV_{act} is the activation voltage losses in volts, and ΔV_{ohm} is the ohmic voltage losses in volts. The losses occur due to a drop in gas absorption at the electrode surface where there is a large current ΔV_{con}. T_s is the stack temperature in k. Co_2 is the amount of liquefied oxygen in mol/cm^3. R_{ion} is the ion flow in the membrane with resistance. I_l is the current limit of the electrode in A. α is a water activity function.

The fuel cell stack's output power is provided as

$$P_{stk} = V_{stk} I_{ext} \tag{18}$$

3. Problem Formulation

The multi-objective problem includes the operation and installation cost of the EVCS and VRP index as objective functions by optimally placing and sizing the PV, wind, and fuel cell. The selected problem formulation considers wind and PV as the primary source and the fuel cell as the secondary source; the size and place of RESs and EVCSs are taken as the objective variables.

3.1. Objective Function and Related Constraints

The objective function and the related constraints of the present work are discussed as follows:

3.1.1. Cost

The objective is to lessen the installation and operation cost of EVCSs, shown in (19)–(24), as illustrated in [3].

Considering both fast and slow charging stations:

$$Cost = C_{installation} + C_{operation} \tag{19}$$

$$C_{installation} = \left\{\sum_{j=1}^{m} F_j \times f_j \times \left(C_{fast}\right)\right\} + \left\{\sum_{j=1}^{m} S_j \times s_j \times (C_{slow})\right\} \tag{20}$$

$$C_{operation} = \left\{\left\{\sum_{j=1}^{m} F_j \times f_j \times \left(CP_{fast}\right)\right\} + \left\{\sum_{j=1}^{m} S_j \times s_j \times (CP_{slow})\right\}\right\} \times P_{elec} \tag{21}$$

Considering only a fast-charging station:

$$Cost = C_{installation} + C_{operation} \tag{22}$$

$$C_{installation} = \left\{ \sum_{j=1}^{m} F_j \times f_j \times \left(C_{fast} \right) \right\} \quad (23)$$

$$C_{operation} = \left\{ \left\{ \sum_{j=1}^{m} F_j \times f_j \times \left(CP_{fast} \right) \right\} \right\} \times P_{elec} \quad (24)$$

where S_j and F_j are the quantity of slow and fast charging infrastructures, and f_j and s_j are the number of fast charging and slow charging points, respectively. C_{slow} and C_{fast} are the installation costs of slow and fast charging infrastructures, respectively. CP_{slow} and CP_{fast} are the power intake of slow and fast charging infrastructures. The cost per unit of electricity is represented as P_{elec}. The cost of installation and operation depends on the quantity of slow and fast charging infrastructures, the quantity of slow and fast charging points, the power intake of chargers, and the electricity cost per unit.

3.1.2. VRP (Voltage Stability, Reliability, Power Loss) Index

The operating considerations of the distribution unit, such as power loss, reliability, and voltage stability, are together considered in one frame. The VRP index has a minimum value of 1 when there is no increase in load, which is considered as an ideal condition. The VRP index is represented mathematically as follows [3].

$$VRP = f\left(P, F_j, S_j, f_j, s_j\right) = w_1 \times VSI + w_2 \times R + w_3 \times P \quad (25)$$

Equation (25) represents a multi-objective function where the decision variables are P, F_j, S_j, f_j, and s_j. P is the placement of the charging infrastructure. w_1, w_2, and w_3 are the weight allotted to VSI, R, and power loss, respectively. VSI represents voltage stability index, R is reliability, and P is power loss, where

$$VSI = \frac{VSI_l}{VSI_{base}} \quad (26)$$

$$R = W_{21}\frac{SAIFI_l}{SAIFI_{base}} + W_{22}\frac{SAIDI_l}{SAIDI_{base}} + W_{23}\frac{CAIDI_l}{CAIDI_{base}} \quad (27)$$

$$P = \frac{P^l_{loss}}{P^{base}_{loss}} \quad (28)$$

Equation (26) represents VSI_{base} and VSI_l as voltage stability index base value and the voltage stability after the placement of charging infrastructure, respectively. $SAIFI_{base}$ and $SAIFI_l$ are the $SAIFI$ base value and the $SAIFI$ value after the employment of charging stations, respectively. The same convention is followed for $SAIDI$ and $CAIDI$. w_{21}, w_{22}, and w_{23} are the weight assigned to $SAIFI$, $SAIDI$, and $CAIDI$, respectively in (27). P^l_{loss} and P^{base}_{loss} are the power loss value after the assignment of the charging station and the power loss value for the base case, as depicted in (28).

The voltage stability index (VSI) before and after employment of the charging infrastructure is given in (29) and (30).

$$VSI_{base} = \sum_{j=1}^{N_D} 2\, V_j^2\, V_{j+1}^2 - 2V_{j+1}^2\left(P_{j+1}r_j + Q_{j+1}x_j\right) - |Z|^2\left(P_{j+1}^2 + Q_{j+1}^2\right) \quad (29)$$

$$VSI_l = \sum_{j=1}^{N_D} 2\, V_j'^2 V_{j+1}'^2 - 2V_{j+1}'^2\left(P'_{j+1}r_j + Q'_{j+1}x_j\right) - |Z|^2\left(P_{j+1}'^2 + Q_{j+1}'^2\right) \quad (30)$$

Here, the quantity of buses in the distribution system is denoted by ND, and Vj and V_j' are the voltage of the jth bus for the base case and the voltage after the employment of charging station (CS), respectively. P_{j+1} and P'_{j+1} are the active power at the (j + 1)th bus before and after the positioning of CSs. The same convention is followed for reactive power (Q). The impedance of the branch is represented as Z, the resistance of the branch is represented as r_j, and reactance of the branch is represented as x_j.

The important reliability indices, i.e., system average interruption duration index (SAIDI), system average interruption frequency index (SAIFI), and consumer average interruption duration index (CAIDI), are expressed in (31) and (32).

$$\left.\begin{aligned} SAIFI_{base} &= \frac{\sum_{j=1}^{N_D} \lambda_j N_j}{\sum_{j=1}^{N_D} N_j} \\ SAIDI_{base} &= \frac{\sum_{j=1}^{N_D} U_j N_j}{\sum_{j=1}^{N_D} N_j} \\ CAIDI_{base} &= \frac{\sum_{j=1}^{N_D} U_j N_j}{\sum_{j=1}^{N_D} \lambda_j N_j} \end{aligned}\right\} \quad (31)$$

$$\left.\begin{aligned} SAIFI_{l} &= \frac{\sum_{j=1}^{N_D} \lambda'_j N_j}{\sum_{j=1}^{N_D} N_j} \\ SAIDI_{l} &= \frac{\sum_{j=1}^{N_D} U'_j N_j}{\sum_{j=1}^{N_D} N_j} \\ CAIDI_{l} &= \frac{\sum_{j=1}^{N_D} U'_j N_j}{\sum_{j=1}^{N_D} \lambda'_j N_j} \end{aligned}\right\} \quad (32)$$

$$\left.\begin{aligned} \lambda'_j &= \frac{\lambda_j}{P_j} \times P'_j \\ U'_j &= \frac{U_j}{P_j} \times P'_j \end{aligned}\right\} \quad (33)$$

where λ_j and λ'_j are the failure rate of bus j before and after placement of CSs, U_j and U'_j are the outage period of bus j before and after employment of CSs, respectively, and N_j is the number of users associated at bus j. P_j and P'_j are the active power at the bus 'j' before and after the employment of charging infrastructures, respectively, as seen in (33).

The computation of power loss before and after the employment of CSs is expressed as:

$$\left.\begin{aligned} P_{loss}^{base} &= \sum_{j=1}^{N_D} I_j^2 r_j \\ P_{loss}^{l} &= \sum_{j=1}^{N_D} I'^2_j r_j \end{aligned}\right\} \quad (34)$$

where I_j and I'_j are the current through branch j before and after employment of CSs, as presented in (34). It can be observed that the VRP index depends on the position of charging infrastructure, the type of CSs, the quantity of CSs, and the quantity of charging points. The SAIFI is considered a frequency-based reliability index that relies on interruption frequency. SAIDI is a time-based dependability file that is affected by the time of the outage. The unitary technique calculates the frequency and duration of interruptions when charging stations are installed [3]. The best locations and sizes of the RESs must be determined centered on the same objective to lessen the VRP index further.

The increase in load by considering both fast and slow charging stations is described in (35) and (36).

$$P'_j = P_j + \{(F_j \times f_j) \times CP_{fast}\} + \{(S_j \times s_j) \times CP_{slow}\} \quad (35)$$

Considering only fast charging station:

$$P_j' = P_j + \{(F_j \times f_j) \times CP_{fast}\} \tag{36}$$

To carry out the sizing of EVCSs and their optimal location, the following constraints are listed in (37)– (41), considering both the highest and lowest amount of fast charging points, slow charging points, true and reactive power of the bus system, the voltage of the bus, and the output power of all the renewable energy sources. Equations (42) and (43) are the demand–supply balance constraint for each bus after counting EVCSs and RESs.

$$0 < F_p \leq F_{max}; \qquad 0 < f_p \leq f_{max} \tag{37}$$

$$0 < S_p \leq S_{max}; \qquad 0 < s_p \leq s_{max} \tag{38}$$

$$Q_j^{min} \leq Q_j \leq Q_j^{max}; \qquad P_j^{min} \leq P_j \leq P_j^{max} \tag{39}$$

$$V_j^{min} \leq V_j \leq V_j^{max} \tag{40}$$

$$P_{min}^{RES} \leq P_j^{RES} \leq P_{max}^{RES} \tag{41}$$

$$P_{gi} - P_{di} - V_i \sum_{j=1}^{N} V_j \, Y_{ij} \, Cos\,(\delta_i - \delta_j - \theta_{ij}) = 0 \tag{42}$$

$$Q_{gi} - Q_{di} - V_i \sum_{j=1}^{N} V_j \, Y_{ij} \, Cos\,(\delta_i - \delta_j - \theta_{ij}) = 0 \tag{43}$$

where P_{gi} and Q_{gi} are the active and reactive power production of the ith bus and P_{di} and Q_{di} are the active and reactive power demand, respectively. V_j is the voltage of jth bus, and Y_{ij} is the magnitude of bus admittance matrix. $\delta_{(i)}$ and $\delta_{(j)}$ are the voltage angle of the ith and jth bus. θ_{ij} is the angle of Y_{ij}.

3.2. Proposed Operational Strategy

It is more economical to indicate the location and sizing of RESs along with EVCSs to minimize the VRP index further. A comprehensive explanation has been stated in a flowchart in Figure 2. The steps followed for the proposed operational strategy are highlighted:

Step 1: Identify the preliminary value $i = 1$.
Step 2: Select the modified IEEE 33- and 123-bus system for the analysis of the VRP index.
Step 3: Figure out the value of wind, PV, and fuel cell output power using (1) to (4).
Step 4: Run the load flow for the base case by the forward-backward technique.
Step 5: Estimate the voltage sensitivity factor (VSF) for all buses using (44) [38].

$$VSF = |dv/dp| \tag{44}$$

Step 6: Based on the VSF, the CS is optimally selected. Increase the charging station load in phases.
Step 7: Run distribution load flow analysis by the forward-backward method.
Step 8: Optimize the VRP index and cost.
Step 9: The wind, PV, and fuel cell output power is integrated into the modified IEEE 33- as well as 123-bus system and the EVCS load.
Step 10: Run a forward-backward load flow method.
Step 11: To minimize the VRP index further, the location and sizing of RESs are optimally selected.

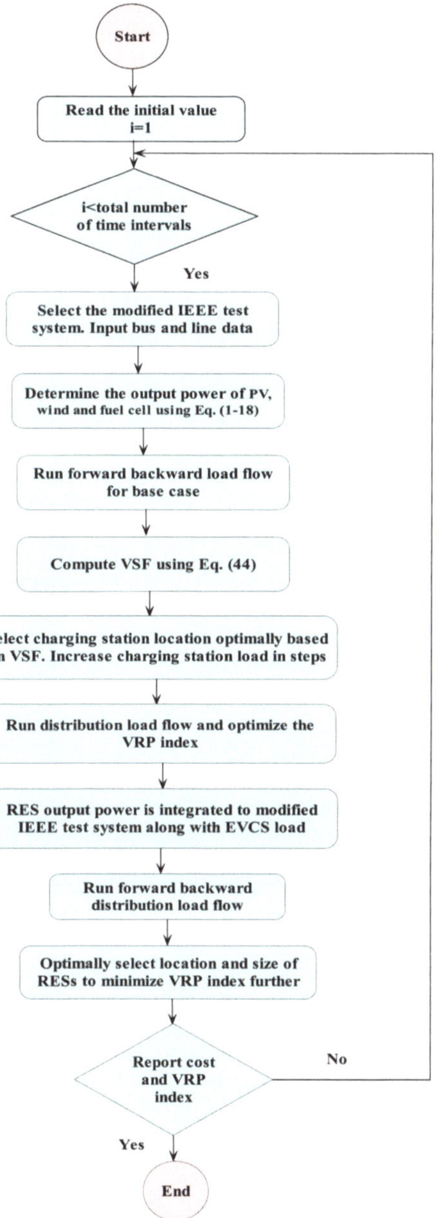

Figure 2. Proposed operation strategy.

4. Optimization Algorithm Based on Modified Teaching-Learning Method

The formulated problem in the preceding section is comprehensive, with multifaceted optimization. Hence, it is crucial to opt for an effective optimization technique. This work uses a modified TLBO to resolve the aforementioned complex problem in two diverse operation cases. The modified TLBO algorithm combines basic TLBO with suitable constraints handling as described in Section 3. The comprehensive discussion of the modified TLBO method is given as follows.

The basic version of the TLBO is a nature-inspired and population-centered method that imitates the teaching method followed by the learning phase. In the learning phase, two approaches are involved. Starting with the teacher phase, interaction takes place between both teacher and student. Then, in the learner phase, communication is established amongst the learners. The population size is selected based on a random number of students. The finest learner P_{best} from the unique population set is obtained and chosen as a teacher factor from the best fitness value. The variance between the P_{best} and the mean result of group, that is, P_{mean}, is applied to sort the newest response as per Equation (45), taken from [39].

$$P_{i,\ new} = P_{i,\ old} + r_i\ (P_{i,best}\ - T_F\ \times\ P_{i,mean}) \tag{45}$$

The arbitrary number 'r_i' is produced in the range [0, 1] in the ith iteration, and teaching factor T_F is selected arbitrarily between 1 and 2. During the learning phase, a learning procedure is followed by a student that includes debate between themselves, and a chance studying companion who is more knowledgeable P_j is picked to apprise the new set of solutions. The newest solution is acceptable if the learner P_i has more awareness than his or her counterpart P_j as per (46), and equally, the solution is updated by (47), taken from [40].

$$P_{i,\ new} = P_{i,\ old} + r_i\ (P_i\ - P_j) \tag{46}$$

$$P_{i,\ new} = P_{i,\ old} + r_i\ (P_{i,best} - T_F\ \times\ P_{i,mean}) \tag{47}$$

The modified TLBO is performed by choosing the size of the population 50 and 500 as the number of generations. Because 'r_i' and '*TF*' are not the specific parameters of TLBO, their values do not need to be tweaked like the mutation and crossover parameters in GA or inertial weight in PSO. The proposed problem formulation is complex and non-linear in nature. Hence, to solve such a complex and multi-objective problem, the modified TLBO technique is employed.

The advantages of TLBO with other population-based techniques such as GA, PSO, Harmony Search (HS), and Artificial Bee Colony (ABC) are three methods that combine a number of alternatives to reach the best one. PSO needs learning elements, weight variation, and maximum velocity, while GA requires crossover probability, selection procedure, and mutation rate. ABC requires the maximum value, while HS demands the rate at which harmony memory, rate of pitch adjustment, and number of improvisations are taken into account. The implementation of TLBO is easier since, unlike other optimization approaches, it does not require any algorithm factors to be tuned. Similar to PSO, TLBO modifies the current solution using the best iteration's outcome for the population, thereby raising the convergence rate. TLBO does not split the population as happened in ABC optimization. The process of GA relies on the three crucial phases, viz., selection, crossover, and mutation. Likewise, in ABC, onlooker and scout bees are employed. On the contrary, TLBO uses dual phases, i.e., the 'teacher phase' and the 'learner phase'. TLBO utilizes the mean value of the population to improvise the solution. The quality solution obtained by TLBO involves the greediness to reach a good candidate solution [41]. The workflow of modified TLBO is depicted in Figure 3. There are no specified control parameters mandatory for this algorithm, so it can be more popularly used compared to other population-based algorithms [42]. For it to work, it only needs a few basic governing parameters like population size and generation number [41]. If the quantity of generations attains the maximum, stop; else, reinitialize from the teacher phase. Table 3 validates the modified TLBO optimization with other procedures well-known in the literature.

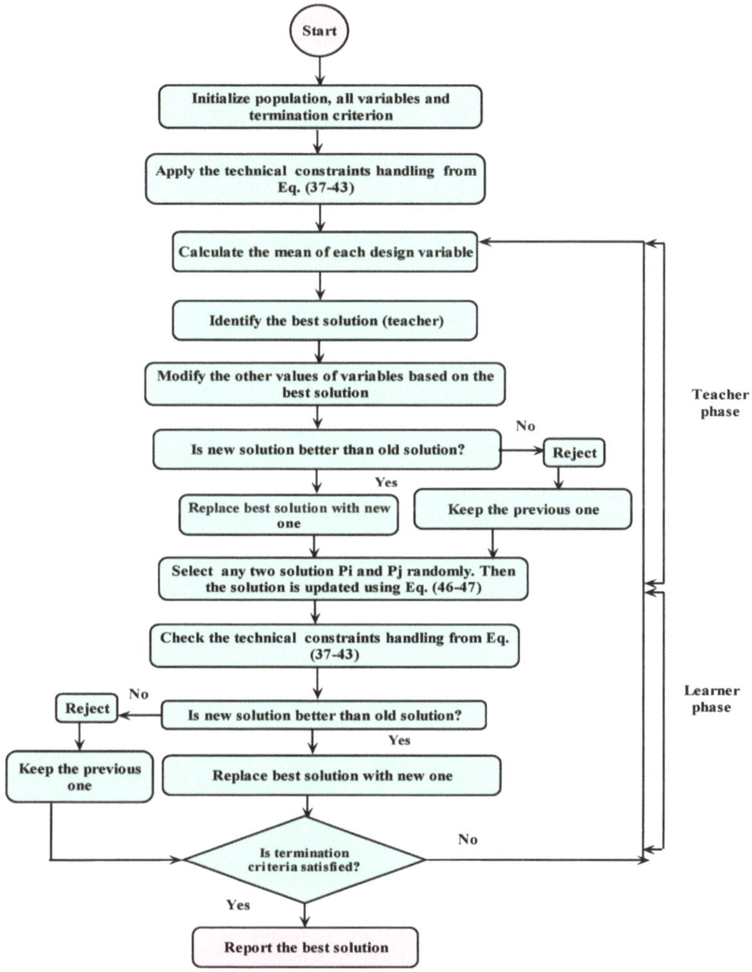

Figure 3. Flowchart of modified TLBO.

Table 3. Comparison of modified TLBO algorithm with other optimization methods.

Reference No.	Type of Optimization Algorithm	Feature of Optimization Algorithm
[4]	CSO—TLBO	Bio-inspired algorithm.
[5]	HGWOPSO	Has better optimization performance.
[9]	GA PSO	Provides better solution with faster convergence.
[16]	BLSA	Performs in terms of search precision and convergence in multidimensional search space.
Present work	Modified TLBO	Controls parameters specific to the algorithm are not required.

5. Results and Discussion

Determination of the optimal sizing and location of EVCSs in a DC MG comprising wind, PV, and fuel cell interlinked with the utility grid is carried out using a modified version of TLBO to lessen the VRP index and cost. The data to IEEE 123-bus test system is taken from [43,44]. The optimization is performed in a MATLAB environment R2019b version of Intel core (TM) i3-2350M CPU clocking at a speed of 2.30 GHz. Figure 4 indicates the optimal location of EVCSs and three different sources on the IEEE 33-bus test system. Similar steps are followed with the extensive IEEE 123-bus test system. The voltage levels of the 33- and 123-bus test system are 12.66 kV and 4.16 kV as the standard requirement. The considered DC MG produces power utilizing solar PV panels and fuel cells followed by wind turbines. The modeling of each source is described in Section 2. The output power of the solar PV and fuel cell is extracted in the form of DC. Hence, the solar PV and fuel cell are optimally placed at bus numbers 2 and 30 with a DC-DC converter. On the contrary, the output power produced from the wind turbine is in the form of AC. Hence, it is connected to bus 18 through a rectifier along with a DC—DC type converter. The utility grid is interlinked to slack bus 1 through a bidirectional AC-DC converter. The sources used in the DC MG store or generate DC power, making it immediately compatible with EVCS load. The EVCSs are linked to the DC bus by exchanging the power on bus numbers 19 and 20 through a bidirectional rectifier along with the DC-DC type converter. At bus 2, one more EVCS is optimally placed through the bidirectional DC-DC converter. The DC bus voltage is controlled by receiving power from the utility grid. When the total output power of sources is less than the total EVCS load, the utility grid injects power to the EVCS load. The DC bus voltage of 400 V with ±5% deviation is considered. All the EV loads are connected to a 400 V DC bus through a converter to balance the power between the DC bus and EVCSs. The DC MG easily integrates with the considered RESs with the smaller number of converters, and it improves the overall efficiency of the system.

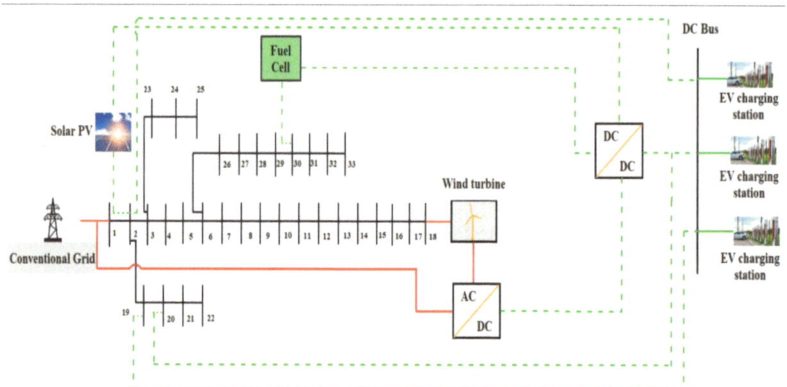

Figure 4. Framework of modified IEEE 33-bus test system.

As the current work is strictly confined to the reduction in VRP index and cost, it requires data related to the IEEE test system. In the present work, results are analyzed on medium as well as extensive networks, such as IEEE 33- and 123-bus test systems. The optimization is performed by considering six non-dominated planning schemes. Each scheme is optimally selected with its own set of fast and slow CSs and fast and slow charging points. The parameters of the PV module utilized in the present work are described in Table 4 [45]. The specifications of a wind turbine (WT) in the selected multi-source MG are described in Table 5 [35]. The number of fast and slow CSs (F_j, S_j) and quantity of slow and fast points of charging (f_j, s_j) are optimized in each scheme, as shown in Table 6. The input factors essential for achieving objective function are taken from [3].

Table 4. Parameters of solar PV module.

Parameter	Value	Parameter	Value
Power	305 W	Converter efficiency	90%
Length	1559 mm	A_{pv}	1.63 m^2
Width	1046 mm	N_{pv}	150
Lifespan	25 years	η_{pv}	18.7%

Table 5. Specifications of WEC system.

Parameter	Value	Parameter	Value
Air density	1.225 kg/m^2	Wind turbine nominal capacity	20 kW
Blade radius	3.7 m	Cut-in speed	3.5 m/s
Number of blades	3	Cut-out speed	18 m/s
C_{pmax}	0.47	Nominal speed	17.5 m/s
η_{con}	90%		

Table 6. The Number of Variables to be Augmented in the Different Schemes of Operation.

Scheme	Fj	fj	Sj	sj
1	1,1,1	1,1,1	1,2,1	6,6,6
2	2,2,2	2,2,2	2,2,2	10,10,10
3	2,1,2	2,1,2	2,2,2	8,8,8
4	2,1,2	2,1,2	2,2,2	10,10,10
5	1,1,1	1,1,1	2,2,2	7,9,16
6	1,1,1	1,1,1	2,1,2	10,7,10

Detailed descriptions of the proposed case are given below.

5.1. Cost Objective Function

The placement and sizing of EVCSs on the modified IEEE 33- and 123-bus distribution system short of RESs are performed to minimize the VRP index and cost. The modified TLBO technique obtains the objective function cost values as described in Table 7. The cost objective function is optimized under six schemes considering a different quantity of fast charging (FCSs) and slow charging stations (SCSs). Table 7 compares the optimized cost value to the most recently published work taken from reference [3]. As cost objective function is mainly reliant on the total number of fast and slow charging stations and charging points, the amount of charging points is less in scheme 1 compared to other schemes of operations.

Table 7. Objective function cost value comparison for the different planning schemes.

Scheme	With FCS and SCS by Modified TLBO (USD × 10^6)	Only FCS by Modified TLBO (USD × 10^6)	With FCS and SCS by CSO-TLBO (USD × 10^6) [3]
1	3.1494	2.0250	3.0206
2	8.6994	7.5750	8.8900
3	7.1994	6.1350	7.1120
4	7.9494	6.8250	7.9977
5	5.0994	3.9750	5.0883
6	5.9094	4.7850	5.9284

It is clear from Table 7 that scheme 1 gives the minimum value of cost when considering both fast and slow CSs and when considering only fast charging stations. The computational time required for the cost analysis is 8.234 s.

5.2. VRP Index

The VRP index minimization tested on IEEE 33- and 123-bus systems is detailed in the following subsections.

5.2.1. IEEE 33-Bus System

The location of CSs in a distribution unit centered on the VRP index is described in Figure 5. The VRP index is minimized to a greater extent by considering only fast charging stations compared to the combination of the FCS and SCS in all the schemes. The lowest value of VRP index is achieved in scheme 2, as the CS is placed at the strongest bus number, 2, of the test system. The optimal location and sizing of the CS centered on the VRP index are given in Tables 8 and 9, respectively. The VSF of bus 2 is the minimum for all the loading factors, which makes it the strongest bus of the system, and bus number 19 is considered as the second strongest bus of the system. The highest capacity of CS is achieved in scheme 2, i.e., 793.87 and 400.92 kW, respectively. The proposed objective function VRP index is a combination of distribution network operating parameters, i.e., voltage stability, systems reliability, and power loss. The reliability indices are further classified as SAIFI, SAIDI, and CAIDI. The indices of reliability are deteriorated due to an increase in EV load. The deteriorated values were fewer than the critical values of the reliability indices. As EV charging load is increased in steps optimally at any bus, the SAIFI, SAIDI, and CAIDI values increase linearly; the system tends to be less reliable and fails to achieve healthy operation, which may lead the system to collapse. Optimal placement of RESs with EVCSs helps to meet the increase in load demand with the healthy operation of the system by maintaining reliability indices within the critical values.

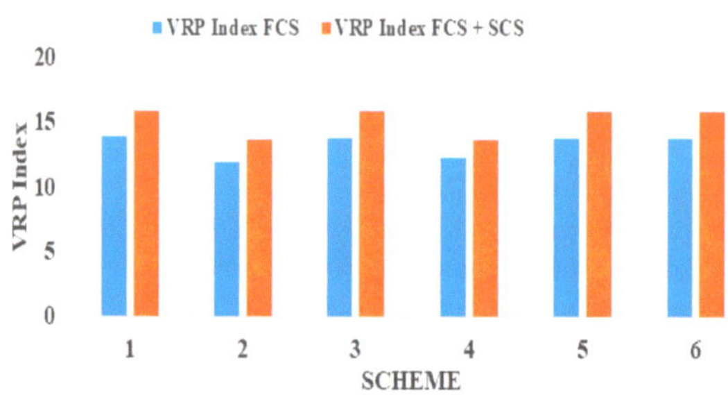

Figure 5. VRP index of modified IEEE 33-bus system considering both FCSs and SCSs.

Table 8. Optimal location of CSs based on VRP index for 33-bus system.

Bus	Number of FCS	Number of SCS
2	2	1
19	2	3
20	1	2

In the proposed case, applying RESs along with EVCS load increases the distribution system's reliability. The power extracted by wind and PV changes considerably because of the variation caused by solar irradiation and wind speed at altered timings due to varying climatic conditions. Hence, the fuel cell is considered as a backup source that supplies only real power. Furthermore, EVs charging with RESs cut down the emission of greenhouse gas. Thus, in the present work, a combination of wind, PV, and fuel-cell-based DC MG is taken for the optimal position and sizing of EVCSs. This work primarily focuses on lessening the VRP index of the distribution network by the best RES placement and sizing with EVCSs. The outcomes obtained for the proposed case are discussed as follows.

Table 9. Optimal sizing of CSs based on VRP index for 33-bus system.

Scheme	Size (kW) (Fast + Slow Charging Station)	Size (kW) Only FCS by Modified TLBO
1	297.6	240
2	793.87	400.92
3	522.61	390.58
4	508.14	378.53
5	297.60	240
6	297.60	240

The value of VSI is computed for each bus, as discussed in Section 3. The stability of the bus is given by the VSI. The VSI value varies between 0 and 1. Any of the buses having a higher value of VSI, close to 1, are considered stable buses, and buses with lower values of VSI, close to zero, are prone to instability and should be taken care of. The placement and sizing of EVCSs and RESs depend on the continuous fall and sudden rise of VSI. The VSI values for the base, after placement of the EVCS load of 1500 kW and the EVCSs considering RESs, are shown in Figure 6. As the EVCS load is installed at the most robust bus, number 2, of the test system, there was less degradation of voltage stability. The stability index of the weakest bus, 14, is improved by 13.13% after installing RESs, considering the base value, and by 15.30% considering only EVCSs. The VSI value improves for all the buses with RES placement compared to EVCS placement alone. Hence, the probability of voltage instability is reduced, and the system becomes more stable with RES placement.

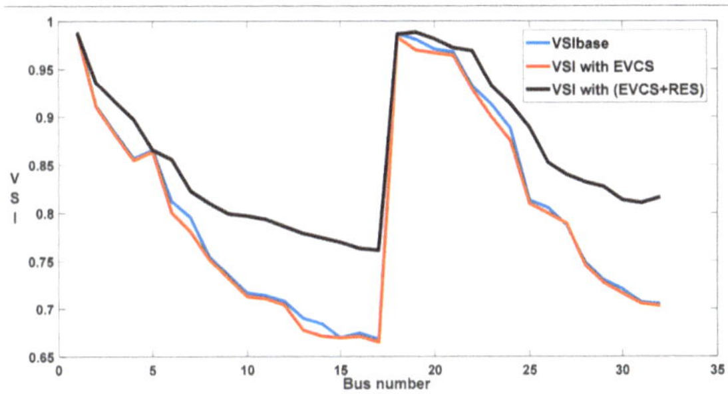

Figure 6. VSI of modified IEEE 33-bus system.

Figure 7 displays the voltage profile of the setup at each bus for the base after installing EVCSs with and without connecting RESs. It indicates that the voltage drop of all the buses after installing EVCSs is much less than the base value, because the CS is positioned at the robust bus of the test system. If the CS is placed at the weakest bus, then the voltage drop at that bus cannot be tolerated. However, the magnitude of voltage at all the buses lies within the acceptable range because of the distribution of CSs between a number of buses instead of concentrating the CSs at a single bus. The lowest value of voltage is obtained at bus number 18, and the maximum voltage appears at bus numbers 2 and 19.

The voltage profile started at 1 p.u. of voltage at bus number 1. It started a slow decrease in voltage up to bus number 18 from bus 1. The voltage profile is significantly and rapidly enhanced from bus number 18 to 19, then gradually decreased until bus 33. The maximum enhancement of voltage is perceived after installing RESs. This could reduce losses and improve the overall system performance. The voltage of the weakest bus, 14, is improved by 3.1% after installing RESs, considering the base value, and 3.4% considering only EVCSs.

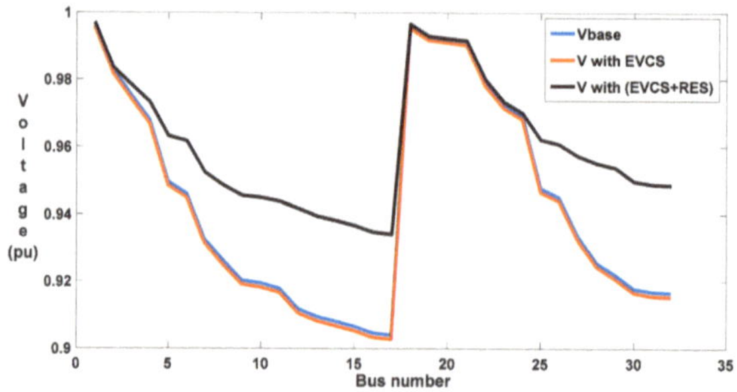

Figure 7. Voltage profile of modified IEEE 33-bus system.

The reliability index of the test system for the proposed case is presented in Figure 8. The SAIFI value for the base load is 0.0890 interruptions/year. When the EVCS of 1500 kW capacity is placed, the SAIFI value is raised to 0.1540 interruptions per year. The value of SAIFI after the placement of RESs is reduced to 0.0587 interruptions/year. A similar implication is perceived in the values of SAIDI and CAIDI. The value of SAIDI raised to 0.4764 h/year with EVCSs and drops to 0.2300 h/year after RES placement. The CAIDI raised to 6.684 h/interruption and drops to 6.002 h/interruption with respect to the base case after installing EVCSs alone and RESs with CSs, respectively.

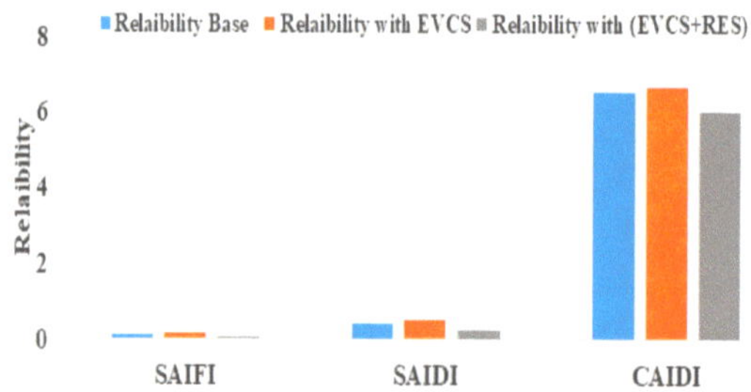

Figure 8. Reliability of modified IEEE 33-bus system.

When a CS is installed, the distribution network power loss will be more than when it is not installed. Figure 9 shows that when RESs are placed on the test system along with EVCSs, there is a significant decrease in the power loss by 9.71%.

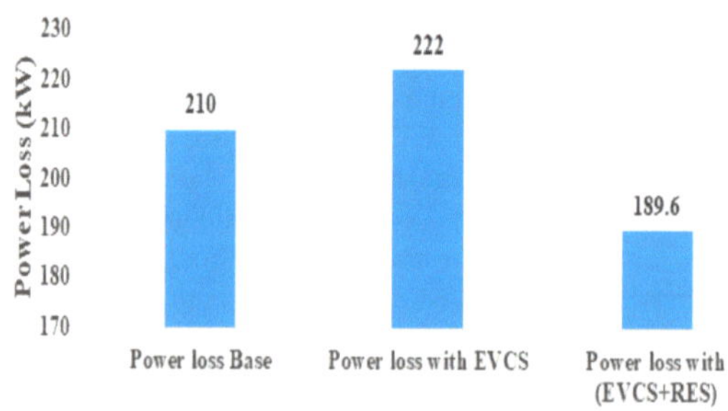

Figure 9. Power loss of the modified IEEE 33-bus system.

The VRP index of a test system in each scheme with and without RESs can be observed in Figure 10. Scheme 2 gives the lowest VRP index of 11.68 for EVCS placement alone. The minimum value of VRP index is obtained with scheme 6 compared to other schemes of operation, i.e., 9.6985 with EVCS and RES placement. The percentage reduction of VRP index on six various schemes for 33 bus system is shown in Figure 11.

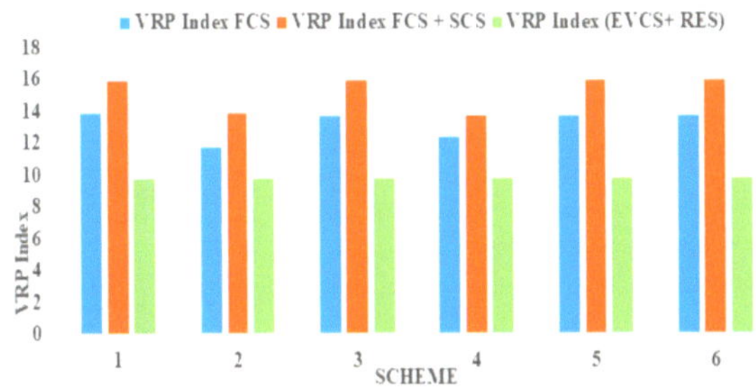

Figure 10. VRP index of the modified IEEE 33-bus system.

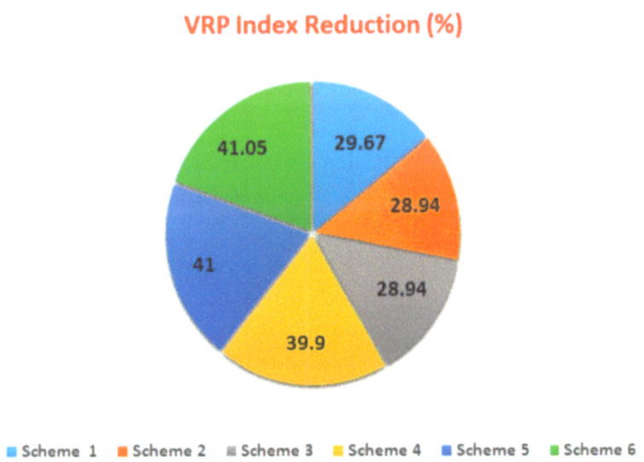

Figure 11. VRP index reduction in the modified IEEE 33-bus system.

The power variation of PV, wind, and EV loading for 24 h is shown in Figure 12, in which time is displayed on the *X*-axis followed by power (kW) represented on the *Y*-axis, concerned with solar irradiation, variation in wind speed, and dynamic behavior of EV load. The output power modeling of sources is described in Section 2. The data associated with the power variation concerned with solar irradiation [45] and wind speed variation are adapted from [33,46,47]. The EV charging and discharging pattern is adapted from [18]. The VRP index for EVCS installation alone and RES along with EVCS installation are 7.8562 and 9.7154, respectively, displayed in Table 10. Figure 13 represents the voltage fluctuation of the 33-bus system for 24 h. The maximum voltage magnitude was obtained from 12 to 1 pm, and the voltage magnitude drops from 1 to 7 am and 6 pm to 12 am for the proposed case, based on the respective charging and discharging of EVs. Less voltage fluctuations can be observed for the base case. The voltage drops at all the hours with EVCS placement alone compared to the base case.

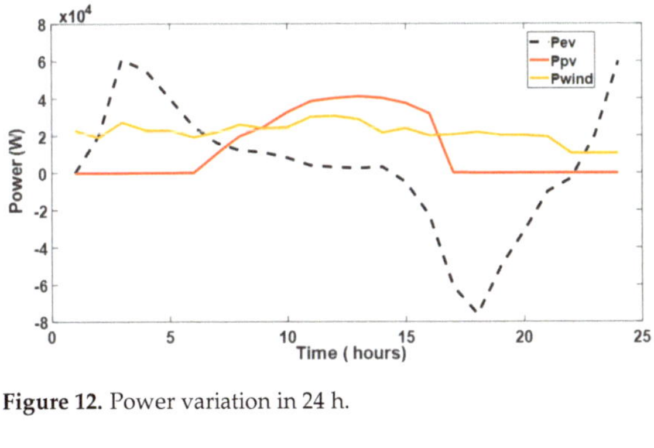

Figure 12. Power variation in 24 h.

Table 10. Objective function value for 24 h variation of power.

Sl.	VRP Index with EVCS	VRP Index with EVCS + RES
1	7.8562	9.7154

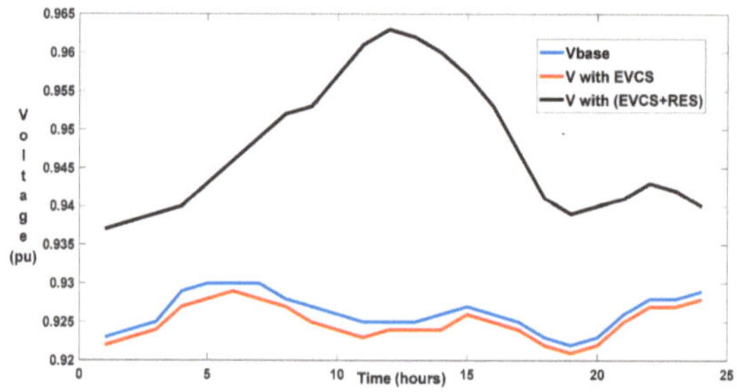

Figure 13. Voltage profile of 33-bus system for 24 h.

5.2.2. IEEE 123-Bus System

The optimal location of CSs centered on the VRP index in the 123-bus test system is given in Table 11. The placement of CSs is performed on the strongest buses of the system, with the major focus of reducing VRP index and enhancing the voltage profile of the system by reducing power losses.

Table 11. Optimal location of CSs based on VRP index on IEEE 123 test system.

Bus	Number of FCS	Number of SCS
16	2	1
43	2	2
107	1	2

The value of VSI on the 123-bus system is determined for each bus as discussed in Section 3. The stability of the bus is given by VSI. The VSI value varies between 0 and 1. The placement and sizing of EVCSs and RESs on the IEEE 123 test system depends on the continuous fall and sudden rise of VSI. Figure 14 shows the VSI value for the base, after placement of the EVCS load of 1500 kW and the EVCSs considering RESs. The EVCS load installed at the most robust bus, 60 of the test system, caused less degradation of voltage stability. The stability index of the weakest bus, 79, is improved by 52.22% after installing RESs, considering the base value, and 74.89% considering only EVCSs. In comparison to EVCS placement alone, the VSI value is increased for all buses with the placement of RESs.

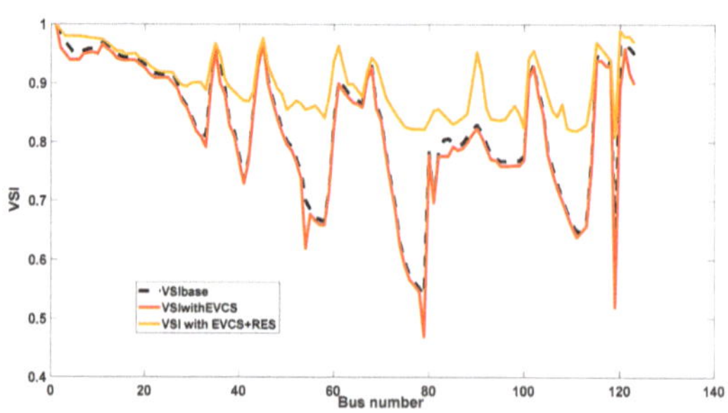

Figure 14. VSI of the IEEE 123-bus system.

Figure 15 displays the voltage profile of the setup at each bus for the base, after installing EVCSs with and without connecting RESs. It indicates that the voltage drop of all the buses after installing EVCSs is much less than the base value because the CS is positioned at the robust bus of the test system. The voltage magnitude of all the buses lies within the acceptable range because of the distribution of CSs between a number of buses instead of concentrating the CSs at a single bus. The lowest value of voltage is obtained at bus number 62, and the maximum voltage is found at bus numbers 83, 84, 85, and 116. The voltage profile started at 1.03 p.u. of voltage at bus number 1. The voltage profile is rapidly and significantly enhanced from bus number 66 to 67 and 115 to 116, then gradually decreased. The significant enhancement of voltage profile at each bus of the system is perceived after installing RESs.

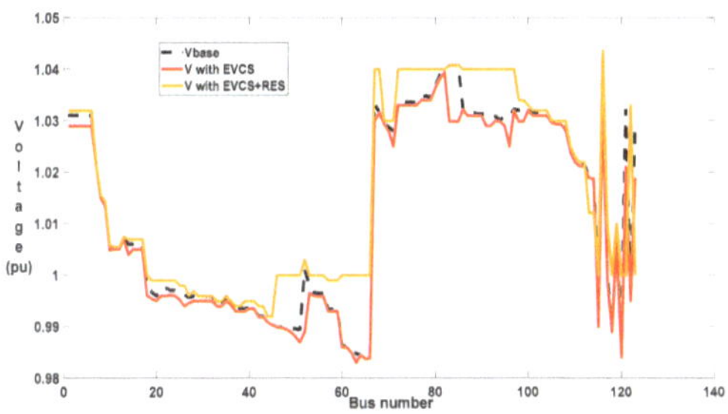

Figure 15. Voltage profile of IEEE 123 bus system.

Figure 16 shows that when RESs are placed on the test system along with EVCSs, there is a significant decrease in the power loss, by 6.09% and 4.44%, with respect to EVCS placement alone and the base value.

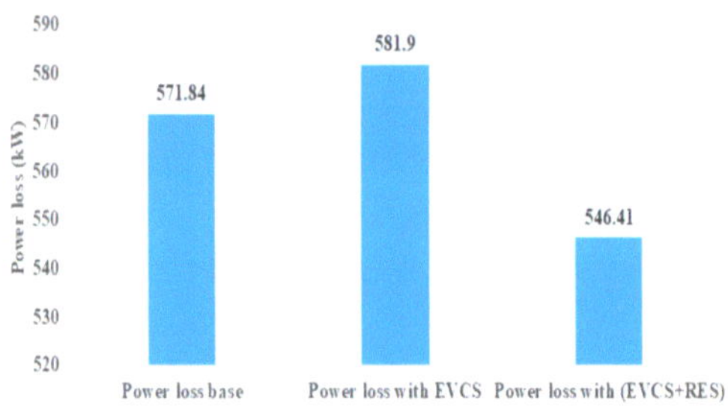

Figure 16. Power loss of the modified IEEE 123-bus system.

The comparison of VRP index in each scheme of operation is observed in Figure 17. Scheme 2 gives the lowest VRP index of 19.99 for EVCS placement alone. The minimum value of VRP index is obtained with scheme 6 in comparison with other schemes of operation, i.e., 17.34 with EVCS and RES placement.

Figure 18 shows the decrease in VRP index for a different operating scheme after installing RESs. Schemes 2 and 4 provide the least VRP index reduction concerning the base, at 13% and 15.36%. The highest reduction in VRP index is achieved in scheme 1 with reference to the base, at 18.27%. The computation time required for the proposed case is 20.89 s.

The power loss reduction and further minimization of the VRP index are achieved by optimal placement of RESs, viz., solar PV, wind, and fuel cells, as described in Table 12. The major RES solar PV of 1080 kW and 1190 kW are placed at bus number 2 and 60 of the 33-bus and 123-bus systems, respectively. The maximum CS load is positioned at bus 2 and 60, as they are the strongest buses of the system, where substantial voltage drop occurs. The wind of 750 kW and 900 kW is placed at bus 18 and 72 because the least value of voltage occurs after placing the CS compared to all other buses. Hence, to maintain the voltage within the tolerable limit, bus numbers 18 and 72 are optimally selected, and the fuel cell of 250 kW and 510 kW is placed at buses 30 and 102, respectively. The values for the objective functions, i.e., cost and VRP index on two different test systems, are provided in Table 13. It validates the proposed case VRP index model with the latest published work derived from reference [3]. Schemes 1 and 6 provide the minimized objective function cost and VRP index values, respectively, for the proposed case.

Figure 19 illustrates the convergence rate of modified TLBO for scheme 1, concerned only with EVCSs, and for the suggested case on both the test systems. The proposed case of the modified TLBO method gives more robust results because it reaches the least possible objective function. Hence, the optimum placement and sizing of EVCSs in a DC MG consisting of wind, solar PV, and fuel cells integrated with the grid are a favorable way out for an optimal reduction in the VRP index of the test system.

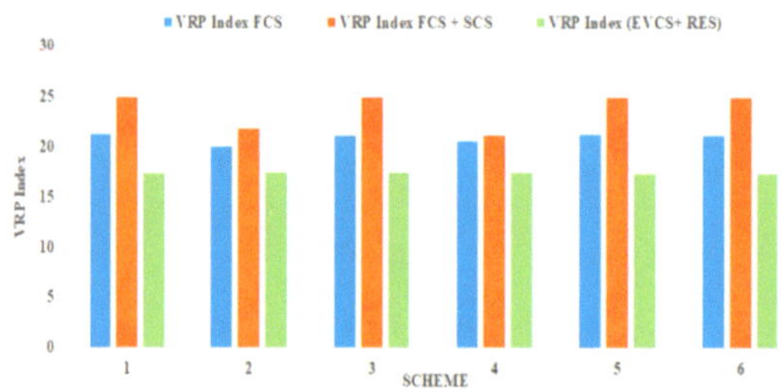

Figure 17. VRP index of the IEEE 123-bus system.

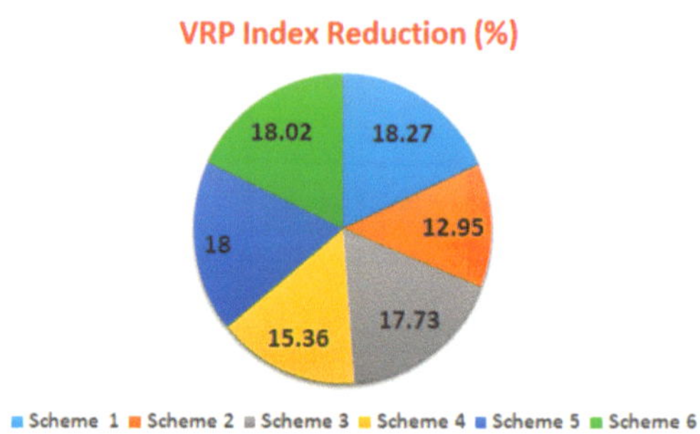

Figure 18. VRP index reduction in the IEEE 123-bus system.

Table 12. Optimal placement and sizing of RESs.

Scheme	Optimal Sizing (kW) on 33-Bus System	Optimal Location on 33-Bus System	Optimal Sizing (kW) on 123-Bus System	Optimal Location on 123-Bus System
PV	1080	2	1190	60
Wind	750	18	900	72
Fuel cell	250	30	510	102

Table 13. Values obtained from objective functions for six innovative schemes for the proposed case (EVCSs with RESs).

Scheme	Cost (USD × 10^6)	VRP Index for the Proposed Case on 123-Bus System by Modified TLBO	VRP Index for the Proposed Case on 33-Bus System by Modified TLBO	VRP Index with only EVCS on 19-Bus System by CSO-TLBO [3]
1	2.0250	17.35	9.7015	11.3011
2	7.5750	17.4	9.7200	11.7391
3	6.1350	17.4	9.7200	11.7018
4	6.8250	17.4	9.7200	11.8097
5	3.9750	17.35	9.6995	10.9291
6	4.7850	17.34	9.6985	11.3119

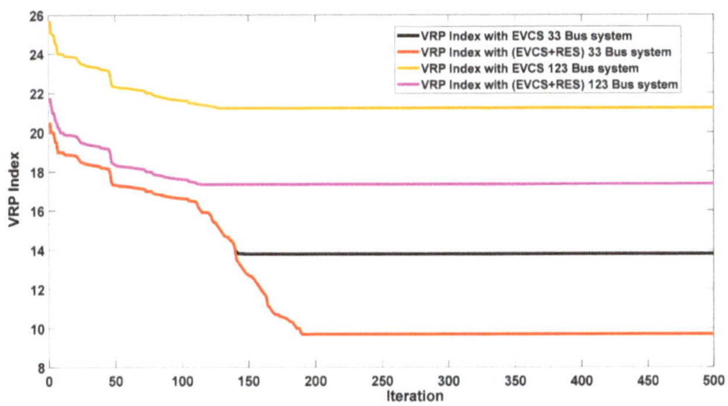

Figure 19. Convergence characteristics comparison of the proposed case on selected test systems.

The pareto graph considering the proposed approach, scheme 6, is given in Figures 20 and 21 for 33- and 123-bus systems, respectively. The pareto optimum solution yields the best compromised objective parameters for six diverse schemes of operation on both the test systems.

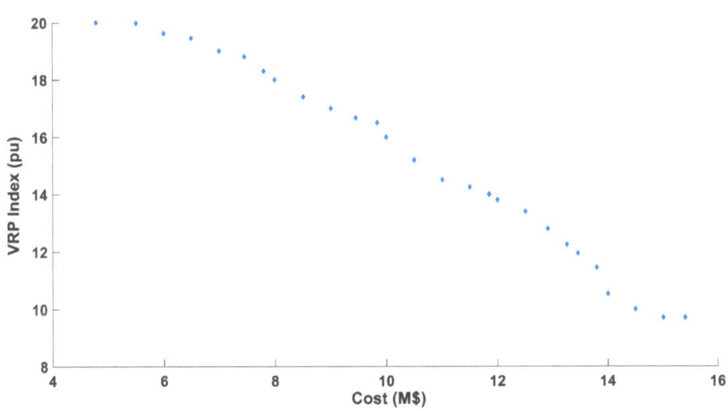

Figure 20. Optimal pareto solution for 33-bus system.

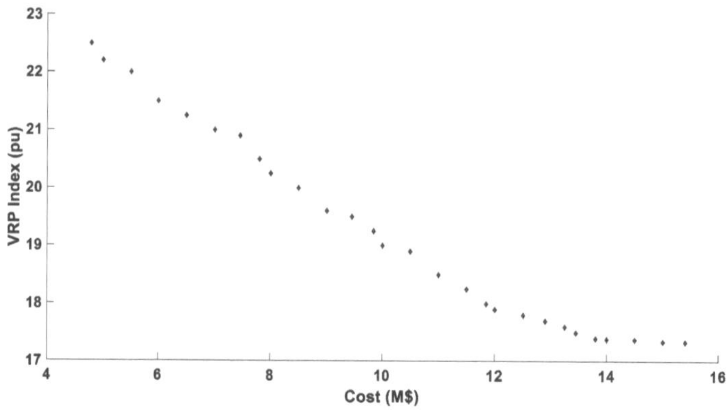

Figure 21. Optimal pareto solution for 123-bus system.

6. Conclusions

The rapid increase in EVCS loads can cause voltage unsteadiness and reliability problems. Due to the large-scale penetration of RESs, the reliable operation of the distribution network may be affected, which can strain the grid network's performance. To address this particular issue, a new operation strategy is employed. This paper gives insight into the optimum sizing and location of EVCSs and RESs in a DC MG integrated with a utility grid. The present work is solved in a multi-objective context including cost and distribution network operational aspects such as voltage stability, power loss, and reliability. The major focus is to decrease the VRP index of the distribution unit under six diverse schemes of operation. For the proposed case, the minimum value of cost obtained is USD 2.0250×10^6. The CSs are optimally placed at bus numbers 2, 19, and 20 and 16, 43, and 107 on 33 and 123 test systems, respectively. The RESs are optimally placed at 2, 18, and 30 and 60, 72, and 102 on 33 and 123 test units, respectively. The power loss on a medium scale and large scale is reduced by 14.59%, 9.714% and 6.09%, 4.44%, respectively, for the proposed case concerning EVCS placement alone and regarding the base case. An efficient meta-heuristic-modified TLBO technique is applied in MATLAB to solve this complex problem. The efficacy of the recommended strategy is evaluated on a modified IEEE 33- and123-bus test system for the proposed case and associated with the base values. The obtained results indicate that the optimal site and sizing of RESs and EVCSs with the proposed operational strategy enhance the voltage profile of a distribution system. The optimal positioning of RESs along with EVCSs reduces the VRP index of a distribution system to a greater extent under scheme 6 compared to other schemes of operation on both test systems. Therefore, the proposed instance is more suitable for minimizing the objective functions. There are numerous possibilities to delve deep into a further study by performing a techno-economic valuation of DC-MG-based EVCSs when linked to the utility grid.

Author Contributions: Conceptualization, N.K.K. and J.N.S.; methodology, N.K.K. and J.N.S.; software, N.K.K. and V.S.R.; validation, N.K.K., V.K.J. and J.N.S.; formal analysis, N.K.K.; investigation, N.K.K., V.K.J. and J.N.S.; resources, N.K.K. and J.N.S.; data curation, N.K.K.; writing—original draft preparation, N.K.K.; writing—review and editing, N.K.K., V.K.J. and J.N.S.; visualization, N.K.K., V.K.J., J.N.S. and A.S.; supervision, V.K.J. and J.N.S.; project administration, V.K.J., D.N.G. and J.N.S.; funding acquisition, V.K.J., G.K., V.S.R. and J.N.S. All authors have read and agreed to the published version of the manuscript.

Funding: This research received no external funding.

Data Availability Statement: Not applicable.

Conflicts of Interest: The authors declare no conflict of interest.

References

1. Fang, S.; Zhao, T.; Xu, Y.; Lu, T. Coordinated Chance-constrained Optimization of Multi-energy Microgrid System for Balancing Operation Efficiency and Quality-of-service. *J. Mod. Power Syst. Clean Energy* **2020**, *8*, 853–862. [CrossRef]
2. Salkuti, S.R. Short-term optimal hydro-thermal scheduling using clustered adaptive teaching learning based optimization. *Int. J. Electr. Comput. Eng.* **2019**, *9*, 3359. [CrossRef]
3. Aluisio, B.; Bruno, S.; De Bellis, L.; Dicorato, M.; Forte, G.; Trovato, M. DC-Microgrid Operation Planning for an Electric Vehicle Supply Infrastructure. *Appl. Sci.* **2019**, *9*, 2687. [CrossRef]
4. Pagidipala, S.; Vuddanti, S. Solving realistic reactive power market clearing problem of wind-thermal power system with system security. *Int. J. Emerg. Electr. Power Syst.* **2021**, *23*, 125–144. [CrossRef]
5. Deb, S.; Tammi, K.; Kalita, K.; Mahanta, P. Charging Station Placement for Electric Vehicles: A Case Study of Guwahati City, India. *IEEE Access* **2019**, *7*, 100270–100282. [CrossRef]
6. Mozafar, M.R.; Moradi, M.H.; Amini, M.H. A simultaneous approach for optimal allocation of renewable energy sources and electric vehicle charging stations in smart grids based on improved GA-PSO algorithm. *Sustain. Cities Soc.* **2017**, *32*, 627–637. [CrossRef]
7. Bilal, M.; Rizwan, M.; Alsaidan, I.; Almasoudi, F.M. AI-Based Approach for Optimal Placement of EVCS and DG With Reliability Analysis. *IEEE Access* **2021**, *9*, 154204–154224. [CrossRef]
8. Dhingra, K.; Singh, M. Smart Charging Station to Cater the Sudden Ingress and Egress of EVs while Supporting the Frequency of Microgrid through VSG. *Arab. J. Sci. Eng.* **2020**, *45*, 6715–6727. [CrossRef]
9. Deb, S.; Tammi, K.; Kalita, K.; Mahanta, P. Impact of electric vehicle charging station load on distribution network. *Energies* **2018**, *11*, 178. [CrossRef]
10. Awasthi, A.; Venkitusamy, K.; Padmanaban, S.; Selvamuthukumaran, R.; Blaabjerg, F.; Singh, A.K. Optimal planning of electric vehicle charging station at the distribution system using hybrid optimization algorithm. *Energy* **2017**, *133*, 70–78. [CrossRef]
11. Cui, Q.; Weng, Y.; Tan, C.-W. Electric Vehicle Charging Station Placement Method for Urban Areas. *IEEE Trans. Smart Grid* **2019**, *10*, 6552–6565. [CrossRef]
12. Jawad, S.; Liu, J. Electrical Vehicle Charging Services Planning and Operation with Interdependent Power Networks and Transportation Networks: A Review of the Current Scenario and Future Trends. *Energies* **2020**, *13*, 3371. [CrossRef]
13. Wang, D.; Locment, F.; Sechilariu, M. Modelling, Simulation, and Management Strategy of an Electric Vehicle Charging Station Based on a DC Microgrid. *Appl. Sci.* **2020**, *10*, 2053. [CrossRef]
14. AbuElrub, A.; Hamed, F.; Saadeh, O. Microgrid integrated electric vehicle charging algorithm with photovoltaic generation. *J. Energy Storage* **2020**, *32*, 101858. [CrossRef]
15. Khaksari, A.; Tsaousoglou, G.; Makris, P.; Steriotis, K.; Efthymiopoulos, N.; Varvarigos, E. Sizing of electric vehicle charging stations with smart charging capabilities and quality of service requirements. *Sustain. Cities Soc.* **2021**, *70*, 102872. [CrossRef]
16. Deb, S.; Tammi, K.; Gao, X.-Z.; Kalita, K.; Mahanta, P.; Cross, S. A Robust Two-Stage Planning Model for the Charging Station Placement Problem Considering Road Traffic Uncertainty. *IEEE Trans. Intell. Transp. Syst.* **2021**, *23*, 6571–6585. [CrossRef]
17. Pal, A.; Bhattacharya, A.; Chakraborty, A.K. Allocation of EV Public Charging Station in Renewable based Distribution Network using HHO Considering Uncertainties and Traffic Congestion. *Res. Sq.* **2021**. *preprint*.
18. Islam, M.; Shareef, H.; Mohamed, A. Improved approach for electric vehicle rapid charging station placement and sizing using Google maps and binary lightning search algorithm. *PLoS ONE* **2017**, *12*, e0189170. [CrossRef]
19. Bhadra, J.; Chattopadhyay, T.K. Analysis of distribution network by reliability indices. In Proceedings of the 2015 International Conference on Energy, Power and Environment: Towards Sustainable Growth (ICEPE), Shillong, India, 12–13 June 2015; pp. 1–5.
20. Jadoun, V.K.; Sharma, N.; Jha, P.; Malik, H.; Garcia Márquez, F.P. Optimal scheduling of dynamic pricing based v2g and g2v operation in microgrid using improved elephant herding optimization. *Sustainability* **2021**, *13*, 7551. [CrossRef]
21. Aluisio, B.; Dicorato, M.; Ferrini, I.; Forte, G.; Sbrizzai, R.; Trovato, M. Optimal Sizing Procedure for Electric Vehicle Supply Infrastructure Based on DC Microgrid with Station Commitment. *Energies* **2019**, *12*, 1901. [CrossRef]
22. Navin, N.K. A Multiagent Fuzzy Reinforcement Learning Approach for Economic Power Dispatch Considering Multiple Plug-In Electric Vehicle Loads. *Arab. J. Sci. Eng.* **2021**, *46*, 1431–1449. [CrossRef]
23. Suresh, V.; Bazmohammadi, N.; Janik, P.; Guerrero, J.M.; Kaczorowska, D.; Rezmer, J.; Jasinski, M.; Leonowicz, Z. Optimal location of an electrical vehicle charging station in a local microgrid using an embedded hybrid optimizer. *Int. J. Electr. Power Energy Syst.* **2021**, *131*, 106979. [CrossRef]
24. Chen, L.; Xu, C.; Song, H.; Jermsittiparsert, K. Optimal sizing and sitting of EVCS in the distribution system using metaheuristics: A case study. *Energy Rep.* **2021**, *7*, 208–217. [CrossRef]
25. Ali, A.; Mahmoud, K.; Lehtonen, M. Optimal planning of inverter-based renewable energy sources towards autonomous microgrids accommodating electric vehicle charging stations. *IET Gener. Transm. Distrib.* **2022**, *16*, 219–232. [CrossRef]
26. Archana, A.N.; Rajeev, T. A Novel Reliability Index Based Approach for EV Charging Station Allocation in Distribution System. *IEEE Trans. Ind. Appl.* **2021**, *57*, 6385–6394. [CrossRef]
27. Shaker, M.H.; Farzin, H.; Mashhour, E. Joint planning of electric vehicle battery swapping stations and distribution grid with centralized charging. *J. Energy Storage* **2023**, *58*, 106455. [CrossRef]
28. Wang, S.; Lu, L.; Han, X.; Ouyang, M.; Feng, X. Virtual-battery based droop control and energy storage system size optimization of a DC microgrid for electric vehicle fast charging station. *Appl. Energy* **2020**, *259*, 114146. [CrossRef]

29. Vinod, K.R.; Singh, S. Solar photovoltaic modeling and simulation: As a renewable energy solution. *Energy Rep.* **2018**, *4*, 701–712. [CrossRef]
30. Shaikh, M.R.S. A Review Paper on Electricity Generation from Solar Energy. *Int. J. Res. Appl. Sci. Eng. Technol.* **2017**, *5*, 1884–1889. [CrossRef]
31. Chedid, R.; Sawwas, A. Optimal placement and sizing of photovoltaics and battery storage in distribution networks. *Energy Storage* **2019**, *1*, e46. [CrossRef]
32. Pandey, A.K.; Jadoun, V.K.; Sabhahit, J.N. Real-Time Peak Valley Pricing Based Multi-Objective Optimal Scheduling of a Virtual Power Plant Considering Renewable Resources. *Energies* **2022**, *15*, 5970. [CrossRef]
33. Djelailia, O.; Kelaiaia, M.S.; Labar, H.; Necaibia, S.; Merad, F. Energy hybridization photovolta-ic/diesel generator/pump storage hydroelectric management based on online optimal fuel consumption per kWh. *Sustain. Cities Soc.* **2019**, *44*, 1–15. [CrossRef]
34. Slootweg, J.G.; de Haan, S.W.H.; Polinder, H.; Kling, W.L. General model for representing variable speed wind turbines in power system dynamics simulations. *IEEE Trans. Power Syst.* **2003**, *18*, 144–151. [CrossRef]
35. Verde, A.; Lastres, O.; Hernández, G.; Ibañez, G.; Verea, L.; Sebastian, P.J. A new method for characterization of small capacity wind turbines with permanent magnet synchronous generator: An experimental study. *Heliyon* **2018**, *4*, e00732. [CrossRef]
36. Wang, C.-N.; Lin, W.-C.; Le, X.-K. Modelling of a PMSG Wind Turbine with Autonomous Control. *Math. Probl. Eng.* **2014**, *2014*, 856173. [CrossRef]
37. Jayalakshmi, N.S.; Gaonkar, D.N. Operation of grid integrated wind/PV hybrid system with grid perturbations. *Int. J. Renew. Energy Res.* **2015**, *5*, 1106–1111.
38. Omran, A.; Lucchesi, A.; Smith, D.; Alaswad, A.; Amiri, A.; Wilberforce, T.; Sodré, J.R.; Olabi, A. Mathematical model of a proton-exchange membrane (PEM) fuel cell. *Int. J. Thermofluids* **2021**, *11*, 100110. [CrossRef]
39. Hailu Kebede, M.; Bekele Beyene, G. Feasibility study of PV-wind-fuel cell hybrid power system for electrification of a rural village in Ethiopia. *J. Electr. Comput. Eng.* **2018**, *2018*, 4015354. [CrossRef]
40. Rahman, M.M.; Barua, S.; Zohora, S.T.; Hasan, K.; Aziz, T. Voltage sensitivity based site selection for PHEV charging station in commercial distribution system. In Proceedings of the 2013 IEEE PES Asia-Pacific Power and Energy Engineering Conference (APPEEC), Hong Kong, China, 8–11 December 2013; pp. 1–6. [CrossRef]
41. Rao, R.V.; Savsani, V.J.; Vakharia, D.P. Teaching–learning-based optimization: A novel method for constrained mechanical design optimization problems. *Comput. Aided Des.* **2011**, *43*, 303–315. [CrossRef]
42. Rao, R. Review of applications of TLBO algorithm and a tutorial for beginners to solve the unconstrained and constrained optimization problems. *Decis. Sci. Lett.* **2016**, *5*, 1–30.
43. Tahmasebi, M.; Pasupuleti, J.; Mohamadian, F.; Shakeri, M.; Guerrero, J.; Khan, M.B.; Nazir, M.; Safari, A.; Bazmohammadi, N. Optimal Operation of Stand-Alone Microgrid Considering Emission Issues and Demand Response Program Using Whale Optimization Algorithm. *Sustainability* **2021**, *13*, 7710. [CrossRef]
44. Feeders, D.T. IEEE PES Distribution System Analysis Subcommittee. 2011. Available online: http://sites.ieee.org/pes-testfeeders (accessed on 22 August 2022).
45. Jiang, Z.; Zou, F.; Chen, D.; Kang, J. An improved teaching–learning-based optimization for multilevel thresholding image segmentation. *Arab. J. Sci. Eng.* **2021**, *46*, 8371–8396. [CrossRef]
46. Zou, F.; Chen, D.; Wang, J. An Improved Teaching-Learning-Based Optimization with the Social Character of PSO for Global Optimization. *Comput. Intell. Neurosci.* **2016**, *2016*, 18. [CrossRef] [PubMed]
47. The Annual Variability of Wind Speed. Available online: https://www.wind-energy-the-facts.org/the-annual-variability-of-wind-speed.html (accessed on 22 August 2022).

energies

MDPI

Article

Analysis of an Urban Grid with High Photovoltaic and e-Mobility Penetration

Florian Maurer *, Christian Rieke, Ralf Schemm and Dominik Stollenwerk

FH Aachen—NOWUM-Energy, University of Applied Sciences Aachen, 52428 Jülich, Germany; rieke@fh-aachen.de (C.R.); schemm@fh-aachen.de (R.S.); stollenwerk@fh-aachen.de (D.S.)
* Correspondence: maurer@fh-aachen.de; Tel.: +49-241-6009-54265

Abstract: This study analyses the expected utilization of an urban distribution grid under high penetration of photovoltaic and e-mobility with charging infrastructure on a residential level. The grid utilization and the corresponding power flow are evaluated, while varying the control strategies and photovoltaic installed capacity in different scenarios. Four scenarios are used to analyze the impact of e-mobility. The individual mobility demand is modelled based on the largest German studies on mobility "Mobilität in Deutschland", which is carried out every 5 years. To estimate the ramp-up of photovoltaic generation, a potential analysis of the roof surfaces in the supply area is carried out via an evaluation of an open solar potential study. The photovoltaic feed-in time series is derived individually for each installed system in a resolution of 15 min. The residential consumption is estimated using historical smart meter data, which are collected in London between 2012 and 2014. For a realistic charging demand, each residential household decides daily on the state of charge if their vehicle requires to be charged. The resulting charging time series depends on the underlying behavior scenario. Market prices and mobility demand are therefore used as scenario input parameters for a utility function based on the current state of charge to model individual behavior. The aggregated electricity demand is the starting point of the power flow calculation. The evaluation is carried out for an urban region with approximately 3100 residents. The analysis shows that increased penetration of photovoltaics combined with a flexible and adaptive charging strategy can maximize PV usage and reduce the need for congestion-related intervention by the grid operator by reducing the amount of kWh charged from the grid by 30% which reduces the average price of a charged kWh by 35% to 14 ct/kWh from 21.8 ct/kWh without PV optimization. The resulting grid congestions are managed by implementing an intelligent price or control signal. The analysis took place using data from a real German grid with 10 subgrids. The entire software can be adapted for the analysis of different distribution grids and is publicly available as an open-source software library on GitHub.

Keywords: e-mobility; smart-charging; distribution grid simulation

Citation: Maurer, F.; Rieke, C.; Schemm, R.; Stollenwerk, D. Analysis of an Urban Grid with High Photovoltaic and e-Mobility Penetration. *Energies* **2023**, *16*, 3380. https://doi.org/10.3390/en16083380

Academic Editors: Evangelos Karfopoulos, Michele Roccotelli and Ioannis Karakitsios

Received: 13 February 2023
Revised: 29 March 2023
Accepted: 4 April 2023
Published: 12 April 2023

1. Introduction

The German government has set itself the goal of reducing greenhouse gas emissions in the transport sector by 48% by 2030 compared with 1990. This requires the electrification of transport and the use of renewable energies. A prerequisite for this is the further expansion of the charging infrastructure. The transformation from fossil energy to renewable energy sources (RES) started in the year 2004 by setting the first novel of the "Erneuerbaren-Energien-Gesetz" (EEG—Law) in charge to meet the EU climate objectives. In 2020, the average yearly share of the electric energy generated by renewables in Germany was 45% [1].

Due to the time-dependent availability of RES and further decentralization of the electricity generation, the power system is changing. On the other hand, the expanding market of Electric Vehicles (EVs) is an additional challenge, which creates a lot of decentralized demand and provides an interesting potential as storage. The German objective of one million EVs on the roads by 2020 was not achieved, but current developments and forecasts indicate a successful ramp-up. In 2020, about 200,000 funding applications are submitted for EVs, and 360,000 EVs were subsidized in 2021 [2]. By 2030, the German government aims for seven to ten million EVs, while the German Association of the Automotive Industry expects 10.5 million EVs [3].

To meet this objective, capable charging infrastructure and control of this infrastructure is required. It is planned to increase the amount of available public charging stations from 21,000 in 2021 to one million in 2030. This plan corresponds with the EU's objective of a ratio of EVs to public charging points of 10:1 [4]. Additionally, 7.1 million charging points in private households are required to ensure the electrification of the transport sector [5]. The share of private charging operations is estimated up to 88%, which results in additional consumption of 10,000 GWh in the low voltage grids, especially in urban regions. In the public domain, 12% of charging operations will be carried out with a corresponding energy volume of 12,000 GWh in 2030 [5].

As grid planning allocates resources for decades of operation, large parts of today's distribution grids were not planned for today's demand of electrification. Multiple studies show that simultaneous charging during peak hours can exhaust and exceed the current grid capacities [6–10].

To avoid congestion, either a grid reinforcement or an intelligent control system can be applied. Therefore, time-dependent charging prices are already considered and tested in some areas [11]. Different simulation tools can model energy system expansion and investments on a larger scale [12], while this study investigates whether congestion in the distribution grid can be reduced with the help of simple controlling or dynamic prices. Individual charging behavior is rarely investigated thoroughly, resulting in high assumptions about concurrent charging processes.

2. Related Work

Distribution grids and possible control signals for EVs have been investigated in many previous studies. The approaches to mapping mobility behavior, the charging strategies, and the modelling approach for the energy supply units and the energy distribution grids vary considerably between the individual studies.

In this section, some studies related to modelling of price-sensitive charging behavior are cited, and their strategies are compared to the ones developed in this paper. The research in this field is conducted worldwide and evolves dynamically, and therefore, the related work is multifaceted. The adapted approaches before the year 2015 are summarized in Gottwalts PhD thesis [13], where he found that Time-of-Use-based pricing can be used to achieve demand response for small shares of flexible loads and investigates group-based pricing in contrast to Realtime pricing. Growing databases and a partial mapping of user behavior were the basis for subsequent works in particular, summarized by Su in the year 2020 [14]. He updates the existing knowledge basis and gives a broad overview of existing EV models, charging strategies and ToU-based vs. group-based pricing approaches for demand response. A high potential for load-shifting EV charging was found in [15], as constantly about 45% of cars are available. However, the most recent studies consider latest EV models, their charging behavior, and the resulting demand response to price signals with different approaches in [16–25]. In [16,17], the EV demand is modeled deterministically considering a non-sensitive approach, such as the first charging strategy *Base Case* chosen in this paper. The first modelling approach presented in this paper is a more flexible adaptive approach with a fixed cap, represented by a maximum willingness to pay for the energy. In the result, the EV users only charge if the SoC is below the required SoC of the coming day. Furthermore, we have derived individual functions of willingness to pay depending on the state of charge based on surveys to further refine the charging strategy.

Many approaches generate scenario data through sampling from existing data to create scenarios with future higher charging demand as seen in [20–22]. A total of 29,262 existing EV charging curves were used in [22] to sample the power demand. A specific case comparing uncontrolled and controlled charging is shown in [26,27], while the economic investment and operation cost were considered in [28].

Yet those approaches do not model price sensitivity and the responding displacement of charging demand. Such factors can be determined by using a stochastic probabilistic simulation with consideration of the grid aspects [18,29–31]. A similar approach to investigate grid planning was done in [32], which investigates four different charging station location cases. The authors in [23] propose a probability-based approach and compare price sensitive and non-sensitive demands with the result of much lower maximum demand and a shift toward the evening hours from 7 and 10 am without sensitivity.

The investigation of PV potentials in the context of E-Mobility was done in [33], where different PV and EV penetration cases were modelled using three different charging behaviors without price sensitivity in a cell-based grid model.

In [24], the EV charging demand is controlled in an indirect manner by adopting a dynamic virtual pricing mechanism according to the forecasted EV charging demand and local renewable generation production. The objective of the EV Agents is to minimize their individual charging cost, according to the price values and therefore schedule the charges at the cheapest hours. This modelling approach is very similar to the second *Case B* and *Case C* charging strategy presented in this paper where the maximization of the PV consumption in the individual households and the maximization of a user satisfying function determine the charging schedule for the EVs. The main difference to the approaches in this paper are the usage of real distribution grid data, PV integration, and energy consumption data for this grid and the benefit function usage as optimization objective.

3. Materials and Methods

In this section, the methods that are used to model the mobility behavior, the resulting charging demand, and the ordinary power demand are explained. In addition, the applied dynamic grid fee and the different scenario cases are derived.

The overall agent-based simulation framework is implemented in Python with common libraries such as *NumPy*, *Pandas* and *GeoPandas*, *PyPSA* and *PVLib*. The results are stored using the timeseries database *TimescaleDB* and are evaluated with the help of a geo referenced data store extension *PostGIS*. To orchestrate the multiple simulation runs across a cluster of HPC servers, the technology *Docker Swarm* is used. Yet, single simulations can also be run on a consumer laptop. To solve the optimization, any solver supported by the solver library pyomo can be used, for example *GLPK*.

3.1. Simulation Framework & Sequence

The framework is designed in such a way that several scenarios/simulations can be calculated in parallel. For this purpose, each simulation is started in a separate Docker container. The containers work on the same database, so that the results can be easily summarized and analyzed.

The scenarios differ depending on the considered user behavior. In the *Base Case*, the user requests a charging process after his journey. A request is always made if the current SoC of the vehicle is not sufficient for the mobility needs of the next day. For example, if the vehicle only has 15% left after the journey, but 45% is needed for the next day, the consumer sends a request to fully charge his EV to 100%. In this case, the charging process is planned so that it starts at the next time step t and continues until the vehicle is fully charged at his nominal power P_{car}.

Depending on the level of the dynamic grid fee G_t, the charging process is accepted and carried out or suspended. If the price level is too high, charging is suspended for 30 min until the user makes another request. A maximum of five rejections are possible until the consumer commits even at higher costs.

This corresponds to the assumption that the need for mobility is always more worth as the total cost of charging, and that the consumer accepts this cost to reach his workplace the next day, for example. The following steps and the corresponding Figure 1 summarize the simulation process in detail for the *Base Case*.

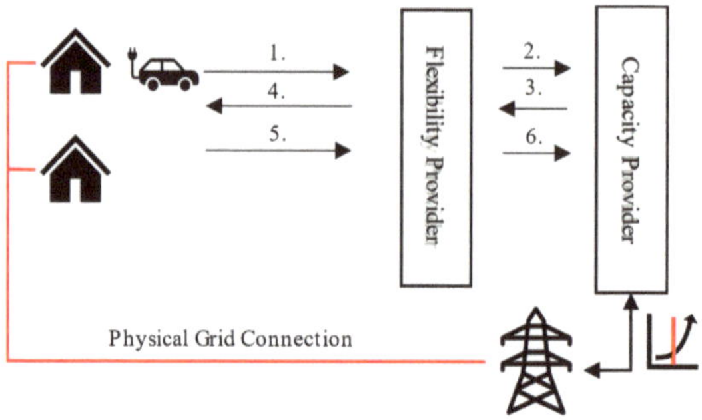

Figure 1. Simulation sequence of the developed simulation framework SmartDSO. The figure shows general simulation steps (1–6) and price communication between price-sensitive mobility demand and the grid fees depending on congestion calculated by the capacity provider.

1. The Consumer sends a charging request for the next simulation period (15 min)
2. The Flexibility Provider maps the demand on the corresponding grid node and sends demand to the Capacity Provider.
3. The Capacity Provider calculates the total grid utilization and answers with a grid fee time series.
4. The Flexibility Provider communicates the Price Level to the Consumer.
5. The Consumer can lock the demand to the current price level and sends a commit or reject message back.
6. Corresponding to the answer, the Flexibility Provider send final demand time series to the Capacity Provider.

In the other scenarios (*Case A*, *Case B*, and *Case C*), each consumer solves an optimization problem at the beginning of each day. Input variables of the optimization are the PV generation and the consumption, resulting from a Standard Load Profile (SLP). The journeys of the day are also considered, as well as the required SoC of the coming day. At the end of the day, the SoC must fulfil the same requirement that is also used in the *Base Case*. In addition, the electricity tariff of the end consumer, including surcharges and taxes, as well as the dynamic grid fee are considered. The grid fee is initially assumed to be 2.5 ct/kWh for the first calculation of the five possible optimization runs.

The Capacity Provider receives a request for the next 96 quarter hours from all consumers who want to charge their EV via the electricity grid. The capacity provider processes the requests in the order in which they were made. Only when the six steps from Figure 1 have been processed for a consumer, the next consumer is handled.

If a consumer rejects the request because, for example, the grid fees exceed the initial 2.5 ct/kWh, it carries out another optimization with the new grid fee information at each time step t. The result of the optimization is either a disclaimer of the recharge because the grid fee is too high in all hours, or a shift to other hours in which the grid fee is lower. The consumer always accepts the last of his five offers, even if the price is higher than the price limit. The objective function of the optimization problem can be formulated as shown in Equation (1).

$$\max\left(B - \sum_{t}^{T} PRequest_t \cdot Tarif_t\right) \tag{1}$$

The users living in a residential area want to maximize their benefit B, meaning to maximize the usage with cheapest cost. For this purpose, a utility function is used, which is defined in *Case A* and *Case B* via a linear function between SoCs of the cars and the price limit. In the last *Case C*, the integral of the demand functions, which are derived in Section 3.3, is used. These are linearized piecewise using the following Equations (2)–(6).

$$\sum_{s} z_s = 1 \tag{2}$$

$$q_s \geq SegVLow_s \cdot z_s \ \forall\ s \tag{3}$$

$$q_s \leq SegVUp_s \cdot z_s \ \forall\ s \tag{4}$$

$$B = \sum_{s} \begin{matrix} SegBLow_s \cdot z_s + \\ SegCoeff_s \cdot (q_s - SegVLow_s \cdot z_s) \end{matrix} \tag{5}$$

$$VTotal = \sum\nolimits_{s} qs \tag{6}$$

The linearization is explained in detail in [34,35]. To increase the benefit B, the car must be charged. This can be done either via a photovoltaic system Ppv_t or by purchasing requested electricity $PRequest_t$ from the grid. If the benefit gain ΔB is lower than the additional cost $Tarif_t$, the car is not charged any further. The Equations (7)–(9) show the corresponding balancing of the charging power $PCharge_t$, the power of the photovoltaic system Ppv_t and the grid purchase $PRequest_t$ at time step t.

$$PRequest_t = PCharge_t - Ppv_t \tag{7}$$

$$PGen_t - Ppv \geq 0 \ \forall\ t \in T \tag{8}$$

$$PRequest_t \geq 0 \ \forall\ t \in T \tag{9}$$

The charging power $PCharge_t$ is the sum of the individual capacities for each car in the household (see Equation (10)). The car can only be charged when it is not in use and when it is at home (see Equation (11)).

$$PCharge_t = \sum\nolimits_{car}^{C} P_{car,t} \tag{10}$$

$$P_{car,t} \leq (1 - U_{car,t}) \cdot PMax_{car} \ \forall\ t \in T,\ car \in C \tag{11}$$

Finally, the SoC level of a car is restricted to the lower limit 0 and the upper limit $EMax_{car}$ as shown in Equations (12)–(14).

$$\begin{gathered} V_{car,t} = V_{car,t-1} + P_{car,t} \cdot dt - EDem_{car,t} \\ \forall\ t \in T,\ car \in C \end{gathered} \tag{12}$$

$$\begin{gathered} 0 \leq V_{car,t} \leq E_{car} \\ \forall\ t \in T,\ car \in C \end{gathered} \tag{13}$$

$$VTotal = \sum\nolimits_{car}^{C} V_{car,T} \tag{14}$$

The study examines four scenarios. For each scenario, the degree of PV expansion is varied between 25% and 100% of the potential maximum capacity in the supply area. The scenarios differ in the modelling of the charging behavior of the consumers and the used electricity tariff. In the first two cases (*Base* and *A*), the end consumers receive a fixed electricity price that corresponds to the average of the day-ahead prices for the analyzed period. In the simulation, the month of May in 2022 is examined. In *Case B* and *C*, the hourly day-ahead price is directly passed on to the consumers. A further 10.8 ct/kWh in surpluses and taxes is added to both the fixed price and the dynamic day-ahead price. Depending on the grid utilization, a dynamic grid fee is added to every requested charging process.

In the *Base Case*, no optimization takes place regarding an objective value. Drivers always request a charging process via the Capacity Provider when they return from a journey and the SoC of their vehicle is no longer sufficient for the journey on the following day. This is described by the SoC limit $SoCLimit_t$. The Consumers have a fixed price limit of 45 ct/kWh. If the charging price is on average more expensive than the limit, the charging process is shifted in hours with less utilization.

In *Case A*, the described optimization takes place at the beginning of each day. The objective value is the maximization of electricity consumption from the own PV system. For this purpose, it is assumed that the PV generation for the coming day is already known. Thus, any uncertainty in PV generation at the time of optimization is not considered. Analogous to the *Base Case*, electricity is only consumed from the grid if the SoC is no longer sufficient for the upcoming mobility demand. Otherwise, only the use of the own PV System is possible. If the average price of a request is again above the limit of 45 ct/kWh, the charging process is shifted again.

In *Case C*, the objective value is the maximization of a benefit function B, which leads to a maximum PV usage and a charging at the cheapest hours. The SoC-dependent benefit function B is described in detail in Section 3.3. In *Case B*, the fixed price cap of 45 ct/kWh is applied. For each scenario case, serval simulations are calculated to get a distribution of possible grid congestions. Table 1 summarize the scenarios.

Table 1. Simulation parameters of investigated scenarios.

Scenario	Charging Strategy	Market	Grid Fee	Mobility Behavior
Base Case	SoC independent	Flat Price	dynamic	random
Case A	maximize PV usage + SoC independent	Flat Price	dynamic	random
Case B	maximize PV usage + SoC independent	Spot Price	dynamic	random
Case C	maximize PV usage + SoC dependent	Spot Price	dynamic	random

3.2. Agents in the Simulation Framework

The agent of the capacity provider calculates the grid utilization and the resulting dynamic grid fee G_t of the requested charging process. The function in Equation (15) is used to calculate the dynamic fee. The function increases monotonically until it reaches its maximum value of 100 ct/kWh at a utilization of 100%. From this limit, the price remains constant. To model the function G_t, the current grid fees in Germany were analyzed. According to [24], the average grid fee for residential areas is approximately 7.5 ct/kWh. A difference of 2.0–3.0 ct/kWh is possible. In this study, a minimum of 2.6 ct/kWh is chosen, and the common 7.5 ct/kWh at a utilization of 40%.

$$G_t = \max_{n \in N}\left(0.15 \cdot \left(0.175 - \ln\left(1 - u_{n,t}^{\frac{2}{3}}\right)\right)\right) \tag{15}$$

To calculate the grid fee G_t, the highest utilization of the component n at the respective time t in the corresponding sub-grid is used. The end consumer or the flexibility provider receives the dynamic grid fee for each requested point in time.

To determine the utilization of each grid component, a calculation of the power flow is carried out using the Python package PyPSA [35]. The power flow corresponds to the solution of a non-linear system of equations, which is solved using the Newton Raphson method. Since the capacity provider cannot directly control the charging of the EVs, the optimal power flow is not calculated, as often in other studies including [9,36]. Therefore, the capacity provider does not solve an optimization task by dispatching the demand and generation, but the dispatch of the consumers takes place via the dynamic grid fee.

In the simulation, a distribution grid on low voltage level (400 V in Germany) with approx. 3100 residential consumers with approx. 3600 EVs is evaluated. The number of consumers is distributed over ten sub grids. The smallest grid has approx. 180 residential consumers, the largest grid approx. 480 residential consumers. The analyzed grid topology is shown in Figure 2. In the 100% PV-Scenario, the total amount of installed photovoltaic capacity is approx. 4.5 MW.

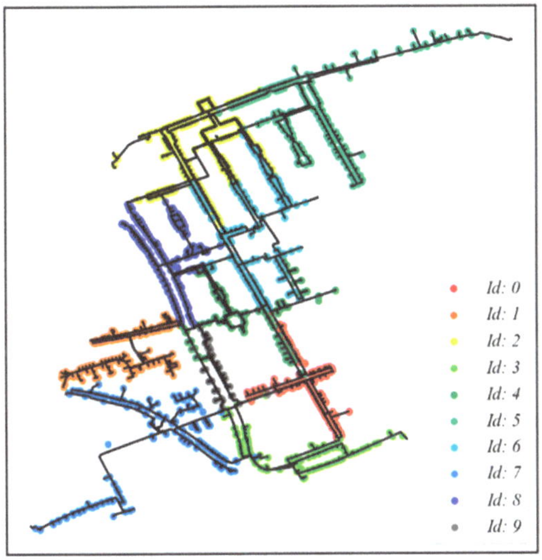

Figure 2. Investigated low-voltage (400 V) grid with 10 subgrids and 3100 consumers. The lines are colored according to the belonging subgrid ID.

The agent of the Flexibility Provider takes over the communication with the Capacity Provider for the consumers. The agent offers different tariffs depending on the analyzed scenario. The varying price signal in *Case B* and C is a composition of the price of the wholesale market (the day-ahead and continuous intraday market). Here, the wholesale market prices are based on the merit order principle and therefore take the scarcity of fluctuating renewable energy into account.

To summarize, the Flexibility Provider takes the role of the energy supplier and is responsible for energy purchase at the wholesale market. In the simulation framework, the agent only aggregates the generation and consumption time series and acts as an intermediary.

3.3. Consumer Behavior

In the simulation framework, three consumer types are implemented. To model the demand time series of each consumer, a SLP is used. Businesses and residential consumers with less than 100,000 kWh per year are forecasted and settled with such a SLP in Germany [37]. Bigger consumers, as the industry sector, are settled using historic meter values. Since no meter values are available for the industry consumers in the analyzed grid, a SLP for industry consumers is derived.

For this purpose, the total demand of residential and commercial customers was subtracted from the time series of the total load in Germany. Next, typical days are summarized. Therefore, all working days (Monday to Friday) are averaged for the annual seasons spring, summer, autumn, and winter. The same applies to Saturdays and Sundays. Therefore, twelve typical profiles are derived in hourly resolution for industry consumers. To estimate the annual consumption in the residential sector per hour, the SLP type H0 is used [37]. The profile is scaled with the total demand of 128 TWh/a. The same applies to commercial consumers, but with profile type G0 [37] and a yearly consumption of 135 TWh/a.

To summarize, all three consumer types are modeled with a SLP. For all residential areas, the H0 profile; for all businesses, the G0 profile; and for all bigger consumers, the derived profile (I0-Profile) is used.

In addition to the static demand, each household is represented by a Consumer-Agent, who has the objective to charge the EV with a specific strategy. The charging process always takes place at home.

To get a well-estimated charging demand, the mobility behavior of a single resident is modelled. The mobility behavior results from a statistic evaluation of the study "Mobilität in Deutschland" [36], where 154,420 households are surveyed. Based on these mobility statistics, different mobility events M_i are implemented. Each mobility event M_i belongs to one of the three categories work, errands, or hobby. A similar classification is carried out in [38]. To define a mobility event M_i, the following variables are used:

- $Ts_{t,j}$ as the start time at the destination in [HH:MM],
- $Td_{t,j}$ as the event duration in [min],
- $T_{t,j}$ as the travel time to the destination and/or back in [min],
- the absence time $Ta_{t,j}$ in [min] and
- $D_{t,j}$ as the distance between the home and destination in [km] for a day t.

To estimate the number of drivers and cars in the supply area, a mean consumption of 1500 kWh per year per person is assumed. With the data of the yearly energy consumption of each residential area, the number of persons per household is calculated and limited to two possible drivers. Each driver has his own EV. The mobility pattern of each driver is drawn as follows:

1. Draw the employment status and the relationship full-time or part-time. $Ta_{t,work}$ is 525 and 240 min for full- and part-time jobs.
2. Determine the mobility weekday $t \in DoW$, five days are drawn for work.
3. Terminate the start time $Ts_{t,j}$ depending on weekdays or weekend.
4. Get the one-way distance $D_{t,j}$.
5. Depending on the distance, determine the travel time $T_{t,j}$ and the means of transport.
6. Repeat step 2–5 for two hobbies with an absence time $Ta_{t,hobby}$ of 90 min and a certain number of errands with an absence time $Ta_{t,errand}$ of 35 min per week.

The maximum number of events is three per day, covering all categories. Next, drivers get a random EV. The vehicle must have at least the range to cover the maximum distance without charging. Equation (16) summarizes the condition for the range D_{EV}.

$$D_{EV} \geq 2 \max_{t \in T} \left(\sum_{j=1}^{N} D_{t,j} \right) \tag{16}$$

A random draw is made from all possible vehicles. The probability that a vehicle is drawn corresponds to the current registration statistics of the Federal Motor Transport Authority in Germany [2] The manufacturer's specification of the battery capacity E_{car}, the charging power P_{car}, and the consumption $EDem_{car}$ are used. The power P_{car} refers to a charging process at an Alternating Current (AC) charging station.

The potential of photovoltaics in the grid area was estimated using the data from source [39]. The portal offers the possibility to query PV data on a building-specific basis. The data include the roof areas of a building with azimuth, tilt, and maximum possible power at an efficiency of 17%. The data were prepared according to the known consumers in the supply area, so that each consumer receives at least one PV system in a 100% scenario. For multi-family houses, the total PV capacity of the building must be distributed among several consumers. The capacity was distributed to the individual residents in proportion to the total annual energy consumption of all consumers in the building.

The weather data from source [40] was used to calculate the final PV generation. The source provides weather data in a square grid of 0.25 degrees in hourly resolution for the whole of Europe. For simplification, the weather data in the investigated supply area were averaged so that every consumer receives the same weather at every hour. The transfer of the weather data into a generation time series is done in Python with the library PVLib [41]. PV feed-in into the public grid is not considered. This means that there are no grid congestion situations due to too high PV feed-in.

Nevertheless, it is possible to quantify the effects of PV on the increasing grid utilization due to e-mobility. To investigate this, different consumer strategies were derived.

In the scenario cases, *Base Case*, *Case A*, and *Case B*, it was assumed that demand for electricity is independent of the dynamic retail price up to a defined threshold C_{Max}. When the threshold is exceeded, the result is a complete demand displacement for the electricity purchase of the EVs. On one hand, this leads to the desired guarantee that not many grid congestions occur. On the other hand, this modeling of the dynamic incentive effect is a clear abstraction of the actual relationship between demand (effect) and retail price (cause). In addition, this modeling technique results in a step function since demand evolves discontinuously at the threshold.

In classical microeconomics, demand is assumed to have a defined price elasticity, which can also reflect the real decision behavior of consumers more realistically. The assumption of an elasticity is also of great importance for the analysis of incentive models for load control, e.g., market- and grid-serving charging [42]. Therefore, a set of price elasticity functions is used in scenario *Case C*.

For the derivation the price elasticity, survey results were used for this study, which were presented in detail in [43]. In the survey design, approximately 90 participants were asked to define three price levels according to the following questions, among others, as part of the online survey: The study provided the following information on price and charging behavior:

- The following total price C_{Min} at AC charging stations is low for me—I would try to charge (Stated in ct/kWh).
- The following total price C_{Max} at AC charging stations is too high for me—I would not charge (Stated in ct/kWh).
- The following total price *PLimit* at AC charging stations is quite high for me—but I would charge if necessary (Stated in ct/kWh).

Figure 3 shows the results for C_{Min} and C_{Max} of the survey in a scatter histogram. Above and to the right of the scatter plot, the absolute frequency distribution (histograms) of C_{Min} and C_{Max} are also plotted. The scatterplot indicates a correlation between C_{Max} and C_{Min}. A higher indication of C_{Min} tends to result in feedback of an increased C_{Max} for the same respondent. This correlation will be analyzed in more detail using a subsequent regression analysis.

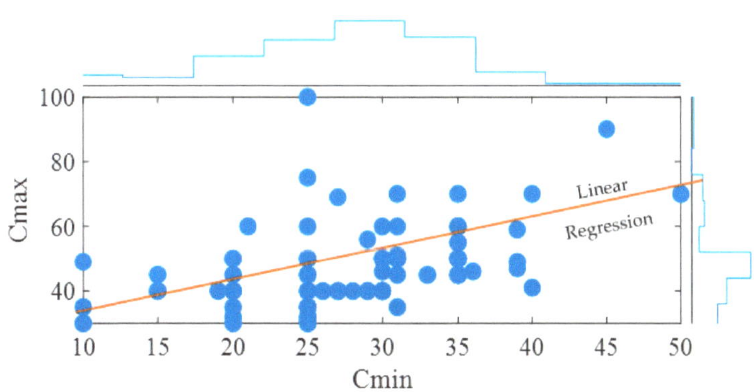

Figure 3. Distribution and correlation between the price levels C_{Min} and C_{Max} in [ct/kWh].

What is striking in the histogram for C_{Min} is the broad distribution of responses. This suggests that end consumers have different price sensitivities, because all customers got the same average price level of 40 ct/kWh in the survey.

In the following, C_{Min} and C_{Max} are to be interpreted as marginal prices at a very high and low SoC. In a next step, the relationship between C_{Min} and C_{Max} is quantified using linear regression and the significance of the slope parameter is assessed. The estimator for the slope parameter m in the amount of 0.75 is significant compared to the H0 hypothesis $m = 0$. The slope of the function between the point C_{Limit} is derived from the difference of the two marginal prices C_{Min} and C_{Max} to the marginal price C_{Limit}. The results of the subsequent regression analysis are shown in Table 2.

Table 2. Regression parameters of price sensitivity curve.

C_{Min} to C_{Max}	Estimate	SE	*p*-Value
Intercept	26.75	4.71	1.94×10^{-7}
m	0.75	0.16	1.31×10^{-5}
C_{Min} to $C_{Limit} - C_{Min}$			
Intercept	26.73	4.70	1.95×10^{-7}
m	−0.25	0.16	0.12
C_{Max} to $C_{Max} - C_{Limit}$			
Intercept	−5.36	1.75	0.003
m	0.27	0.035	2.65×10^{-11}

A positive correlation between the difference of C_{Min} and C_{Max} to C_{Limit} could also be demonstrated, indicating a differentiation of the slopes of the demand functions.

For the concrete demand function, a simplified assumption was made that the points C_{Max}, C_{Limit} and C_{Min} are associated with a respective SoC of 20%, 40% and 90%. The values 90% and 20% correspond to a battery-saving charging behavior. [N6] The third interpolation point of 40% was derived from the fact that at a SOC of 40%, over 90% of all motion profiles can be mapped.

From the regression analysis, the support points for the continuous demand function F(SoC) were derived: $C_{Max} = (C_{Max}; 15\%)$, $C_{Limit} = (C_{Max} - 0.27{\cdot}C_{Max}; 40\%)$, $C_{Min} = (C_{Min}; 85\%)$. The H0-Hypthesis for the slope between min and $C_{Max} - C_{Min}$ is rejected in case of $\alpha < 12\%$. The lower value of m underlines the existence of a different slope along the curve, but a future survey with a larger number of participants should be conducted to further investigate the relationship and decrease the standard deviation of the estimator. The course of the demand function between the grid points was interpolated with the help of a spline. In a final step, it was considered that, according to the survey results, individual demand curves can be assumed (see histogram of C_{Min} in Figure 3). This is particularly important when considering the simultaneity of the reaction to a price change.

Based on the survey results, it can be expected that a defined price signal results in a demand-specific reduction in demand. In terms of content, this reflects a different and individual "range anxiety" of individual residents, since the demand restraint leads to a reduced SoC. Residents will judge the "value" of the SoC differently.

The graph in Figure 4 shows thirteen exemplary trajectories of each demand function F(SoC). Each curve represents a corresponding risk preference. This becomes comprehensible when considering a defined end customer price with the associated advised SoC. For example, at a price of approx. 62 ct/kWh, a very risk-averse household will achieve a minimum SoC of about 30%, while a very price-sensitive household will not seek charging at this price and will charge its EV solely from its home PV system. Conversely, when the retail price is below the median price level of 40 ct/kWh, the individual profiles also aim for a different SoC state. This corresponds to the fact that price-sensitive users want to maintain charging flexibility for their own PV electricity or for future hours with even significantly lower electricity prices and thus do not aim for a SoC of 100%.

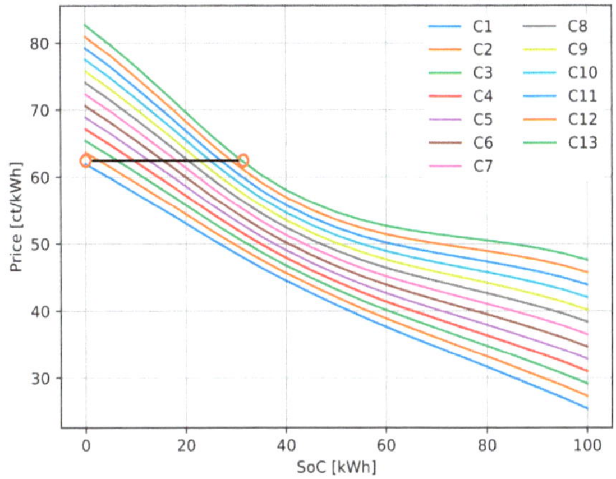

Figure 4. 13 Trajectories of each demand function depending on the State of charge—drawn from the survey. The trajectories show different exemplary price sensitivity behavior.

The assignment of end users to one of the mapped demand functions F(SoC) is done with a random variable that picks up the relative frequency distribution of C_{Min}. This ensures that extreme manifestations of risk aversion occur less frequently in the total consumer base than demand functions located in the middle of the set of curves.

Exemplary behavior curves of a specific car are shown in Figure 5. The usage behavior is the same through all scenarios since the visited car stays the same. In the *Base Case*, the car is only charged when it becomes available without optimizing, using most of the energy from the grid. The second Case shows how optimizing the PV usage increases the SoC and reduces charging from the grid. As the market price is constant, the grid charging is optimized according to the grid fee signal, resulting in many short charging periods. In *Case B*, a dynamic spot market price is added, which has a much larger impact than the grid fee did, resulting in a preference to charge when the market prices are low. Often, market prices are low during the day when cars are unavailable (see Figure 6), creating a preference to charge shortly before or after the unavailability of the car. Finally, *Case C* introduces the derived demand function F(SoC), which is based on the current SoC and results in a lower average SoC.

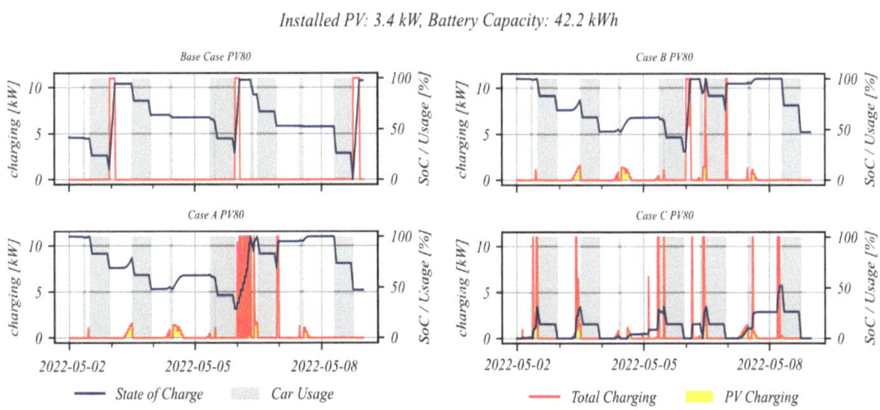

Figure 5. Exemplary resulting charging curves of the same EV visualized for the *Base Case*, *Case A*, *Case B*, and *Case C* with 80% installed PV potential on buildings in the simulation.

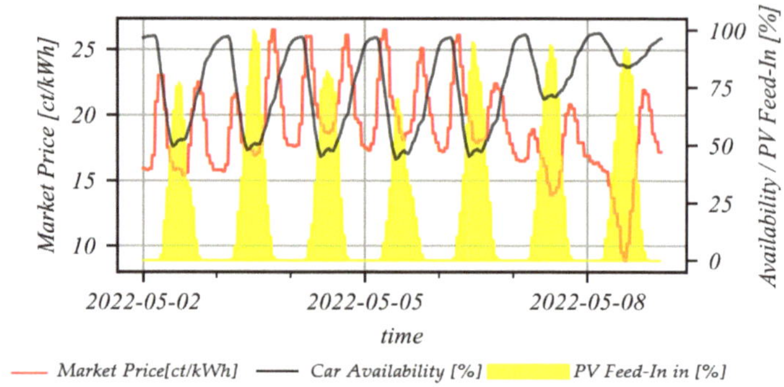

Figure 6. Spot market prices + percentage of PV feed-in in contrast to the car availability.

4. Results

In this section, the main results of the evaluation are presented. For this purpose, the analyses are divided into a comparison of the *Base Case* with *Case A* (increase of PV self-consumption) and an analysis of the strategies regarding a more realistic representation of user behavior in the case of dynamic market prices (*Case B* and *Case C*).

For the following explanations, it is useful to remember Figure 6. It shows the market price, the normalized PV-feed in, and the average availability of the EVs at home. The figure illustrates the challenge of synchronizing the charging of EVs with periods of low prices or high PV feed-in in the different use cases. Times of high PV-feed in and low market prices are often attended to low availability of EVs for charging operations.

4.1. Analysis of a Self-Supply-Based Charging Strategy

In a first step of the analysis, it is analyzed whether an increased PV self-consumption can significantly contribute to the reduction of the grid load. Figure 7 shows the sorted duration line of the average transformer utilization. In all simulations of different PV capacities in *Case A*, an average transformer utilization of >50% is observed in less than 2% of the simulation hours. An increase in installed PV capacity results in a more noticeable reduction in average utilization, especially in the middle and at the right edge of the duration line.

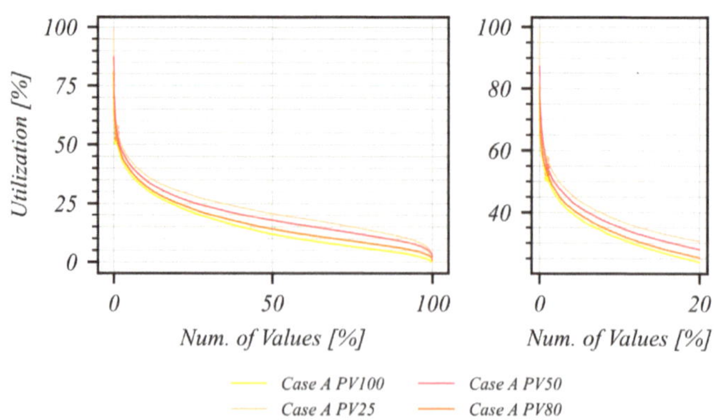

Figure 7. Impact of the PV expansion on the main transformer utilization.

When considering the 1% quantile, which shows an average utilization between 50–55%, the reduction effect is less noticeable with an increase in installed PV capacity. This effect is understandable since the PV feed-in is based on a fluctuating supply (solar radiation) and thus cannot be controlled, and in addition, there is a natural asynchrony between PV feed-in and charging demand (see Figure 6). A closer look at the values in the 1% quantile shows that these occur in hours with low/no solar radiation where the PV supply cannot act grid serving.

A more detailed breakdown of the utilization analysis for the respective 10 transformers of the distribution grid shows that the PV effect is also differentiated depending on the area. The type of use in the respective area, and thus the ratio of installed PV capacity to the number of EVs, have a decisive influence on the reduction potential. The upper graph of Figure 8 shows the number of EVs and the potential of PV generation. Basically, a correlation with the transformer utilization is shown here in the second graph. The different reduction potential of the second graph can be explained in connection with the different shares of PV to electromobility in the individual areas. Therefore, the most significant impact can be seen in subgrid 2. Figure 8 shows that a detailed evaluation is useful, especially for inhomogeneous grid areas, to determine the influence of different PV-installed capacity precisely.

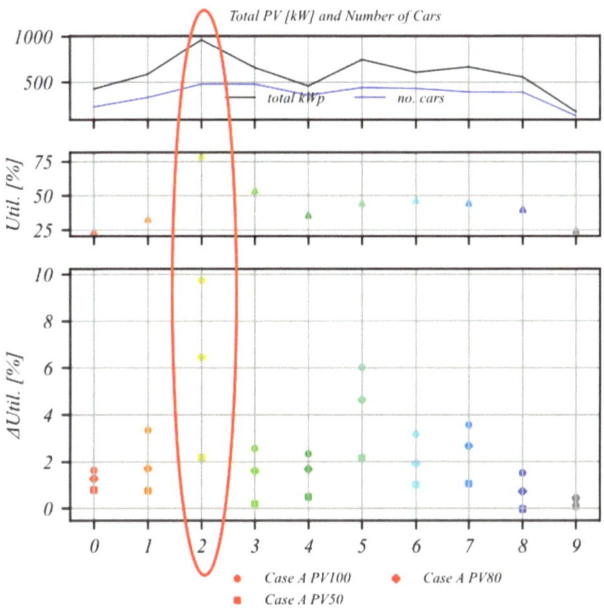

Figure 8. Impact of the demand function on the utilization on grid level for all 10 transformers (0–9 and colored) with its PV installation size and number of cars in its subgrid.

4.2. Analysis of More Complex Derivation of the Charging Behavior

In the following, the effects of the more realistic mapping of the demand function F(SoC) and the consideration of a dynamic market price on the grid utilization are evaluated. The individual utilization under a high penetration of installed PV capacity of 80% of the potential is shown in the right panel of Figure 9. The 1% quantile as a risk measure for the grid utilization is clearly differentiated over the four considered cases. It is understandable that the introduction of a self-consumption optimizing (red) and additionally to a dynamic market price aligned charging behavior (green) lead to a clear increase in the utilization factor in the 1% quantile. However, from the authors' point of view, a grid planning based on these values would lead to an inefficient grid planning. This is shown by the scenario of an additional consideration of a differentiated demand elasticity (red) that is closer to reality. The higher simultaneity in the charging behavior caused by the dynamic market price component does not represent individual risk preferences—but creates a swing effect for all simulation households. The demand functions parameterized based on the survey mitigate this effect of the dynamic market price component in the residential tariff. This result demonstrates the need for a more in-depth analysis of user behavior in further research to map this aspect more precisely and to refine the impact analysis on grid utilization.

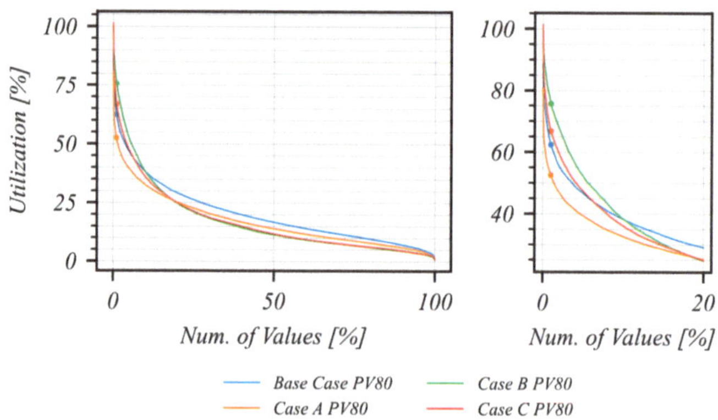

Figure 9. Impact of the demand function on the mean transformer utilization.

4.3. Grid and Economic Impact Analyses

In the following analyses, the grid utilization as well as the economic implications of the individual cases are considered from the end customer's perspective.

The individual cases require a corresponding regulatory framework. It is of decisive importance to know about the individual benefit function of the user's behavior. Assuming an economically rational agent, the more complex user behavior depicted in *Case C* should therefore be compatible with an associated economic incentive.

In Table 3, the consumption (grid- and PV-side) as well as the associated costs for a detailed analysis (split market- and grid-side) are presented.

Table 3. Result overview for the scenarios with a used PV potential of 80%.

	Unit	*Base Case*	*Case A*	*Case B*	*Case C*
Charging					
Grid	%	98	71	71	69
PV	%	2	29	29	31
Shifted	%	<0.3	<0.1	4.9	38.9
Pricing					
Average Cost	ct/kWh	21.8	15.6	15.2	14.0
Market	%	78	79	72	73
Grid	%	22	21	28	27
ø Grid Fee	ct/kWh	4.3	3.9	3.4	3.0
95% quantile	ct/kWh	6.4	5.8	7.4	6.6
ø 95–100%	ct/kWh	6.8	6.2	8.8	7.5
Utilization					
ø Util.	%	20	17	17	17
95%-quantile	%	47	40	52	47
ø 95–100%	%	56	47	67	59

Figure 10 shows the utilization for a given transformer throughout all scenarios for one week of simulation time. Comparing the *Base Case* to *Case A*, one can see that a reduction in grid utilization is performed through the optimization of the PV charging. Furthermore, adding the price signal without differentiation in the user behavior creates large utilization spikes (*Case B*), which are reduced if the benefit function F(SoC) is added (*Case* C).

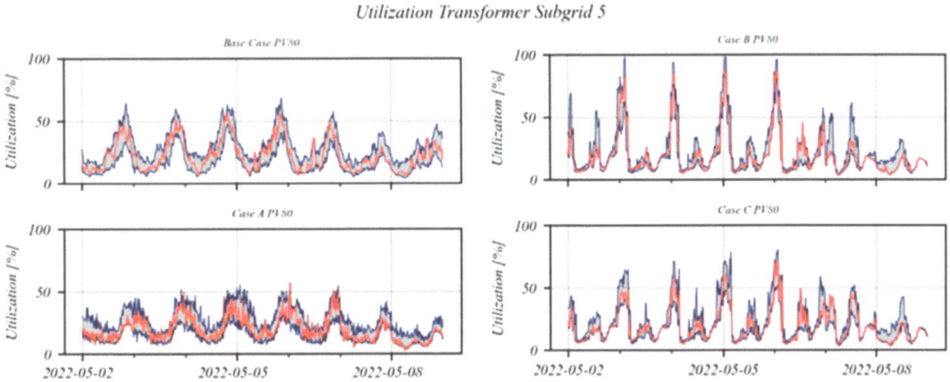

Figure 10. Utilization in percent on an exemplary subgrid throughout the different scenarios analyzed in this paper—the minimum, mean, and maximum utilization of the given transformer are shown as lower and upper band of this.

On one hand, looking at consumption shows the increase in PV self-consumption in *Case A–C* described earlier. The largest cost reduction is achieved by the significant increase in PV self-consumption from *Case A* onwards due to better synchronization of charging and PV feed-in.

The further cost reduction in *Case B* and *Case C* is largely based on an additional response to the dynamic price signal. It is noticeable that the differentiation of the demand function can significantly reduce the peak grid congestions as also seen in Figure 9. The level in this range almost reaches the utilization level of the *Base Case* again (without dynamic market price signal). However, this must be contrasted with the fact that a correspondingly mapped user behavior, in contrast to PV synchronization, is only provided with a relatively low economic incentive in the model evaluation (reduction of the market-side costs from 15.2 to 14.0 ct/kWh).

Optimizing PV usage reduces the cost by 6.2 ct/kWh, which shows the greatest potential for cost reduction from the users' point of view.

5. Conclusions

In this paper, we presented a multi-agent-based simulation for a realistic view of possible grid congestions due to the increase in e-mobility. This model is open-source available on GitHub at https://github.com/NOWUM/smartdso/. Furthermore, we investigated the effect of dynamic grid and market prices. To summarize, the dynamic market prices result in high simultaneity factors, which can be handled by the provided dynamic grid fee and differentiated price sensitivity curves.

In the future, it can be assumed that due to increased fluctuating feed-in, the day-ahead and intraday prices will be subject to higher fluctuations, and thus, the incentive for dynamic tariffs should increase. Nevertheless, it can be assumed that a very high degree of automation and a low restriction of the quality of life are required to observe the user behavior depicted here in practice.

The presented methodology focuses on the agent-based modeling of individual mobility behavior where each mobility demand is priced by the grid operator/capacity provider.

Modeling the user's behavior independently of the grid operator prices allows simulating heterogenous individual charging behavior of the agents.

The assigned user behavior in *Case C* leads to a significant change in grid utilization, which highlights the requirement for more realistic modelling of charging behavior when analyzing future grid development plans. As a next step, the user behavior and scope of functions of the market actors should be improved. Moreover, other sector-coupled technologies such as power-to-heat and battery storages should be introduced to cover the total scope of energy demand. This allows to further investigate a more realistic view of the impact of flexibility from e-Mobility and PV systems on the energy system.

Author Contributions: Conceptualization, F.M., C.R., R.S. and D.S.; methodology, R.S., F.M. and C.R.; software, F.M. and C.R.; validation, F.M., C.R. and R.S.; formal analysis, R.S. and C.R.; investigation, F.M. and C.R.; resources, D.S.; data curation, F.M. and C.R..; writing—original draft preparation, F.M., C.R., R.S. and D.S.; writing—review and editing, F.M., C.R., R.S. and D.S.; visualization, C.R. and R.S.; supervision, R.S.; project administration, D.S.; funding acquisition, D.S. All authors have read and agreed to the published version of the manuscript.

Funding: This research was partially financed by the European Fund for Regional Development (EFRE), Grant Nr. EFRE-0801805.

Data Availability Statement: Results can be reproduced using the open-source software provided. Unfortunately, we do not have permission to publish the grid data to the public as it was only made available for research.

Conflicts of Interest: The authors declare no conflict of interest. The funders had no role in the design of the study; in the collection, analyses, or interpretation of data; in the writing of the manuscript; or in the decision to publish the results.

Nomenclature

RES	Renewable energy sources
EEG	Erneuerbare-Energien-Gesetz
EV	Electric Vehicle
PV	Photovoltaic
PSA	Power System Analysis
SLP	Standard Load Profile
AC	Alternating Current
SoC	State of Charge
T	Set of time steps
t	time step
C	Set of EV/cars
c	car
D	[km] Travel Distance
D_{EV}	[km] range of the EV
P_{car}	[kW] maximum charging power of a car
E_{car}	[kWh] maximum capacity of a car
$EDem_{car,t}$	[kW] demand of a car at time step t
$V_{car,t}$	[kWh] capacity level of a car at time step t
G_t	[ct/kWh] dynamic grid fee at time step t
B	benefit, which depends on $V_{car,t}$
S	set of segments
s	segment of linearized B function
z_s	variable to indicate the active segment
$SegVLow_s$	lower capacity of linearized B function
$SegVUp_s$	upper capacity of linearized B function
$SegCoef_s$	slope of the linearized B function
$VTotal$	[kWh] sum of single $V_{car,t}$
Ppv_t	[kW] power of the attached PV system at t
$PCharge_t$	[kW] charged power at timestep t
C_{Limit}	[ct/kWh] mean price for SoC 40%
C_{Max}	[ct/kWh] upper price, charge when SoC $<$ 20%
C_{Min}	[ct/kWh] lower price, charge when SoC $<$ 90%

References

1. Arbeitsgruppe Energie-Monitoring. *Monitoringbericht 2021*; Bundesnetzagentur für Elektrizität, Gas, Telekommunikation, Post und Eisenbahnen: Berlin, Germany, 2021; p. 540.
2. Statistik Kraftfahrtbundesamt. Available online: https://www.kba.de/DE/Statistik/Fahrzeuge/fahrzeuge_node.html;jsessionid=6437EAC3E3F71138982F44CD869FEFEC.live21324 (accessed on 27 October 2022).
3. Elektromobilität in Deutschland. Available online: https://www.bmwk.de/Redaktion/DE/Dossier/elektromobilitaet.html415 (accessed on 3 April 2023).
4. Robinson, J.; Erickson, L.E. Infrastructure for Charging Electric Vehicles. In *Solar Powered Charging Infrastructure for Electric Vehicles*; Erickson, L.E., Robinson, J., Brase, G., Cutsor, J., Eds.; CRC Press: Boca Raton, FL, USA, 2016; pp. 35–51. ISBN 978-1-4987-3156-0.
5. Kasten, P.; Minnich, L.; Randrianarisoa, J.; Ritter, D.; Vogel, M.; Dünzen, K.; Just, L.; Novirdoust, A.A.; Diers, H.; Niesler, N.; et al. Szenarien Und Regulatorische Herausforderungen Für Den Aufbau Der Ladeinfrastruktur Für Elektrische Pkw Und Lkw. Available online: https://www.oeko.de/fileadmin/oekodoc/ENSURE-II_Ladeinfrastruktur.pdf (accessed on 3 April 2023).
6. Lopes, J.A.P.; Soares, F.J.; Almeida, P.M.R. Integration of Electric Vehicles in the Electric Power System. *Proc. IEEE* **2011**, *99*, 168–183. [CrossRef]
7. Green, R.C.; Wang, L.; Alam, M. The Impact of Plug-in Hybrid Electric Vehicles on Distribution Networks: A Review and Outlook. *Renew. Sustain. Energy Rev.* **2011**, *15*, 544–553. [CrossRef]
8. Brinkel, N.; Alskaif, T.; van Sark, W. The Impact of Transitioning to Shared Electric Vehicles on Grid Congestion and Management. In Proceedings of the 2020 International Conference on Smart Energy Systems and Technologies (SEST), Istanbul, Turkey, 7–9 September 2020.
9. Resch, M.; Buhler, J.; Schachler, B.; Sumper, A. Techno-Economic Assessment of Flexibility Options Versus Grid Expansion in Distribution Grids. *IEEE Trans. Power Syst.* **2021**, *36*, 3830–3839. [CrossRef]
10. Brinkel, N.B.G.; Schram, W.L.; AlSkaif, T.A.; Lampropoulos, I.; van Sark, W.G.J.H.M. Should We Reinforce the Grid? Cost and Emission Optimization of Electric Vehicle Charging under Different Transformer Limits. *Appl. Energy* **2020**, *276*, 115285. [CrossRef]
11. Zethmayr, J.; Kolata, D. Charge for Less: An Analysis of Hourly Electricity Pricing for Electric Vehicles. *World Electr. Veh. J.* **2019**, *10*, 6. [CrossRef]
12. Sinha, S.; Chandel, S.S. Review of Software Tools for Hybrid Renewable Energy Systems. *Renew. Sustain. Energy Rev.* **2014**, *32*, 192–205. [CrossRef]
13. Gottwalt, S. Managing Flexible Loads in Residential Areas. 2015. Available online: https://publikationen.bibliothek.kit.edu/1000048803 (accessed on 3 April 2023).
14. Su, J. Smart Management of Electric Vehicles Charging in Distribution Networks. Ph.D. Thesis, Auckland University of Technology, Auckland, New Zealand, 2020.
15. Babrowski, S.; Heinrichs, H.; Jochem, P.; Fichtner, W. Load Shift Potential of Electric Vehicles in Europe. *J. Power Sources* **2014**, *255*, 283–293. [CrossRef]
16. Su, W.; Wang, J.; Roh, J. Stochastic Energy Scheduling in Microgrids With Intermittent Renewable Energy Resources. *IEEE Trans. Smart Grid* **2014**, *5*, 1876–1883. [CrossRef]
17. Diaz-Cachinero, P.; Munoz-Hernandez, J.I.; Contreras, J. A Linear Model for Operating Microgrids with Renewable Resources, Battery Degradation Costs and Electric Vehicles. In Proceedings of the 2018 15th International Conference on the European Energy Market (EEM), Lodz, Poland, 27–29 June 2018; IEEE: Lodz, Poland, 2018; pp. 1–5.
18. Soares, J.; Borges, N.; Fotouhi Ghazvini, M.A.; Vale, Z.; de Moura Oliveira, P.B. Scenario Generation for Electric Vehicles' Uncertain Behavior in a Smart City Environment. *Energy* **2016**, *111*, 664–675. [CrossRef]
19. Schäuble, J.; Kaschub, T.; Ensslen, A.; Jochem, P.; Fichtner, W. Generating Electric Vehicle Load Profiles from Empirical Data of Three EV Fleets in Southwest Germany. *J. Clean. Prod.* **2017**, *150*, 253–266. [CrossRef]
20. Pflaum, P.; Alamir, M.; Lamoudi, M.Y. Probabilistic Energy Management Strategy for EV Charging Stations Using Randomized Algorithms. *IEEE Trans. Contr. Syst. Technol.* **2018**, *26*, 1099–1106. [CrossRef]
21. Gerritsma, M.K.; AlSkaif, T.A.; Fidder, H.A.; Sark, W.G.J.H.M. van Flexibility of Electric Vehicle Demand: Analysis of Measured Charging Data and Simulation for the Future. *WEVJ* **2019**, *10*, 14. [CrossRef]
22. Wang, B.; Zhao, D.; Dehghanian, P.; Tian, Y.; Hong, T. Aggregated Electric Vehicle Load Modeling in Large-Scale Electric Power Systems. *IEEE Trans. Ind. Applicat.* **2020**, *56*, 5796–5810. [CrossRef]
23. Diaz-Cachinero, P.; Munoz-Hernandez, J.I.; Contreras, J. A Microgrid Model With EV Demand Uncertainty and Detailed Operation of Storage Systems. *IEEE Trans. Ind. Applicat.* **2022**, *58*, 2497–2511. [CrossRef]
24. Xydas, E.; Marmaras, C.; Cipcigan, L.M. A Multi-Agent Based Scheduling Algorithm for Adaptive Electric Vehicles Charging. *Appl. Energy* **2016**, *177*, 354–365. [CrossRef]
25. Zhang, J.; Yan, J.; Liu, Y.; Zhang, H.; Lv, G. Daily Electric Vehicle Charging Load Profiles Considering Demographics of Vehicle Users. *Appl. Energy* **2020**, *274*, 115063. [CrossRef]
26. Karapidakis, E.; Tsikalakis, A.; Paspatis, A.; Fotakis, E.; Stavrakakis, G.; Chatzipoulka, C.; Zervas, P. Grid Operation Assessment under a Specific EV Chargers Deployment Plan in the City of Heraklion. *Electronics* **2021**, *10*, 2831. [CrossRef]
27. Konstantinidis, G.; Karapidakis, E.; Paspatis, A. Mitigating the Impact of an Official PEV Charger Deployment Plan on an Urban Grid. *Energies* **2022**, *15*, 1321. [CrossRef]

28. Paspatis, A.; Fiorentzis, K.; Katsigiannis, Y.; Karapidakis, E. Smart Campus Microgrids towards a Sustainable Energy Transition—The Case Study of the Hellenic Mediterranean University in Crete. *Mathematics* **2022**, *10*, 1065. [CrossRef]
29. Pagani, M.; Korosec, W.; Chokani, N.; Abhari, R.S. User Behaviour and Electric Vehicle Charging Infrastructure: An Agent-Based Model Assessment. *Appl. Energy* **2019**, *254*, 113680. [CrossRef]
30. Sundstrom, O.; Binding, C. Flexible Charging Optimization for Electric Vehicles Considering Distribution Grid Constraints. *IEEE Trans. Smart Grid* **2012**, *3*, 26–37. [CrossRef]
31. Sundström, O.; Binding, C. Planning Electric-Drive Vehicle Charging under Constrained Grid Conditions. In Proceedings of the 2010 International Conference on Power System Technology, Hangzhou, China, 24–28 October 2010; pp. 1–6.
32. Sun, L.; Lubkeman, D. Agent-Based Modeling of Feeder-Level Electric Vehicle Diffusion for Distribution Planning. *IEEE Trans. Smart Grid* **2021**, *12*, 751–760. [CrossRef]
33. Vopava, J.; Bergmann, U.; Kienberger, T. Synergies between E-Mobility and Photovoltaic Potentials—A Case Study on an Urban Medium Voltage Grid. *Energies* **2020**, *13*, 3795. [CrossRef]
34. Tsiakis, P.; Shah, N.; Pantelides, C.C. Design of Multi-Echelon Supply Chain Networks under Demand Uncertainty. *Ind. Eng. Chem. Res.* **2001**, *40*, 3585–3604. [CrossRef]
35. Brown, T.; Hörsch, J.; Schlachtberger, D. PyPSA: Python for Power System Analysis. *JORS* **2018**, *6*, 4. [CrossRef]
36. infas, DLR, IVT und infas 360 Mobilität in Deutschland (on behalf of BMVI). 2018. Available online: https://www.mobilitaet-in-deutschland.de/archive/pdf/MiD2017_Ergebnisbericht.pdf (accessed on 10 April 2023).
37. Standardlastprofile-Strom. Available online: https://www.bdew.de/energie/standardlastprofile-strom/ (accessed on 27 October 2022).
38. Jazaeri, J.; Gordon, R.L.; Alpcan, T. Influence of Building Envelopes, Climates, and Occupancy Patterns on Residential HVAC Demand. *J. Build. Eng.* **2019**, *22*, 33–47. [CrossRef]
39. Solarkataster. Available online: https://www.energiedaten.nrw.de (accessed on 27 October 2022).
40. Muñoz Sabater, J. ERA5-Land Hourly Data from 1950 to Present. Copernicus Climate Change Service (C3S) Climate Data Store (CDS). 2019. Available online: https://cds.climate.copernicus.eu/cdsapp#!/dataset/10.24381/cds.e2161bac?tab=overview (accessed on 10 April 2023).
41. Holmgren, F.; Hansen, W.; Mikofski, A. Pvlib Python: A Python Package for Modeling Solar Energy Systems. *JOSS* **2018**, *3*, 884. [CrossRef]
42. Frank, D.; Kleis, K.; Dr., Schmid, E.; Schrader, T.-S.; Schwarze, E. Hendrik Zimmermann Effektive Stromnetzentgelte Für Die Transformation. Available online: https://www.germanwatch.org/de/85071 (accessed on 3 April 2023).
43. Georg, J.H. "Smartes" Laden an Öffentlichen Ladesäulen: Anreizmodell Für Markt- Und Netzdienliches Laden, Energiewirtschaftliche Tagesfragen 72. 2022. Available online: http://www.jhc-energie.de/.cm4all/uproc.php/0/et-Artikel-29072022-PDF_18115_et_7-8_2022_S.37-41.pdf (accessed on 3 April 2023).

energies

Article

An Enhanced Path Planner for Electric Vehicles Considering User-Defined Time Windows and Preferences

Maximiliano Cubillos [1,*], Mauro Dell'Amico [2], Ola Jabali [1], Federico Malucelli [1] and Emanuele Tresoldi [3]

[1] Dipartimento di Elettronica, Informazione e Bioingegneria, Politecnico di Milano, Piazza Leonardo da Vinci 32, 20133 Milano, Italy; ola.jabali@polimi.it (O.J.); federico.malucelli@polimi.it (F.M.)
[2] Department of Sciences and Methods for Engineering, University of Modena and Reggio Emilia, Via Amendola 2, 42122 Reggio Emilia, Italy; mauro.dellamico@unimore.it
[3] Dipartimento di Informatica, Università degli Studi di Milano, Via Celoria 18, 20133 Milano, Italy; emanuele.tresoldi@unimi.it
* Correspondence: maximiliano.cubillos@polimi.it

Abstract: A number of decision support tools facilitating the use of Electric Vehicles (EVs) have been recently developed. Due to the EVs' limited autonomy, routing and path planning are the main challenges treated in such tools. Specifically, determining at which Charging Stations (CSs) to stop, and how much the EV should charge at them is complex. This complexity is further compounded by the fact that charging times depend on the CS technology, the EV characteristics, and follow a nonlinear function. Considering these factors, we propose a path-planning methodology for EVs with user preferences, where charging is performed at public CSs. To achieve this, we introduce the Electric Vehicle Shortest Path Problem with time windows and user preferences (EVSPPWP) and propose an efficient heuristic algorithm for it. Given an origin and a destination, the algorithm prioritizes CSs close to Points of Interest (POIs) that match user inputted preferences, and user-defined time windows are considered for activities such as lunch and spending the night at hotels. The algorithm produces flexible solutions by considering clusters of charging points (CPs) as separate CSs. Furthermore, the algorithm yields resilient paths by ensuring that recommended paths have a minimum number of CSs in their vicinity. The main contributions of our methodology are the following: modeling user-defined time windows, including user-defined weights for different POI categories, creating CSs based on clusters of CPs with sufficient proximity, using resilient paths, and proposing an efficient algorithm for solving the EVSPPWP. To facilitate the use of our methodology, the algorithm was integrated into a web interface. We demonstrate the use of the web interface, giving usage examples and comparing different settings.

Keywords: electric vehicles; shortest path; points of interest; path planner

Citation: Cubillos, M.; Dell'Amico, M.; Jabali, O.; Malucelli, F.; Tresoldi, E. An Enhanced Path Planner for Electric Vehicles Considering User-Defined Time Windows and Preferences. *Energies* **2023**, *16*, 4173. https://doi.org/10.3390/en16104173

Academic Editor: Javier Contreras

Received: 18 April 2023
Revised: 15 May 2023
Accepted: 16 May 2023
Published: 18 May 2023

1. Introduction

Electric vehicle (EV) sales have significantly grown in the recent years. In Europe, for example, EV registrations have tripled from 3.5% in 2019 to 11.6% in 2020. This includes a 6.2% increase in fully electric vehicles [1]. Indeed, EVs are perceived to be a sustainable alternative to traditional combustion engine vehicles, offering advantages in terms of performance and reduced emissions [2]. However, a number of barriers impede a wider EV uptake. These include high purchase costs, limited autonomy, scarce charging infrastructure and lengthy charging times. To counter some of these barriers, an increasing number of decision support tools facilitating the use of EVs are being deployed. Such tools mainly rely on optimizing operational issues related EVs. In this vein, several contributions from the transportation science community have developed models and solution methods to EV routing problems (see Kucukoglu et al. [3] and Froger et al. [4] for examples).

We propose a methodology for planning EV paths considering user preferences, where charging is performed at public Charging Stations (CSs). The need for such a methodology for private EV users was identified through the eCharge4Drivers project, which is funded by the European Union Horizon 2020 programme. The proposed methodology was integrated in a web interface enabling a large audience to use it. The core optimization problem we deal with relates to planning the shortest path (in terms of time) for an EV needing to travel between a given origin and a given destination. This problem is particularly relevant for long-distance trips, where the shortest path between the origin and the destination exceeds the EV's autonomy. The key decisions to be made in such problems are where to charge and how much to charge. The latter entails an amount of time which depends on the CS technology, the EV characteristics and follows a non-linear function. Moreover, as the charging infrastructure in road networks outside cities is fairly sparse, EVs often need to deviate from their shortest paths to reach CSs. The intertwined relationship between the time spent in detouring and the time spent charging is the main complicating feature of the Electric Vehicle Shortest Path Problem (EVSPP).

As EV charging times may take several hours, users (i.e., drivers) may prefer having charging stops at locations that match their interests. We consider that such interests may be captured via Points of Interests (POIs) that include categories such as restaurants, shopping areas, and nature sites. Moreover, we also consider that the user may want to perform some particular activity during specific time windows, like having lunch or spending the night at a hotel. Such user-defined time windows are also matched with relevant POIs locations. Thus, our overarching goal is to propose an efficient offline methodology integrating user preferences for POIs and user-specific time window into an EVSSP setting. To do so, we introduce the Electric Vehicle Shortest Path Problem with time windows and user preferences (EVSPPWP).

We model the EVSPPWP and propose an efficient heuristic algorithm to solve it. In particular, the algorithm prioritizes CSs that are close to POIs matching the user preferences. Furthermore, while we provide planned (offline) paths, we propose a number of algorithmic enhancements aimed at producing solutions that are practically executable in real time on large road networks. Specifically, the algorithm provides the user with robust and flexible solutions. To incorporate robustness, we only consider paths that have at least a given number CSs in their vicinity. This ensures that the EV could always reach a CS, in case its predicted energy consumption does not match the actual one. Additionally, it may occur that the EV encounters an occupied CS, which was planned in its path. To counter such situations, our algorithm produces flexible solutions by considering a CS as a cluster of charging points (CPs). While previous studies have used clustering for districting to reduce problem sizes in EVRPs [5], we focus on providing solutions so that users can find alternative CPs in cases where selected CPs are not available. Specifically, we produce small sized clusters of CPs that we treat as separate CSs. We plan the path based on those CSs. This entails having a number of CPs in the vicinity of a planned charging stop. Thus, if a CP is busy upon arrival, the user would have multiple alternatives in its vicinity.

Considering the previously discussed enhancements, we develop a heuristic algorithm for the EVSPPWP. We first include weights for each CS according to its vicinity to POIs that match the user-defined preferences. We then use a modified A^* algorithm to establish the nodes where the user-defined time windows are going to be performed. Considering these nodes along with the user-defined origin and destination, we sequentially solve a series of modified EVSPPs (using the algorithm of Kullman et al. [6]), which we then aggregate into a complete solution. Lastly, the algorithm provides up to three alternative paths in a web interface where the users can visualize the path details.

To summarize, the main contributions of this papers are as follows.

1. We incorporate time windows for activities during the trip that are defined by the user, such as taking lunch breaks, visiting POIs, or spending the night at a hotel.
2. We account for user preferences by using weights for different POI categories.
3. We consider CSs based on clusters of CPs with sufficient proximity, thus allowing the user flexibility in the choice of a CP upon arrival.
4. We ensure resilient paths that have a minimum number of CSs within a threshold distance

5. We propose an efficient heuristic algorithm to solve the resulting EVSPPWP.
6. We demonstrate our methodology on a series of real usage examples.

The remainder of this paper is organized as follows. In Section 2, we present a brief literature review on related work. In Section 3, we describe the EVSPPWP and present a Mixed Integer Linear Programming (MILP) model for it. In Section 4, we present our solution methodology. In Section 5, we illustrate the usage of the algorithm, giving usage examples and comparing different settings in the area of northern Italy. Finally, we provide conclusions in Section 6.

2. Related Work

Our core problem falls into the category of EVSPPs, which mainly consists of shortest path problems in which the objective is to minimize the travel time between a given origin and a given destination accounting for the EV's limited autonomy and charging constraints. The main decisions in the EVSPP relate to where and how much to charge at a given set of available CSs. Our algorithm for solving the EVSPP is embedded in a decision support tool for planning EV paths. There are existing decision tools for planning variants of the EVSPP, including A Better Route Planner [7], Zap-Map EV [8], and PlugShare [9]. These tools allow the user to select an origin and a destination, including user-defined waypoint stops. As these tools are commercial, we are not able to verify their modeling assumptions, nor certify the quality of their solutions. Therefore, in the following we focus on academic contributions on the EVSPP.

The EVSPP is strongly related to the Electric Vehicle Routing Problem (EVRP). The main decisions in this latter problem relate to routing a set of EVs from a central depot to serve a set of customers. The limited autonomy of the EV is typically handled by allowing it to detour to CSs and recharge. There has been a rapid growth in studies addressing the EVRP in recent years, with several variants being proposed (see Schniffer et al. [10] and Kucukoglu et al. [3] for comprehensive reviews). From the early formulations of the problem [11], variants of the EVRP have evolved to include features such as time windows and cargo capacity constraints [12]; heterogeneous fleets and charging infrastructure [13,14]; energy consumption functions and charging profiles [15–17]; energy consumption uncertainty [18]; and limited charging capacity [4].

A critical modeling assumption in the EVRP regards the amount of energy to charge at each recharging stop. While some studies assume the EV must completely recharge before leaving a CS [11,19,20], other works consider the charging amount to be a decision of the model [12,17,21]. Several studies have assumed linear charging functions [13,22]. In practice, the charging function of an EV is nonlinear with respect to time, and depends on the charging technology of the CS. The nonlinear charging functions are modeled as piece-wise linear concave functions by Montoya et al. [17] and Froger et al. [21]. We adopt similar assumptions in this paper. For a repository with such functions for approximately 300 EV models, used in our tests, the reader is referred to OpenEV [23].

Different variants and methodologies have been proposed to solve the EVSPP. These variants use different assumptions, objective functions, and solution methods. Zündorf [24] studies the EVSPP considering different types of charging stations having distinct technologies with variable charging quantities. The author proposes heuristics to solve large EVSPP instances based on a multi-objective search and obtains a fast and feasible path on continental graphs. Sweda et al. [25] focus on the EVSPP considering that the availability of CSs is uncertain. They present two heuristic methods for finding adaptive policies considering both adaptive recharging decisions only, and adaptive routing and recharging decisions. Baum et al. [26] study the EVSPP with varying charging power and battery-swapping stations. The authors propose a combination of algorithmic techniques to achieve good performance using realistic instances. Baum et al. [27] propose a functional representation of the optimal energy consumption between two locations, which subsequently led to the development of an efficient heuristic algorithm for computing energy-optimal paths. Recent studies have focused on the computation of energy consumption of EV based on practical driving conditions, including driving patterns, road conditions, and other restrictions. Methodologies to include such vehicle dynamics include Markov Chain models with reinforcement learning [28] and deep reinforcement learning [29].

An important subproblem of the EVRP is the Fixed Route Vehicle Charging Problem (FRVCP), which was first introduced by Montoya et al. [19]. Given a fixed sequence of customers, the FRVCP determines which CSs to visit and the amount of energy to charge in them. The objective is to minimize the completion time of the path, while respecting the battery capacity constraints. Roberti and Wen [30] propose a labeling algorithm for the FRVCP considering a linear charging function and partial charging. Montoya et al. [17] propose a mixed integer linear programming model and a heuristic considering piece-wise linear charging functions and partial charging. Froger et al. [21] propose an exact labeling algorithm for the same problem considered by Montoya et al. [17]. The exact labeling algorithm is available in the Python `frvcpy` package [6] and used in the implementation of our algorithm. Specifically, we solve FRVCP instances in the second step of our algorithm (see Section 4.2.2).

The addition of POIs into vehicle routing decisions has been studied in the literature mainly in tourist routing applications [31]. One of the main treated problems in this context is the Tourist Trip Design Problem (TTDP) [32]. The objective of this problem is to define paths that maximize the number of visited attractions or POIs, while minimizing the total travel costs or meet time constraints. An extension of the TTDP including EVs is known as the Electric Vehicle Routing Tour Planning (EVRTP) [33]. In this problem, the maximization of the tourist satisfaction is considered while meeting path constraints such as total length, visiting times at each POI, or budget constraints. Extensions of the TTDP include the use of uncertain scenarios [34], hybrid fleets [2], and multi-period fleets [35]. To the best of our knowledge, Cassia et al. [36] is the first study to consider the EVSPP with both user preferences as POIs and time windows. However, the proposed heuristic in that paper considers a limited set of charging levels, and thus greatly simplifies the problem.

3. The Electric Vehicle Shortest Path Problem with Time Windows and User Preferences

We define the EVSPPWP as follows. Let $\mathcal{G} = \langle \mathcal{S}_{\mathcal{O},\mathcal{D}}, \mathcal{A} \rangle$ be a directed graph, where $\mathcal{S}_{\mathcal{O},\mathcal{D}} = \{\mathcal{O} \cup \mathcal{D} \cup \mathcal{S}\}$ is a set of nodes containing the origin $\mathcal{O}$, destination $\mathcal{D}$, and the set $\mathcal{S}$ of CSs at which recharging may take place. $\mathcal{A}$ is the set of arcs connecting each pair of nodes $i, j \in \mathcal{S}_{\mathcal{O},\mathcal{D}}, i \neq j$. Each arc is associated with a driving time $t_{ij} \geq 0$, a distance $d_{ij} \geq 0$, and an energy consumption $e_{ij} \geq 0$. These parameters are computed based on the fastest path on the real road network. We assume that the EV has an average consumption of η (kWh/km). To account for the EV's limited autonomy, we discard all arcs with a distance larger than $d_{\max}$. The path between the $\mathcal{O}$ and $\mathcal{D}$ nodes is performed by a single EV with a battery capacity of Q that may be partially recharged at CSs. The EV starts at node $\mathcal{O}$ at t_{start} with an initial state of charge (SoC) (i.e., percentage of the level of the battery with respect to its capacity), equal to q_{start}, and it must arrive to the destination node $\mathcal{D}$ with a final SoC greater than or equal than q_{end}. The EV must arrive at each CS with an amount of energy which is greater than or equal to $q_{\min}$. Both q_{start} and q_{end} are user-defined parameters.

Each CS, $i \in \mathcal{S}$, has a piece-wise linear concave charging function $\Phi_i(\Delta)$, where Δ is the time spent charging. If q is the SoC of the EV when it arrives at the charging station i, then the SoC when it leaves is $\Phi_i(\Delta + \Phi_i^{-1}(q))$. Let $B_i = \{0, b_1, \ldots, b_{m_i}\}$ be the ordered set of breakpoints of the piece-wise linear approximation of the charging curve of CS i. Let c_{ik} and a_{ik} be the charging time and SoC of breakpoint $k \in B_i$. Each breakpoint connects $(c_{i,k-1}, a_{i,k-1})$ and (c_{ik}, a_{ik}) with a linear function with slope equal to ρ_{ik} (see Figure 1).

Let $\mathcal{P}$ be the set of categories of POIs. Each CS is associated with an importance value for each category $p \in \mathcal{P}$ (see Section 4.1.2 for their computation procedure), which represents the accessibility of POIs (relevant to p) from the CS. As minimizing total travel times conflicts with maximizing the total collected importance when visiting nodes, we use a modified arc weight $\tilde{s}_{ij}$ defined as:

$$\tilde{s}_{ij} = t_{ij} + \Delta_j - \mu \sum_{p \in \mathcal{P}} \sigma_{jp}, \tag{1}$$

where σ_{jp} represents the importance value of $j \in \mathcal{S}$ for POI category $p \in \mathcal{P}$, μ is a weight on the total importance value of the POIs with respect to travel time, and Δ_j is the time spent charging at CS j.

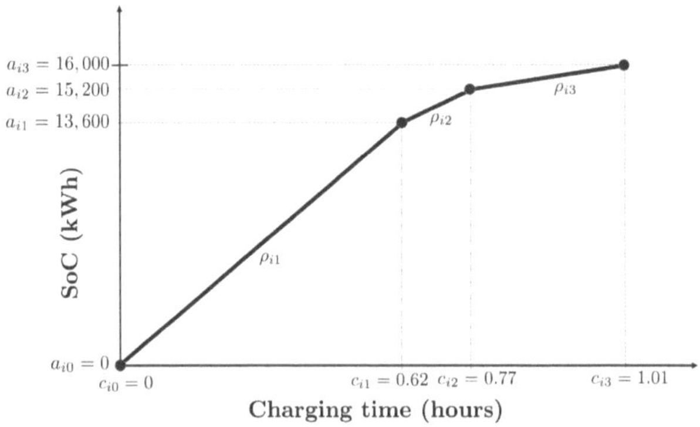

Figure 1. Example of a piece-wise linear approximation for a CS $i \in \mathcal{S}$ with a power of 22 kWh adapted from Montoya et al. [17].

We consider that the user can impose certain types of stops during the path, and use the time spent at the stops to charge the EV. We model this feature by a set of ordered time windows $\mathcal{W}$. Each time window requires a specific POI category, and the EV must stop during all the time windows in their given order. A time window $w \in \mathcal{W}$ is defined by a minimum start time γ_w^L, a maximum start time γ_w^U, and a minimum time that the EV needs to stop $t_w^{\min}$. We define the set $\mathcal{S}_w \subset \mathcal{S}$ as the set of CSs with POIs in their neighborhood that match the POI specified for $w \in W$. The objective is to find the path that minimizes the total modified weights (defined in Equation (1)) of the selected arcs between $\mathcal{O}$ and $\mathcal{D}$, while satisfying all the charging and time windows constraints.

A Mixed-Integer Linear Programming Formulation

We now present a MILP formulation for the EVSPPWP. This formulation is an adaptation of the one presented by Cassia et al. [36]. Let $\mathcal{S}_{\mathcal{O}} = \mathcal{S} \cup \{\mathcal{O}\}$, $\mathcal{S}_{\mathcal{D}} = \mathcal{S} \cup \{\mathcal{D}\}$. Let $\underline{q}_i$ and $\overline{q}_i$ be the SoC when the EV arrives and departs from CS i. The variables $\underline{h}_i$ and $\overline{h}_i$ are respectively the start and end time for charging an EV. The variables $\underline{\lambda}_{ik}$ and $\overline{\lambda}_{ik}$ represent the coefficients associated with the breakpoint (c_{ik}, a_{ik}) in the linear approximation, when the EV enters and leaves CS i. Let $\underline{w}_{ik}$ and $\overline{w}_{ik}$ be binary variables equal to 1 when the SoC is in the interval $[a_{i,k-1}, a_{ik}]$, when respectively the EV enters and leaves the CS i, and 0 otherwise. Variables $\underline{\tau}_i$ and $\overline{\tau}_i$ track respectively the time when the EV arrives and leaves the CS $i \in \mathcal{S}$.

The binary variable x_{ij} equals 1 if the EV uses arc (i,j), and 0 otherwise. Binary variable y_{jw} equals 1 if the EV stops in j in time window w, and 0 otherwise. The model is defined as follows:

$$\min \quad \sum_{(i,j) \in \mathcal{A}} \tilde{s}_{ij} x_{ij} \tag{2}$$

$$\text{s.t.} \quad \sum_{(i,j) \in \mathcal{A}} x_{ij} \leq 1 \qquad \forall i \in \mathcal{S}_{\mathcal{O}} \tag{3}$$

$$\sum_{(i,j) \in \mathcal{A}} x_{ij} - \sum_{(j,i) \in \mathcal{A}} x_{ji} = \begin{cases} 1 & \text{if } i = \mathcal{O} \\ -1 & \text{if } i = \mathcal{D} \\ 0 & \text{otherwise} \end{cases} \qquad \forall i \in \mathcal{S}_{\mathcal{O},\mathcal{D}} \tag{4}$$

$$\text{s.t.}\quad e_{ij}x_{ij} - (1 - x_{ij})Q \leq \overline{q}_i - \underline{q}_j \leq e_{ij}x_{ij} + (1 - x_{ij})Q \qquad \forall (i,j) \in \mathcal{A} \tag{5}$$

$$\overline{q}_{\mathcal{O}} = q_{\text{start}} \tag{6}$$

$$\underline{q}_{\mathcal{D}} \geq q_{\min} \tag{7}$$

$$q_{\min} \sum_{(i,j)\in\mathcal{A}} x_{ij} \leq \underline{q}_i \leq \overline{q}_i \leq Q \sum_{(i,j)\in\mathcal{A}} x_{ij} \qquad \forall i \in \mathcal{S} \tag{8}$$

$$\underline{q}_i = \sum_{k\in B_i} \underline{\lambda}_{ik} a_{ik} \qquad \forall i \in \mathcal{S} \tag{9}$$

$$\underline{h}_i = \sum_{k\in B_i} \underline{\lambda}_{ik} c_{ik} \qquad \forall i \in \mathcal{S} \tag{10}$$

$$\sum_{k\in B_i} \underline{\lambda}_{ik} = \sum_{k\in B_i\setminus\{0\}} \underline{w}_{ik} \qquad \forall i \in \mathcal{S} \tag{11}$$

$$\sum_{k\in B_i\setminus\{0\}} \underline{w}_{ik} = \sum_{(i,j)\in\mathcal{A}} x_{ij} \qquad \forall i \in \mathcal{S} \tag{12}$$

$$\underline{\lambda}_{i0} \leq \underline{w}_{i1} \qquad \forall i \in \mathcal{S} \tag{13}$$

$$\underline{\lambda}_{ik} \leq \underline{w}_{ik} + \underline{w}_{i,k+1} \qquad \forall i \in \mathcal{S}, \forall k \in \mathcal{B}_i\setminus\{0, b_{m_i}\} \tag{14}$$

$$\underline{\lambda}_{i,b_i} \leq \underline{w}_{i,b_i} \qquad \forall i \in \mathcal{S} \tag{15}$$

$$\overline{q}_i = \sum_{k\in B_i} \overline{\lambda}_{ik} a_{ik} \qquad \forall i \in \mathcal{S} \tag{16}$$

$$\overline{h}_i = \sum_{k\in B_i} \overline{\lambda}_{ik} c_{ik} \qquad \forall i \in \mathcal{S} \tag{17}$$

$$\sum_{k\in B_i} \overline{\lambda}_{ik} = \sum_{k\in B_i\setminus\{0\}} \overline{w}_{ik} \qquad \forall i \in \mathcal{S} \tag{18}$$

$$\sum_{k\in B_i\setminus\{0\}} \overline{w}_{ik} = \sum_{(i,j)\in\mathcal{A}} x_{ij} \qquad \forall i \in \mathcal{S} \tag{19}$$

$$\overline{\lambda}_{i0} \leq \overline{w}_{i1} \qquad \forall i \in \mathcal{S} \tag{20}$$

$$\overline{\lambda}_{ik} \leq \overline{w}_{ik} + \overline{w}_{i,k+1} \qquad \forall i \in \mathcal{S}, \forall k \in \mathcal{B}_i\setminus\{0, b_{m_i}\} \tag{21}$$

$$\overline{\lambda}_{i,b_i} \leq \overline{w}_{i,b_i} \qquad \forall i \in \mathcal{S} \tag{22}$$

$$\Delta_i = \overline{h}_i - \underline{h}_i \qquad \forall i \in \mathcal{S} \tag{23}$$

$$\overline{\tau}_{\mathcal{O}} = t_{\text{start}} \tag{24}$$

$$t_{\text{start}} \sum_{(i,j)\in\mathcal{A}} x_{ij} \leq \underline{\tau}_i \leq \overline{\tau}_i \qquad \forall i \in \mathcal{S} \tag{25}$$

$$\underline{\tau}_i + \Delta_i \leq \overline{\tau}_i \qquad \forall i \in \mathcal{S} \tag{26}$$

$$\sum_{w\in\mathcal{W}} y_{jw} \leq 1 \qquad \forall j \in \mathcal{S} \tag{27}$$

$$y_{jw} \leq \sum_{(i,j)\in\mathcal{A}} x_{ij} \qquad \forall w \in \mathcal{W}, \forall j \in \mathcal{S}_w \tag{28}$$

$$\underline{\tau}_j \geq \gamma_w^L y_{jw} \qquad \forall w \in \mathcal{W}, \forall j \in \mathcal{S}_w \tag{29}$$

$$\overline{\tau}_j \leq \gamma_w^U y_{jw} \qquad \forall w \in \mathcal{W}, \forall j \in \mathcal{S}_w \tag{30}$$

$$\sum_{j\in\mathcal{S}_w} y_{jw} = 1 \qquad \forall w \in \mathcal{W} \tag{31}$$

$$\Delta_i \geq \sum_{w\in\mathcal{W}} t_w^{\min} y_{iw} \qquad \forall i \in \mathcal{S} \tag{32}$$

$$x_{ij} \in \{0,1\} \qquad \forall (i,j) \in \mathcal{A} \tag{33}$$

$$y_{jw} \in \{0,1\} \qquad \forall j \in \mathcal{S}_w, \forall w \in \mathcal{W} \tag{34}$$

$$z_w \in \{0,1\} \qquad \forall w \in \mathcal{W} \tag{35}$$

$$\underline{q}_i \geq 0,\ \underline{\tau}_i \geq 0, \qquad \forall i \in \mathcal{S}_{\mathcal{D}} \tag{36}$$

$$\overline{q}_i \geq 0,\ \overline{\tau}_i \geq 0 \qquad \forall i \in \mathcal{S}_{\mathcal{O}} \tag{37}$$

$$\underline{\lambda}_{ik} \geq 0,\ \overline{\lambda}_{ik} \geq 0 \qquad \forall i \in \mathcal{S}, \forall k \in \mathcal{B}_i \tag{38}$$

$$\underline{w}_{ik},\ \overline{w}_{ik} \in \{0,1\} \qquad \forall i \in \mathcal{S}, \forall k \in B_i\setminus\{0\} \tag{39}$$

$$\underline{h}_i \geq 0,\ \overline{h}_i \geq 0,\ \Delta_i \geq 0 \qquad \forall i \in \mathcal{S} \tag{40}$$

The objective function (2) minimizes the total modified weights of the used arcs. Constraints (3) require a CS to be visited once at most. Constraints (4) impose the flow conservation conditions. Constraints (5) track the SoC of the EV for each pair of nodes. Constraint (6) imposes that at the beginning to the trip the EV has a charge equal to q_{start}. Similarly, constraint (7) imposes the minimum charge at destination. Constraints (8) require that the SoC of the EV when leaving an SC is greater than the SoC when it arrives. These constraints also impose that the EV must arrive at the CS with a minimum residual energy, and leave with less than the EV battery capacity.

Constraints (9)–(15) define the SoC and the charging time based on linear approximation of the charging function of a CS. In the same fashion, constraints (16)–(22) define the SoC and the charging time upon departure from a CS. Constraints (23) define the time spent at CS i. Constraint (24) imposes the starting time. Constraints (25) impose that the arrival time has to be lower than the departure, and both must be greater than t_{start}. Constraints (26) link arrival, departure and waiting times. Constraints (27) assure that every CS $j \in \mathcal{S}$ must be used for at most one time window $w \in \mathcal{W}$, while constraints (28) link x_{ij} and y_{jw} variables. Constraints (29) and (30) impose that, for every $w \in \mathcal{W}$, the arrival time is between γ_w^L and γ_w^U. Constraints (31) ensure that every required time window is served. Constraints (32) impose a minimum waiting time. Finally, (33)–(40) define the domains of the variables.

4. Solution Method

In this section, we present our methodology to solve the EVSPPWP. The main innovations of this methodology are as follows: (1) we create CSs composed of clusters of CPs with sufficient proximity, thus providing the user with alternative CPs within a cluster; (2) we account for user preferences by assigning weights for different POI categories and attribute weights to each CS according to its proximity to POIs; (3) we use resilient paths, that is, using only arcs that only have a minimum number of CSs within a distance of d_r to ensure access to CSs in case of emergencies; and (4) we consider user-defined time windows for activities during the trip such as having lunch, visit POIs, or spending a night at a hotel. The proposed methodology has been successfully applied to five geographical regions (see Section 5 for details).

We divide this section into two parts. First, in Section 4.1 we present a pre-processing step to generate the graph which is used for planning the path. This pre-processing step is used to create a more compact graph in the backend. In Section 4.2, we describe the two-step algorithm heuristic that is used to produce up to three paths for a given user request, i.e., an origin and a destination along with other user preference specifications.

4.1. Graph Generation

In this pre-processing step we generate a graph $\mathcal{G} = \langle \mathcal{S}_{\mathcal{O},\mathcal{D}}, \mathcal{A} \rangle$ based on CP data and POI locations, which were provided by the project partners. To generate this graph, we create a set of CSs by clustering CPs, compute the modified weight of the arcs $\tilde{s}_{ij}$, including the importance value of the CSs, and filter arcs by computing the resilience value for each arc in the graph. We describe each of these steps in the subsequent sections.

4.1.1. Clustering

We consider public charging interoperable CSs. Each such CS may group several CPs. At the same time, each CP may have different plug types. The data are typically provided in terms of CPs and not CSs. We generate CSs by clustering the CP data using the following procedure. We consider a maximum number $n_{\max}$ of CPs per cluster, a minimum number $n_{\min}$ of CPs per cluster, and the maximum geodesic distance d_c between CPs in a cluster. These parameters are selected to provide the user with alternative CPs that are close enough to each other. The purpose of this procedure is to cover the case when a selected CP is not available at the moment of arrival, by providing alternatives to the user. Thus, the selection of the parameters depends on the characteristics of the geographical area, such as the total number of CPs and their sparsity.

Let $\mathcal{C}$ be the set of original CP locations, and $\bar{\mathcal{C}}$ be the set of clustered CPs, initially $\bar{\mathcal{C}} = \{\emptyset\}$. The clustering algorithm starts by selecting a random CP $i \in \mathcal{C}$, which initializes the first cluster. The coordinates of the center of the cluster are set to be the coordinates of CP i. Sequentially, we add CPs that are within a distance d_c to the center of the cluster Each time a CP is added, the coordinates of the center of the cluster are updated to be the average between the previous center and the coordinates of the new CP. A cluster is closed when either n_{max} is reached or there are no other CPs available within d_c. When a cluster is closed, a new random CP is selected, and the procedure is repeated with the remaining CPs in the dataset. The algorithm terminates when there are no other CPs available. If the final size of a cluster is lower than n_{min} the cluster is removed from $\bar{\mathcal{C}}$. At the end of the procedure, each cluster is characterized by the coordinates of its center, and the information of each of the CPs in the cluster. In what follows we refer to a cluster of CPs as a CS and all distances are computed based on the center of the clusters.

4.1.2. Modified Arc Weights

In order to compute the modified weights of the arcs $\tilde{s}_{ij}$ as defined in Equation (1), we first compute the importance value σ_{jp} associated with CS j and POI category p. To do so, we divide the POIs into three categories: (1) food (the user prefers to stop in clusters with restaurants), (2) nature (the user prefers to stop in clusters close to nature sites), and (3) shop (the user prefers to stop in places with several shops like malls, department stores, etc.). For each such category p, the importance value σ_{jp} is computed as the number of available POIs within a distance d_{POI} from the CS $j \in \mathcal{S}$ in the given category $p \in \mathcal{P}$.

4.1.3. Resilience Value

For each arc $(i, j) \in \mathcal{A}$ we define a resilience value r_{ij} to represent the availability of CSs within a critical distance from the actual road path represented by the arc. The resilience value ensures a minimum number of CSs are within a critical distance d_r from the path in case of charging emergencies. To compute the resilience value, we divide the path between nodes i and j into sections of length d_t. For each section k, we define the number of CSs that are within a distance of d_r as $C_k = |\{j \in \mathcal{S} | d(j,k) \leq d_r\}|$, where $d()$ represents the shortest geodesic distance between CS j and section k. The resilience value r_{ij} for the path between nodes i and j is defined as:

$$r_{ij} = med(\{C_k, \forall k \subset (i,j) \in \mathcal{A}\}), \quad (41)$$

where $med()$ is the median value of the set. In our methodology, we consider a modified set of arcs $\bar{\mathcal{A}} = \{(i,j) \in \mathcal{A} | r_{ij} > r_{\text{min}}, d_{ij} > d_r\}$, where r_{min} is the minimum resilience value allowed for each arc. The later condition (i.e., $d_{ij} > d_r$) implies that if two CSs are close enough to each other, the arc connecting them is considered to be resilient.

4.1.4. User Input Parameters

A user request is an origin–destination pair of locations. The user must also specify the EV type from a preset list. This choice establishes EV related parameters (i.e., the capacity Q of the battery, the average consumption of η, and the breaks in the charging curve B_i). Such parameters are obtained from the OpenEV repository [23]. In addition, the user may specify the starting SoC (q_{start}), desired SoC at destination (q_{end}), start time at the origin, and a rating for each POI category. These ratings are established with a slidebar that defines a weight parameter ranging from 0 to 100%. These user-defined weights are included in the model by directly multiplying them with the corresponding importance value σ_{jp}. The user may also specify two types of time windows, one for lunch and one for night stays. Lunch stops are clusters that have at least one restaurant, whereas night stops consider hotels with CSs on their premises. Each optional intermediate stop is defined by a minimum start time γ_w^L, a maximum start time γ_w^U, and a minimum time that the EV needs to stop t_w^{min}. Finally, the user may select the plug types ("Type 2" and "Ccs"). CPs in CSs that do not match the selected plug types are discarded. Subsequently, the compatible CP with the maximum charging power in each CS is considered, ties are broken arbitrarily. Thus, the final address and CP specifications in the solution corresponds to the fastest compatible CP in the cluster.

4.1.5. Graph Reduction

In order to reduce computational run-times, we apply another filtering strategy using non EV-paths. In this step, we compute up to three alternative shortest paths between the origin and destination without considering CSs. These three paths are computed using mapbox [37]. We filter the graph $\mathcal{G}$ by only considering the CSs in $\mathcal{S}$ that are within a threshold of d_f kilometers around the three non EV-paths. This step defines a new reduced graph $\tilde{\mathcal{G}} = \langle \tilde{\mathcal{S}}_{\mathcal{O},\mathcal{D}}, \tilde{\mathcal{A}} \rangle$, where $\tilde{\mathcal{S}}_{\mathcal{O},\mathcal{D}}$ is the filtered set of nodes and $\tilde{\mathcal{A}} \subset \mathcal{A}$ is the set of considered arcs between $\tilde{\mathcal{S}}_{\mathcal{O},\mathcal{D}}$. In Figure 2, we illustrate the graph in this reduction step. The three solid lines show the three non EV-paths between the origin and destination. The colored areas show an approximation of the threshold from each path using $d_f = 50$ km. All arcs and nodes that do not fall into the colored area are discarded.

Figure 2. Illustration of the graph reduction phase for three non EV-paths from Florence to Milan, using a threshold of $d_f = 50$ km.

4.2. Two-Step Algorithm

We solve the EVSPPWP on the modified graph $\tilde{\mathcal{G}}$ using a two-step algorithm. In the first step, we use a modified A^* algorithm to establish the nodes where the user-defined time windows are going to be performed. The result of this step is a set of consecutive nodes, each one representing the location at which each time window is performed. Let $\mathcal{N}_\mathcal{W} \in \tilde{\mathcal{S}}_{\mathcal{O},\mathcal{D}}$ be the ordered set of nodes selected by the algorithm to perform time windows in $\mathcal{W}$. The output of the algorithm defines then $|\mathcal{N}_\mathcal{W}| + 1$ legs. In what follows, we refer to legs as the paths to be performed from $\mathcal{O}$ to the first node in $\mathcal{N}_\mathcal{W}$, the path between the ordered nodes in $\mathcal{N}_\mathcal{W}$, and the path between the last node in $\mathcal{N}_\mathcal{W}$ and $\mathcal{D}$. In the second step, we solve the FRVCP for each leg defined in the first step using the labeling algorithm proposed by Kullman et al. [6]. The final solution of the algorithm is the merged solution of each FRVCP instance defined by the legs in the first step.

4.2.1. A^* Algorithm

We implement a modified A^* search algorithm to find the nodes in which the user-specified time windows are performed in order to minimize the total travel time. For this search, we assume that recharging stops are only performed when the remaining energy in the battery is less than the energy needed to reach the next node. At each recharging stop, we assume that the EV is fully charged. The algorithm starts at node $\mathcal{O}$ and sequentially generates the tree of all paths to reach the next node using a best-first search until the destination node is reached. The best-first search is conducted by selecting the next node that minimizes the following objective:

$$f(i) = g(i) + h(i) \tag{42}$$

where i is the current node, $g(i)$ is the duration of the path from $\mathcal{O}$ to i, and $h(i)$ is an estimate of the duration of the shortest path from i to $\mathcal{D}$. Duration computed in $g(i)$ and $h(i)$ include both travel and charging times. We use the modified travel times $\tilde{s}_{ij}$ as defined in Equation (1). By the use of a labeling search, the modified A^* algorithm finds the nodes in $\tilde{\mathcal{G}}$ where times windows $k \in \mathcal{W}$ are performed, which define the legs in the following step.

4.2.2. FRVCP

We solve the FRVCP for each leg (i.e., considering the two extreme nodes of the legs as the fixed nodes) to determine the decisions on where and how much to charge at each recharging stop while minimizing the total travel time. To ensure the final SoC requirement q_{end} is fulfilled at the destination, we add a dummy node at the end of each leg with a sufficient distance from the leg destination node, which is removed at the end of this step. We solve the problem for each leg exactly by the labeling algorithm of Kullman et al. [6] using the Python library `frvcpy`. The labeling algorithm returns the optimal energy-feasible path for each leg, i.e., the location of the stops, the amount to be charged at each stop, and the total duration. After computing the optimal solution for each leg, we merge together all solutions to obtain a single complete EV-path from $\mathcal{O}$ to $\mathcal{D}$. The heuristic solution of the merged legs is the solution of the EVSPPWP.

5. Case Study

Within the eCharge4Drivers project and with the support of project partner Route 220 S.p.A. (Evway), we implemented our path planning methodology in the webpage available at https://planner.evway.net/, accessed on 15 May 2023. Specifically, the tool was implemented in five regions: Northern Italy and Regions North of Italy (NIRNI), Puglia, Greece, Turkey, and Spain. In Figure 3 we show the original CPs used in each of the five regions. For each region, the algorithm was tested using several combinations of origins and destinations. We implemented all parts of the algorithm in Python 3.8. In our tests, all run times for a single path were observed to be under 30 s. In the following, we focus on examples in the NIRNI area. In Section 5.1, we demonstrate how the graph is generated. In Section 5.2, we provide examples on various user inputs, and in Section 5.3, we demonstrate the results of our path planner on various origin–destination examples.

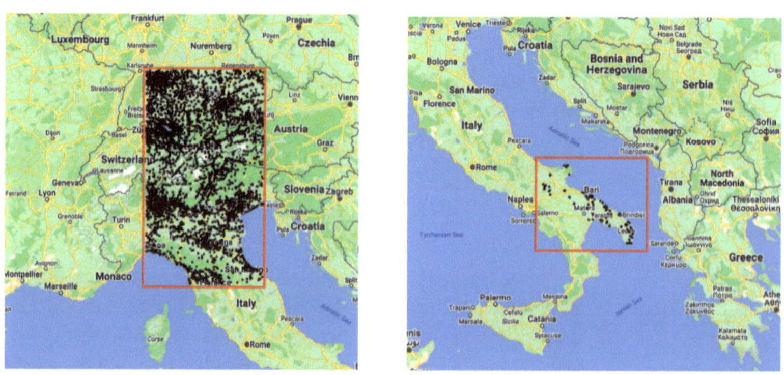

(a) Northern Italy and North of Italy (b) Puglia

Figure 3. *Cont.*

(**c**) Greece (**d**) Spain

(**e**) Turkey

Figure 3. The five regions in which the path-planning methodology was tested.

5.1. Graph Generation

The original dataset provided by the project partners contains 14,767 CPs, providing the location, address, and plug types. A second dataset contains the information on the POIs. This dataset was obtained by both the partner company and OpenStreetMap. Given the high density of CPs in the NIRNI area, we calibrated the three parameters of our clustering algorithm to have a balance between the number of clusters and the availability of CPs outside urban areas. We selected the following parameters in the graph generation step: $n_{\max} = 15$, $n_{\min} = 3$, $d_c = d_{\text{POI}} = d_t = l_{\min} = 1$ km, $d_r = 10$ km, $d_{\max} = 300$ km, $d_f = 50$ km, $\mu = 0.5$, $r_{\min} = 0.1$, and $q_{\min} = 0.2Q$.

In Figure 4, we show the original CP locations and the final CPs clustered in CSs using the procedure described in Section 4.1.1. The resulting graph contains a total number of 3223 CSs.

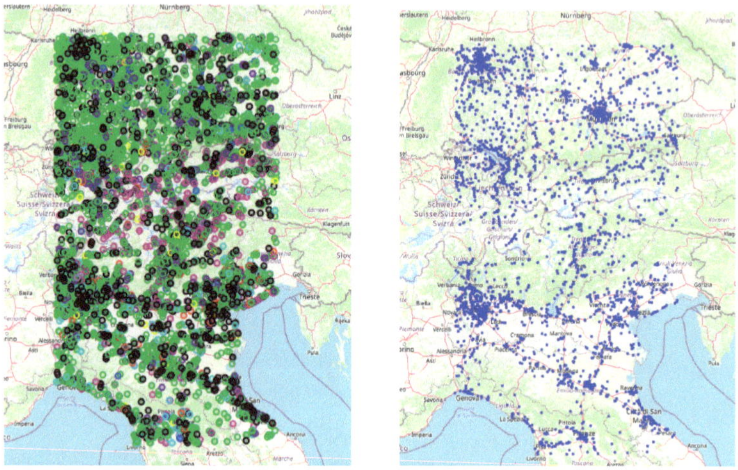

(**a**) Original CPs (**b**) Clusters (CSs)

Figure 4. (**a**) Original 14,767 CPs colored by plug type, and (**b**) resulting 3223 clusters of CPs in the NIRNI area.

5.2. *User Input*

The web interface provides users with the ability to select several inputs directly. These inputs include the start point, destination, and any intermediate waypoints for the trip. In addition, the user can choose the type of EV they will be using, selecting from available options in the OpenEV repository. This choice sets the parameters on battery capacity, average consumption, and charging curve breaks. The user can also set the battery SoC at the start of the trip, and desired SoC at the end of the trip. Additionally, the user can specify preferences for the three categories of stops, using sliders to set the level of importance for each category. The weights are directly multiplied by the importance value for each category. The interface also includes options for setting up lunch and hotel stops during the trip, and specifying their time windows. Finally, the user can choose the plug type they require, with options for "Type 2" and/or "Ccs" plugs.

As previously mentioned, we assume that the EV has an average consumption of η (kWh/km), which is multiplied by d_{ij} to yield e_{ij}. While there may be tools that allow a more accurate calculation of energy between nodes (e.g., accounting for vehicles dynamics according to geophysical properties of the traversed segments), such tools were not available to us. Specifically, we were able to query the systems of our project partners and obtain aggregate path information on a path between two addresses (e.g., distance, travel time). Therefore, as in most of the EVRP literature (see Kucukoglu et al. [3] for a classification of studies by their assumptions on energy consumption), we opted to estimate the energy consumption as a linear factor of the distance. We recognize that this is a shortcoming; yet, we note that, from an algorithmic perspective, energy consumption between nodes is an input. Therefore, the methodology will not change by more accurate energy consumption estimates.

The use of vehicle dynamics to obtain more accurate energy estimates entails the use of a more elaborate set of parameters related to each vehicle. To account for the geographical properties of the roads, one would need to retrieve detailed intersection by intersection road information between every pair of nodes. For example, in the NIRNI area, this corresponds to retrieving detailed intersection by intersection information for approximately 10,380,000 paths corresponding to the pairs of CSs. While such an additional step is computationally intensive, it would belong to the preprocessing stage since energy consumption is an input to our algorithm. From a computational perspective, we expect the run times not to be affected by changing the algorithm's input.

After setting the user preferences on the web interface, the algorithm computes up to three path options between the origin and the destination. These three paths are based on the non EV paths as described in Section 4.1.5. Then the algorithm described in Section 4.2 is executed on each of the three paths establishing where to stop and how much to charge. In Figure 5a we show the three path options given by the algorithm for a trip between Milan, Italy, and Stuttgart, Germany. The figure consists of a left panel displaying the main results for each path, and a right panel displaying a map with the paths in different colors. Using the left panel, the user can select one of the paths for further details. After selecting the first path, as shown in Figure 5a, details on the place and duration of the recharging stops are shown in the left panel, while in the right panel, the selected path is displayed along with grey points representing alternative CPs. In Figure 5b, we compare the original path with a path using the option of adding a waypoint. Figure 6a shows the recommended path, whereas Figure 6b shows the path after the user selects a waypoint in Munich, Germany. The actual path and the charging stops are shown in green, and alternative CPs are shown in grey. The user may change a proposed CP by selecting an alternative CP. If an alternative CP is selected, the algorithm is re-run while fixing the selected CP.

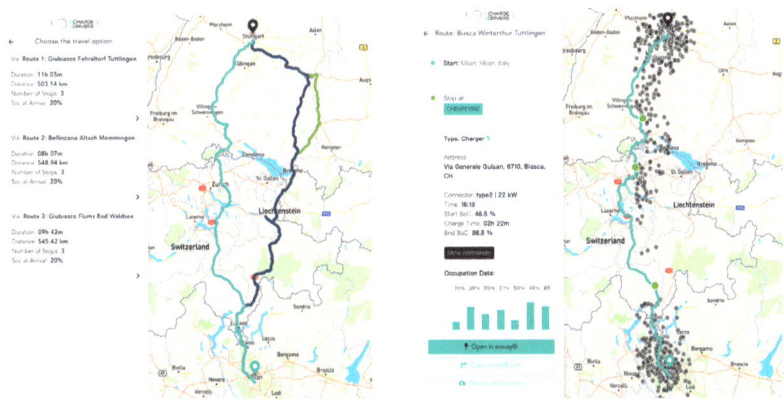

(**a**) Three path options (**b**) Details of a selected path

Figure 5. Example of three path options for a trip from Milan to Stuttgart on the web interface.

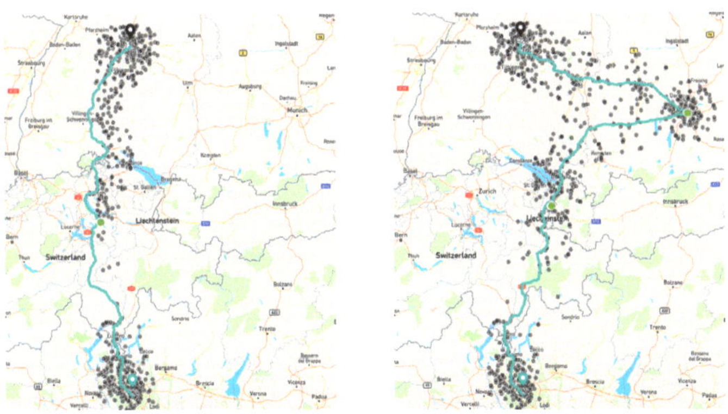

(**a**) Original path (**b**) Path with a waypoint

Figure 6. Example of the addition of one waypoint in Munich, for a trip between Milan and Stuttgart.

5.3. Origin–Destination Examples

We illustrate the use of our path planner methodology using three origin–destination pairs accounting for short, medium, and long-distance trips in the NIRNI area. For each trip, we compute the path using four configurations of user preferences: a base case without preferences, one assigning 100% to the shopping POI weight, one having a lunch stop, and one having a night stop. From the three possible path recommendations, we selected the fastest path. In all cases, we set the EV type to be BWM i3S 120Ah, a start SoC of 80%, and a desired end SoC of 20%. For the lunch break, the earliest arrival is set to 12:30, latest arrival is set to 13:30, and a duration of two hours is selected. For the night rest, the earliest desired check-in is 20:00, the latest check-in is 22:00, and the check-out is at 8:30. Table 1 reports the results for the three origin and destinations: Venice–Milan, Trento–Stuttgart, and Florence–Munich. All paths are set to start at 10:00, except from Venice–Milan and Trento–Stuttgart with night stops, which are set to start at 20:00 for its duration to match a night stop. For each path, we report the start time, end time, duration, distance, and number of recharging stops. We observe that the duration and distance of the trips vary depending on the preference selection. These differences are due to different selected stops and/or different charging times at the stops given the preferences. For example, trips with the night stop option have a notable increase in the total duration since this preference has a fixed check out at 8:30.

In Table 2, we show the address, start time, duration, SoC at start, and SoC at end for all four recharging stops in the path between Florence and Munich, for the four user preferences combinations. Except for the first stop, the addresses for the charging stops differ depending on the preference. A similar behavior can be observed in terms of the start time and duration of the charging stops. In Figure 7, we show a comparison of the paths without preferences and with lunch break. We can observe in the map that the first and last charging stops (1 and 4) are the same; however, they have different charging times. The intermediate charging stops (2 and 3) both show different locations and charging times.

Table 1. Path details for three origin–destination pairs. For each pair, four different user preferences are selected.

Origin	Destination	Preference	Start Time	End Time	Duration	Distance (km)	Charging Stops
Venice	Milan	-	10:00	15:17	05 h 17 min	268.58	1
		100% weight for shopping	10:00	13:36	03 h 36 min	269.87	1
		Lunch	10:00	15:16	05 h 16 min	274.32	2
		Night	20:00	10:31	14 h 31 min	268.58	1
Trento	Stuttgart	-	10:00	17:24	7 h 24 min	499.84	3
		100% weight for shopping	10:00	18:15	08 h 15 min	506.25	3
		Lunch	10:00	18:16	08 h 16 min	500.40	3
		Night	20:00	13:10	17 h 10 min	500.40	3
Florence	Munich	-	10:00	21:54	11 h 54 min	649.15	4
		100% weight for shopping	10:00	22:51	12 h 51 min	651.75	4
		Lunch	10:00	21:56	11 h 56 min	639.94	4
		Night	10:00	10:09	24 h 09 min	649.15	4

Table 2. Recharging stop decisions details for a trip from Florence to Munich. Results are shown for four different user preferences.

Preference	Stop	Address	Start Time	Duration	SoC Start (%)	SoC End (%)
-	1	Via Isonzo 16, 40033, IT	11:14	01 h 43 min	36.4	100.0
	2	Via Dante 14, 38063, IT	14:48	01 h 50 min	23.1	76.5
	3	Schindergries Parking, 39043, IT	18:05	00 h 25 min	20.0	75.0
	4	Roßhütte 419, 6100, AT	19:54	00 h 19 min	27.8	71.5
100% weight for shopping	1	Via Isonzo 16, 40033, IT	11:59	01 h 36 min	36.4	100.0
	2	Via Dante 14, 38063, IT	15:33	01 h 50 min	23.1	76.5
	3	Schindergries Parking, 39043, IT	18:50	01 h 02 min	20.0	100.0
	4	Marienpl. 17, 82467, DE	21:49	00 h 39 min	39.0	58.0
Lunc h	1	Via Isonzo 16, 40033, IT	11:14	00 h 43 min	36.4	63.4
	2	Via del Commercio 33, 46030, IT	13:03	02 h 54 min	20.0	94.8
	3	Via Lancia - Lanciastraße 14, 39100, IT	17:42	00 h 30 min	20.0	79.0
	4	Roßhütte 419, 6100, AT	19:53	00 h 23 min	20.0	71.5
Night	1	Via Isonzo 16, 40033, IT	11:14	01 h 43 min	36.4	100.0
	2	Via Dante 14, 38063, IT	14:48	01 h 50 min	23.1	76.5
	3	Schindergries Parking, 39043, IT	18:05	00 h 21 min	20.0	67.2
	4	Roßhütte 419, 6100, AT	19:51	12 h 38 min	20.0	100.0

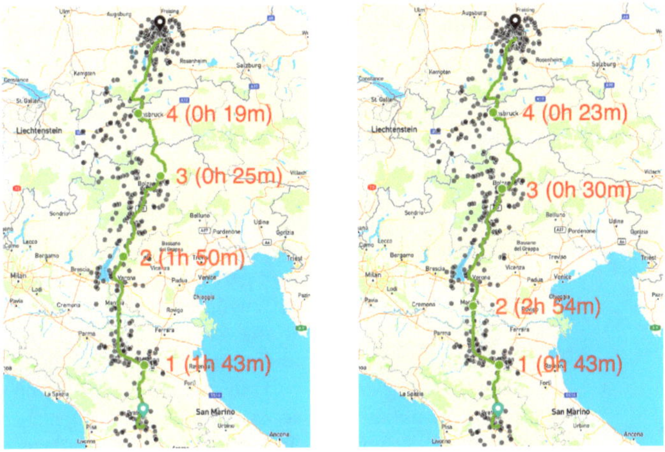

(**a**) Without preferences (**b**) With lunch break

Figure 7. Comparison of paths between Florence and Munich using (**a**) no user preferences and (**b**) selecting a lunch break of two hours, with charging stop number and duration.

In Table 3, we compare the duration, distance (km), and number of stops in a trip from Florence and Munich for four EV models. We show results using different starting SoCs, ranging from 60% to 100%, and the same desired SoC at arrival of 10%. We observe clear differences in the duration and number of stops in the trip depending on the selected EV type. Considering a starting SoC of 100%, the Tesla Roadster 2022 shows the fastest path with a duration of seven hours and five minutes with no recharging stops, while the slowest trip with Kia EV6 2WD 2021 lasts twelve hours and twenty-one minutes with two recharging stops. The same pattern can be seen when comparing EV models with lower starting SoC. As expected, we observe for all EV models a decrease in the number of stops from the starting SoC. In all cases, decreasing this value from 100% to 60% of SoC results in one additional recharging stop.

Table 3. Duration, distance (km), and number of stops for a trip between Florence and Munich for different EV models and SoCs at start.

EV Model		**SoC at Start (%)**				
		60	**70**	**80**	**90**	**100**
Tesla Roadster 2022	Duration	09 h 45 min	08 h 51 min	07 h 53 min	07 h 05 min	07 h 05 min
	Distance	654.94	652.11	652.11	648.42	648.42
	Stops	1	1	1	0	0
Tesla Model Y Long Range 2020	Duration	12 h 05 min	11 h 20 min	10 h 40 min	10 h 06 min	9 h 24 min
	Distance	659.21	655.19	655.19	658.57	652.04
	Stops	2	2	2	2	1
BMW - i3S - 120 A h 2020	Duration	11 h 16 min	12 h 10 min	11 h 54 min	12 h 10 min	11 h 33 min
	Distance	643.4	649.15	649.15	638.2	639.6
	Stops	4	4	4	3	3
Kia EV6 2WD 2021	Duration	12 h 12 min	15 h 05 min	14 h 33 min	12 h 15 min	12 h 21 min
	Distance	662.49	662.63	662.62	663.13	658.61
	Stops	3	2	2	2	2

6. Conclusions

Given an origin and a destination, we proposed a path-planning methodology for EVs, considering user preferences. The main innovations in our methodology are the use of weights for three categories of POIs to reflect user preferences in the selection of CSs, the use of clustered CSs to ensure alternatives in the vicinity of CSs, the use of resilient paths, and allowing the user to define time windows for activities such as lunch and hotel stays. We divided our methodology into two parts. We first performed a pre-processing step in which we generate a graph with clustered CSs and resilient paths from raw CP and POI data. Then, we proposed a two-step algorithm to solve the shortest path problem The final solution of our methodology is a heuristic solution for the EVSPPWP.

We conducted experiments for three origin–destination pairs in the NIRNI area, and for each trip, the fastest path is selected from four configurations of user preferences. We compared the duration, distance, and number of charging stops to demonstrate the usage of our methodology. In addition, we compared four EV models for a trip from Florence to Munich, which showed clear differences in duration and the number of stops, with the Tesla Roadster 2022 having the fastest path at seven hours and five minutes with no recharging stops when starting at 100% SoC, while a Kia EV6 2WD 2021 lasts twelve hours and twenty-one minutes with two recharging stops. Decreasing the starting SoC from 100% to 60% resulted in one additional recharging stop for all EV models.

An interesting extension to our methodology is to include the geophysical properties of the roads to obtain more accurate estimations of the energy consumption between CSs. Also, future extensions include considering dynamic planning, which accounts for dynamic energy consumption based on vehicle dynamics, driving conditions, traffic, road conditions, and other real-time restrictions.

Author Contributions: Conceptualization, M.C., M.D., O.J., F.M. and E.T.; methodology, M.C., M.D., O.J., F.M. and E.T.; software, M.C. and E.T.; validation, M.C., M.D., O.J., F.M. and E.T.; formal analysis, M.C., M.D., O.J., F.M. and E.T.; writing—original draft preparation, M.C. and O.J.; writing—review and editing, M.C. and O.J. All authors have read and agreed to the published version of the manuscript.

Funding: This work is a part of the eCharge4Drivers project. This project has received funding from the European Union Horizon 2020 research and innovation programme under grant agreement No. 875131. This content reflects only the author's view and the European Commission is not responsible for any use that may be made of the information this publication contains.

Data Availability Statement: Not applicable.

Conflicts of Interest: The authors declare no conflict of interest.

Nomenclature

The following nomenclature is used in this manuscript:

EV	Electric Vehicle
CP	Charging Point
CS	Charging Station
POI	Point of Interest
SoC	State of Charge
kWh	Kilowatt-hour
EVSPP	Electric Vehicle Shortest Path Problem
EVSPPWP	Electric Vehicle Shortest Path Problem with time windows and user preferences
MILP	Mixed Integer Linear Programming
EVRP	Electric Vehicle Routing Problem
FRVCP	Fixed Route Vehicle Charging Problem
TTDP	Tourist Trip Design Problem
EVRTP	Electric Vehicle Routing Tour Planning
NIRNI	Northern Italy and Regions North of Italy
$\mathcal{S}$	Set of charging stations
$\mathcal{O}$	Origin node
$\mathcal{D}$	Destination node
$\mathcal{S}_{\mathcal{O},\mathcal{D}}$	Set of nodes containing the origin $\mathcal{O}$, destination $\mathcal{D}$, and the set $\mathcal{S}$

t_{start}	Start time at $\mathcal{O}$
q_{start}	Initial SoC
q_{end}	Final SoC
$q_{\min}$	Minimal energy at arrival
Δ	Time spent charging at a CS
$\Phi_i(\Delta)$	Piece-wise linear concave charging function of charging station i
$\mathcal{G}$	Graph
$\mathcal{A}$	Set of arcs in the graph $\mathcal{G}$
t_{ij}	Driving time between nodes i and j
d_{ij}	Distance between nodes i and j
e_{ij}	Energy consumption between nodes i and j
η	EV average consumption
Q	Battery capacity
B_i	Ordered set of breakpoints of the charging curve of CS i
c_{ik}	Charging time of breakpoint $k \in B_i$
a_{ik}	SoC of breakpoint $k \in B_i$
ρ_{ik}	Slope of the linear function that connects breakpoints $(c_{i,k-1}, a_{i,k-1})$ and (c_{ik}, a_{ik})
σ_{jp}	Importance value of $j \in \mathcal{S}$ for POI category $p \in \mathcal{P}$
$d_{\max}$	Maximum distance between arcs
μ	Weight on the total importance value of the POIs with respect to travel time
$\mathcal{P}$	Set of POI categories
Δ_j	Time spent charging at CS j
$\mathcal{W}$	Set of ordered time windows that require a specific POI category
γ_w^L	Minimum start time of time window w
γ_w^U	Maximum start time of time window w
$t_w^{\min}$	Minimum time that the EV needs to stop for time window w
$\mathcal{S}_w$	Set of CSs with POIs in their neighborhood that match the POI specified for $w \in W$
$\tilde{s}_{ij}$	Modified weight of the arc between nodes i and j
$\underline{q}_i$	SoC when the EV arrives at i
$\overline{q}_i$	SoC when the EV departs from i
$\underline{h}_i$	Start time for charging an EV at i
$\overline{h}_i$	End time for charging an EV at i
$\underline{\lambda}_{ik}$	Coefficients associated with the breakpoint (c_{ik}, a_{ik}) when the EV enters i
$\overline{\lambda}_{ik}$	Coefficients associated with the breakpoint (c_{ik}, a_{ik}) when the EV leaves i
$\underline{w}_{ik}$	Binary variable equal to 1 when the SoC is in $[a_{i,k-1}, a_{ik}]$, when the EV enters i
$\overline{w}_{ik}$	Binary variable equal to 1 when the SoC is in $[a_{i,k-1}, a_{ik}]$, when the EV leaves i
$\underline{\tau}_i$	Time when the EV arrives at i
$\overline{\tau}_i$	Time when the EV departs from i
x_{ij}	Binary variable equal to 1 if the EV uses arc (i, j)
y_{jw}	Binary variable equal to 1 if the EV stops in j in time window w
d_{POI}	Distance threshold for considering POIs around a CS
r_{ij}	Resilience value of the arc between nodes i and j
d_r	Critical distance for ensuring availability of CSs
d_t	Length of a section in which the resilience value is computed
C_k	Number of CSs within a distance of d_r from section k
d_f	Threshold distance around the non EV-paths
$\tilde{\mathcal{G}}$	Reduced graph
$\tilde{\mathcal{S}}_{\mathcal{O},\mathcal{D}}$	Filtered set of nodes
$\tilde{\mathcal{A}}$	Set of considered arcs between $\tilde{\mathcal{S}}_{\mathcal{O},\mathcal{D}}$
$\mathcal{N}_{\mathcal{W}}$	Ordered set of nodes selected by the algorithm to perform time windows in $\mathcal{W}$

References

1. EEA. Average Carbon Dioxide Emissions from New Cars Registered in EUROPE Decreased by 12% in 2020. 2022. Available online: https://www.eea.europa.eu/highlights/average-carbon-dioxide-emissions-from (accessed on 13 April 2023).
2. Martins, L.d.C.; Tordecilla, R.D.; Castaneda, J.; Juan, A.A.; Faulin, J. Electric vehicle routing, arc routing, and team orienteering problems in sustainable transportation. *Energies* **2021**, *14*, 5131. [CrossRef]
3. Kucukoglu, I.; Dewil, R.; Cattrysse, D. The electric vehicle routing problem and its variations: A literature review. *Comput. Ind. Eng.* **2021**, *161*, 107650. [CrossRef]

4. Froger, A.; Jabali, O.; Mendoza, J.E.; Laporte, G. The electric vehicle routing problem with capacitated charging stations. *Transp. Sci.* **2022**, *56*, 460–482. [CrossRef]
5. Sánchez, D.G.; Tabares, A.; Faria, L.T.; Rivera, J.C.; Franco, J.F. A Clustering Approach for the Optimal Siting of Recharging Stations in the Electric Vehicle Routing Problem with Time Windows. *Energies* **2022**, *15*, 2372. [CrossRef]
6. Kullman, N.D.; Froger, A.; Mendoza, J.E.; Goodson, J.C. frvcpy: An open-source solver for the fixed route vehicle charging problem. *INFORMS J. Comput.* **2021**, *33*, 1277–1283. [CrossRef]
7. A Better Route Planner. 2023. Available online: https://abetterrouteplanner.com (accessed on 15 March 2023).
8. ZapMapEV. 2023. Available online: https://www.zap-map.com (accessed on 15 March 2023).
9. PlugShare. 2023. Available online: https://www.plugshare.com (accessed on 15 March 2023).
10. Schiffer, M.; Schneider, M.; Walther, G.; Laporte, G. Vehicle routing and location routing with intermediate stops: A review. *Transp. Sci.* **2019**, *53*, 319–343. [CrossRef]
11. Erdoğan, S.; Miller-Hooks, E. A green vehicle routing problem. *Transp. Res. Part E Logist. Transp. Rev.* **2012**, *48*, 100–114. [CrossRef]
12. Schneider, M.; Stenger, A.; Goeke, D. The electric vehicle-routing problem with time windows and recharging stations. *Transp. Sci.* **2014**, *48*, 500–520. [CrossRef]
13. Felipe, Á.; Ortuño, M.T.; Righini, G.; Tirado, G. A heuristic approach for the green vehicle routing problem with multiple technologies and partial recharges. *Transp. Res. Part E Logist. Transp. Rev.* **2014**, *71*, 111–128. [CrossRef]
14. Hiermann, G.; Hartl, R.F.; Puchinger, J.; Vidal, T. Routing a mix of conventional, plug-in hybrid, and electric vehicles. *Eur. J. Oper. Res.* **2019**, *272*, 235–248. [CrossRef]
15. Goeke, D.; Schneider, M. Routing a mixed fleet of electric and conventional vehicles. *Eur. J. Oper. Res.* **2015**, *245*, 81–99. [CrossRef]
16. Keskin, M.; Çatay, B. Partial recharge strategies for the electric vehicle routing problem with time windows. *Transp. Res. Part C Emerg. Technol.* **2016**, *65*, 111–127. [CrossRef]
17. Montoya, A.; Guéret, C.; Mendoza, J.E.; Villegas, J.G. The electric vehicle routing problem with nonlinear charging function. *Transp. Res. Part B Methodol.* **2017**, *103*, 87–110. [CrossRef]
18. Pelletier, S.; Jabali, O.; Laporte, G. The electric vehicle routing problem with energy consumption uncertainty. *Transp. Res. Part B Methodol.* **2019**, *126*, 225–255. [CrossRef]
19. Montoya, A.; Guéret, C.; Mendoza, J.E.; Villegas, J.G. A multi-space sampling heuristic for the green vehicle routing problem. *Transp. Res. Part C Emerg. Technol.* **2016**, *70*, 113–128. [CrossRef]
20. Andelmin, J.; Bartolini, E. An exact algorithm for the green vehicle routing problem. *Transp. Sci.* **2017**, *51*, 1288–1303. [CrossRef]
21. Froger, A.; Mendoza, J.E.; Jabali, O.; Laporte, G. Improved formulations and algorithmic components for the electric vehicle routing problem with nonlinear charging functions. *Comput. Oper. Res.* **2019**, *104*, 256–294. [CrossRef]
22. Desaulniers, G.; Errico, F.; Irnich, S.; Schneider, M. Exact algorithms for electric vehicle-routing problems with time windows. *Oper. Res.* **2016**, *64*, 1388–1405. [CrossRef]
23. OpenEV. GitHub-Chargeprice/Open-ev-Data. 2022. Available online: https://github.com/chargeprice/open-ev-data (accessed on 15 March 2023).
24. Zündorf, T. Electric Vehicle Routing with Realistic Recharging Models. Master's Thesis, Karlsruhe Institute of Technology, Karlsruhe, Germany, 2014, *unpublished*.
25. Sweda, T.M.; Dolinskaya, I.S.; Klabjan, D. Adaptive routing and recharging policies for electric vehicles. *Transp. Sci.* **2017**, *51*, 1326–1348. [CrossRef]
26. Baum, M.; Dibbelt, J.; Gemsa, A.; Wagner, D.; Zündorf, T. Shortest feasible paths with charging stops for battery electric vehicles. *Transp. Sci.* **2019**, *53*, 1627–1655. [CrossRef]
27. Baum, M.; Dibbelt, J.; Pajor, T.; Sauer, J.; Wagner, D.; Zündorf, T. Energy-optimal routes for battery electric vehicles. *Algorithmica* **2020**, *82*, 1490–1546. [CrossRef]
28. Aljohani, T.M.; Mohammed, O. A Real-Time Energy Consumption Minimization Framework for Electric Vehicles Routing Optimization Based on SARSA Reinforcement Learning. *Vehicles* **2022**, *4*, 1176–1194. [CrossRef]
29. Aljohani, T.M.; Ebrahim, A.; Mohammed, O. Real-Time metadata-driven routing optimization for electric vehicle energy consumption minimization using deep reinforcement learning and Markov chain model. *Electr. Power Syst. Res.* **2021**, *192*, 106962. [CrossRef]
30. Roberti, R.; Wen, M. The electric traveling salesman problem with time windows. *Transp. Res. Part E Logist. Transp. Rev.* **2016**, *89*, 32–52. [CrossRef]
31. Ruiz-Meza, J.; Montoya-Torres, J.R. A systematic literature review for the tourist trip design problem: Extensions, solution techniques and future research lines. *Oper. Res. Perspect.* **2022**, *9*, 100228. [CrossRef]
32. Gavalas, D.; Konstantopoulos, C.; Mastakas, K.; Pantziou, G. A survey on algorithmic approaches for solving tourist trip design problems. *J. Heuristics* **2014**, *20*, 291–328. [CrossRef]
33. Wang, Y.W.; Lin, C.C.; Lee, T.J. Electric vehicle tour planning. *Transp. Res. Part D Transp. Environ.* **2018**, *63*, 121–136. [CrossRef]
34. Karbowska-Chilinska, J.; Chociej, K. Genetic algorithm for generation multistage tourist route of electrical vehicle. In Proceedings of the International Conference on Computer Information Systems and Industrial Management, Bialystok, Poland, 16–18 October 2020; Springer: Berlin, Germany, 2020; pp. 366–376.

35. Cortés-Murcia, D.L.; Afsar, H.M.; Prodhon, C. Multi-period profitable tour problem with electric vehicles and mandatory stops. *Int. J. Sustain. Transp.* **2022**, *17*, 473–489. [CrossRef]
36. Cassia, A.; Jabali, O.; Malucelli, F.; Pascoal, M. The electric vehicle shortest path problem with time windows and prize collection. In Proceedings of the 2022 17th Conference on Computer Science and Intelligence Systems (FedCSIS), Sofia, Bulgaria, 4–7 September 2022; IEEE: Piscataway, NJ, USA, 2022; pp. 313–322.
37. Mapbox. Mapbox Optimization Tool. 2023. Available online: https://www.mapbox.com (accessed on 15 March 2023).

energies

MDPI

Article

Locating and Sizing Electric Vehicle Chargers Considering Multiple Technologies

Tommaso Schettini [1,*], Mauro dell'Amico [2], Francesca Fumero [3,*], Ola Jabali [4] and Federico Malucelli [4]

1 GERAD, École de Technologie Supérieure, HEC Montréal, Montréal, QC H3T 2A7, Canada
2 Department of Sciences and Methods for Engineering, Università di Modena e Reggio Emilia, 42122 Modena, Italy; mauro.dellamico@unimore.it
3 Dipartimento di Ingegneria Gestionale, Politecnico di Milano, 20133 Milano, Italy
4 Dipartimento di Elettronica, Informazione e Bioingegneria, Politecnico di Milano, 20133 Milano, Italy; ola.jabali@polimi.it (O.J.); federico.malucelli@polimi.it (F.M.)
* Correspondence: tommaso.schettini@hec.ca (T.S.); francesca.fumero@polimi.it (F.F.)

Abstract: In order to foster electric vehicle (EV) adoption rates, the availability of a pervasive and efficient charging network is a crucial requirement. In this paper, we provide a decision support tool for helping policymakers to locate and size EV charging stations. We consider a multi-year planning horizon, taking into account different charging technologies and different time periods (day and night). Accounting for these features, we propose an optimization model that minimizes total investment costs while ensuring a predetermined adequate level of demand coverage. In particular, the setup of charging stations is optimized every year, allowing for an increase in the number of chargers installed at charging stations set up in previous years. We have developed a tailored heuristic algorithm for the resulting problem. We validated our algorithm using case study instances based on the village of Gardone Val Trompia (Italy), the city of Barcelona (Spain), and the country of Luxembourg. Despite the variability in the sizes of the considered instances, our algorithm consistently provided high-quality results in short computational times, when compared to a commercial MILP solver. Produced solutions achieved optimality gaps within 7.5% in less than 90 s, often achieving computational times of less than 5 s.

Keywords: charging station location; charging infrastructure planning; electric vehicles; facility location

Citation: Schettini, T.; dell'Amico, M.; Fumero, F.; Jabali, O.; Malucelli, F. Locating and Sizing Electric Vehicle Chargers Considering Multiple Technologies. *Energies* **2023**, *16*, 4186. https://doi.org/10.3390/en16104186

Academic Editors: Michele Roccotelli, Evangelos Karfopoulos and Ioannis Karakitsios

Received: 30 April 2023
Revised: 11 May 2023
Accepted: 12 May 2023
Published: 18 May 2023

1. Introduction

Due to environmental concerns, electric vehicles (EVs) are increasingly gaining importance as efficient and clean transport modes. Indeed, the widespread adoption of EVs is prioritized by various policymakers to mitigate climate change impacts and foster sustainability. However, the diffusion of these vehicles is challenged by several adoption barriers, such as relatively high prices, safety concerns, and relatively poor performance [1]. Additionally, sparse public charging infrastructure is one of the most significant adoption barriers for private EV users [2]. Therefore, a well-planned design of a pervasive charging network is crucial to facilitating users' recharging access and stimulating EV adoption. Furthermore, Sustainable Development Goal 7 (SDG7) calls for ensuring "universal access to affordable, reliable and modern energy services", expanding infrastructure, and upgrading suitable technologies, so as to make energy supply available to everyone.

Planning the locations and capacities of public charging infrastructure is a complex task as it encompasses various intertwined issues. Namely, EV chargers should be spread out in a way that facilitates easy access to users. However, the construction and maintenance of charging infrastructure can be costly. The magnitude of such investments, coupled with the projected growth in EV adoption, suggests that the strategic planning of charging infrastructure should be done by considering a multi-year planning horizon. Furthermore, different charging technologies should be evaluated, according to user preferences and needs.

We investigate the problem of locating and sizing public charging stations (CSs) for private EV users; we consider a multi-year planning horizon as well as multiple charging technologies. In particular, we consider slow and fast charging technologies. The former is more suited for EVs parked for long durations, such as at home or work [2], whereas the latter is suited for major urban motorways and dense road traffic areas. The specifications of our problem were identified through the eCharge4Drivers project, which is funded by the European Union Horizon 2020 program.

The design process of an adequate network of public CSs can be decomposed into three major conceptual steps. The first step consists of estimating the charging demand for different technologies in the considered area. Two main modeling approaches are used for this purpose [3]: (a) estimating origin–destination traffic flows, with the aim of intercepting recharging needs during trips, and (b) aggregating demand in zones (e.g., neighborhoods) considered as nodes of a network. We use demand estimates derived from the second approach, which is suitable for vehicles parked for long durations (e.g., at home or work) in urban and semi-urban areas. The second step identifies a set of potential locations for CSs. This is done on the basis of land-use planning, environmental effects, and safety requirements. Furthermore, the compatibility of locations with charging technologies (e.g., fast chargers) is identified. This typically includes the consideration of energy grid capabilities and possible upgrades. Given the input from the first two steps, the third step consists of optimizing the locations of CSs along with their sizing. The latter entails establishing the number of chargers of each technology to install at a CS. This paper is centered on modeling and developing an efficient solution algorithm for the third step. However, we validate our model using input from the first two steps, which is derived from the eCharge4Drivers project. We were provided with demand estimates for each technology for two daily periods (i.e., daytime and nighttime), for a set of zones. As we investigate the problem of locating and sizing public CSs, we presume that the demand (used as input to our problem) already discounts the charging demand that is satisfied by home charging. Moreover, zones were also classified as potential CS locations in terms of their ability to accommodate fast and slow charging technologies. Finally, the existing CSs were also provided as input.

We modeled a multi-year charger location and sizing problem. We simultaneously considered expanding the charging capacities of existing CSs by installing more chargers, setting up new CSs, and determining their number of chargers. To this end, we considered costs related to setting up a new CS and the costs of installing chargers in a setup CS. These costs depend on the installed technology. Moreover, the capacity of each charging technology was modeled via an input parameter expressed as the amount of kWh it could provide in a period. As such, our model is general and can work with any charging technology, which is specified through input parameter values. For more details concerning the latest developments in charging technologies, we refer the reader to the following surveys [4–7]. The objective of the problem is to minimize the total investment costs while guaranteeing a target level of demand converge, expressed as a percentage of total demand. Given the multi-year aspect of the problem, we assume that location and sizing decisions are performed at the beginning of each year, and that installed chargers at a given CS remain operational for the subsequent years. We denote the resulting problem as the incremental charger location problem (ICLP).

We propose a mixed-integer linear programming (MILP) model for the ICLP. We derive a maximum flow formulation for a subproblem of ICPL, in which the charging infrastructure is fixed. Such a subproblem can efficiently be solved in polynomial time using well-established algorithms. We use this result in developing an efficient heuristic algorithm for the ICLP based on iteratively installing CSs and chargers at promising locations until the desired level of coverage is reached. Finally, we accelerate our proposed algorithm by using a priority-based strategy, which significantly reduces the computational effort associated with the identification of promising locations at each iteration of the algorithm. We validate the algorithm on three case study instances obtained through the eCharge4Drivers project. Specifically, we show that our algorithm consistently produces high-quality results in a fraction of the time required by a state-of-the-art commercial MILP solver. Finally, we note that maximum flow problems can be very efficiently solved by existing open-source algorithms. Thus, our algorithm can be easily implemented by a broad audience, as it does not necessitate the use of commercial MILP solvers that often require costly software licenses.

To summarize, the contributions of this paper are as follows.

- We model a multi-year charger location and sizing problem by considering multiple charging technologies and periods (ICLPs). Specifically, we consider that the demand for each year, technology, and period is provided for a set of zones.
- We reformulate the single-year ICLP with a fixed charger infrastructure as a maximum flow problem.
- We utilize the maximum flow subproblem in developing an efficient heuristic algorithm for the ICLP.
- We validate the proposed heuristic on three realistic case studies.

The rest of the paper is organized as follows. In Section 2, we briefly outline related works on the charging location problem. In Section 3, we present MILP models for the ICPL and several of its subproblems. We describe our proposed algorithm in Section 4, and validate its performance through computational experiments in Section 5. Finally, we discuss our conclusions and possible future research directions in Section 6.

2. Literature Review

Optimizing the location of CSs has received considerable attention in recent years (see Kchaou-Boujelben [3] for a comprehensive survey). An important aspect of such problems is related to the estimation and modeling of EV charging demand. In particular, the models proposed in the literature often fall into one of two categories: flow-based models and node-based models. In the former case, the demand is modeled through a set of origin–destination trips, whose charging requirements are satisfied at possible locations that intercept the traffic flow. This is suitable for locating CSs on motorways, where the objective is to capture EV flows. In the latter case, the demand is assumed to be located on a set of given nodes. Such demand models are suitable for locating CSs in inhabited areas (e.g., neighborhoods) where EV users live or work. In this paper, we adopt a node-based demand model. This choice stems from the addressed cases in the eCharge4Drivers project. Furthermore, we assume that the energy demand is given as an input to the problem. Thus, the problem of estimating energy demand is beyond the scope of this paper. However, we remark that different methodologies can be applied to estimate the demand for EV charging, such as data-driven techniques, machine learning, spatial statistical analysis (e.g., [8–12]). In what follows, we focus on contributions related to optimizing CS locations considering node-based demand.

A variety of modeling techniques has been employed in CS location optimization, such as fixed-charge models [13], set covering models [8], and p-median models [14]. Those models have been solved using a wide range of approaches, including exact methods [8,14], approximate methods [15], spatial analysis techniques [10], evolutionary algorithms [16,17], and neural networks [11]. The extensive survey by Kchaou-Boujelben [3] provides details on the used solution methods.

An important distinction between models pertains to the choice of objective functions used in the optimization. Several types of objective functions have been considered in the literature, such as minimizing investment costs [15], maximizing covered demands [10], and maximizing service levels. This last objective is typically achieved by minimizing the users' traveling costs/distances to installed chargers [14]. Several authors have considered multiple objectives in the optimization simultaneously. Bai et al. [16] introduced a bi-objective function by accounting for investment costs and service levels, determined by the users' driving time to the chargers as well as recharging times. Given a fixed budget, Cavadas et al. [9] maximized the satisfied demand, which was weighted to prevent users from having to travel long distances for recharging. Analogously, Vazifeh et al. [12] ensured demand satisfaction, minimized the user's total excess driving distance to reach installed CSs, and minimized the number of CSs. We consider minimizing investment costs while guaranteeing that a given percentage of the demand is satisfied.

Two main cost components may be considered when minimizing investment costs. Firstly, a setup cost, comprising land, construction, and grid upgrades; secondly, an installation cost, associated with each installed charger. The setup cost may vary depending on the specific location, e.g., setting up a CS in the city center may be expensive, due to the cost of renting/purchasing land in that area. Conversely, the installation cost is roughly proportional to the number of installed chargers. When considering expanding the number of chargers at an existing CS, the setup cost may be ignored. This possibility was explicitly modeled by Bai et al. [16], who included a "saving cost" in their objective function, which occurs whenever a CS is based at an existing gas station. We note that the installation cost is meaningfully distinct from the setup cost when the optimization problem also considers the sizing of chargers (i.e., determining the number of chargers to be installed at each CS); thus, the problem explicitly models the charging capacity of the infrastructure.

The enforcement of capacity constraints is often done through an allocation-based approach, where the charging demand is directly allocated to CSs to measure capacity consumption. This is often done through a system optimum approach, where the model is free to determine the allocation of demand to chargers as part of the optimization, e.g., [9,13,18]. Ko and Shim [14] considered the nearest allocation rule, where demand is allocated to the nearest open charger, whereas Xie et al. [15] and Hu et al. [11] considered a simulation-based approach to allocate the demand. Furthermore, demand is usually allowed to be allocated to any CS within a maximum distance [9,13], with some authors opting to limit the amount of demand that can be allocated to distant chargers [18].

In some cases, locating and sizing decisions are not made simultaneously but are addressed sequentially. For example, Xie et al. [15] proposed a two-stage approach for locating renewable-powered charging stations on a highway network, where the locations of CSs are first determined and then the capacities of the CSs and energy storage units are established. Several authors simultaneously optimized the setup costs and the installation costs (e.g., [19]). Sadeghi-Barzani et al. [20] minimized these costs by considering fast chargers. They proposed a mixed-integer non-linear model for the resulting problem. Frade et al. [18] defined a maximum covering model that establishes the number and capacities of CSs, in order to satisfy morning and peak hour requests within acceptable levels of service. Zhu et al. [17] considered setup costs and installation costs while accounting for user access costs. The authors formulated an integer programming model for the resulting problem and solved the model via a genetic algorithm.

Considering the previously discussed literature, three additional features are fundamental to optimizing CS locations. First, the demand volume depends on different periods of the day, e.g., day and night. Such a distinction was made in models by Frade et al. [18], Dong et al. [10], and Cavadas et al. [9]. We considered distinct demand input values for day and night periods. Second, several charging technologies exist (e.g., slow and fast), inducing different costs and demand coverage characteristics. This was considered in [13,16]. We consider the option of simultaneously installing fast and slow chargers at a single location. Third, given the strategic nature of the problem, a multi-year planning horizon should be considered. A multi-year planning horizon was considered by Hu et al. [11]. We adopt a similar setting multi-year setting. Furthermore, we allow setting up new locations every year, as well as expanding the capacities of locations installed in previous years.

We summarize the main features of the surveyed literature review in Table 1. The last line of the table positions our paper with respect to the literature.

Table 1. Comparison with representative papers.

Paper	Application	Coverage Requirement	Coverage Condition	Demand Allocation	Existing Infrastructure	Capacity	Objective	Problem Characteristics
Frade et al. (2011) [18]	city block	max	distance dependent	optimal allocation		sizing	maximize coverage	P
Baouche et al. (2014) [13]	city	full	max range	optimal allocation		fixed capacity	minimize setup plus travel costs	T
Sadeghi-Barzani et al. (2014) [20]	city block	full	max range	minimum distance		sizing	minimize setup plus energy loss(grid plus EVs)	
Cavadas et al. (2015) [9]	city	max	max range	optimal allocation		fixed capacity	maximize coverage	P
Asamer et al. (2016) [8]	city	max	max range	no	yes	no	maximize coverage	
Ko and Shim (2016) [14]	city (taxi network)	full	none	nearest		sizing (capacity scenarios)	minimize travel cost	
Zhu et al. (2016) [17]	metropolitan area	full	max range	optimal allocation		sizing	minimize installation plus travel costs	
Xie et al. (2018) [15]	highway network	full	max range	simulation		sizing (2-step)	minimize setup plus installation costs	
Bai et al. (2019) [16]	city	full	none	optimal allocation	yes	sizing	minimize setup plus installation plus travel costs	T
Bouguerra and Layeb (2019) [19]	city	full	max range	optimal allocation		sizing	minimize setup plus installation plus travel costs	
Dong et al. (2019) [10]	city	max	max range	no		no	maximize coverage	P
Vazifeh et al. (2019) [12]	metropolitan area	full	distance dependent	no		fixed capacity	minimize travel cost	
Hu et al. (2020) [11]	metropolitan area	max	max range	simulation		sizing	maximize coverage	Y
This paper	metropolitan area, city, village	full/partial	max range	optimal allocation	yes	sizing	minimize setup plus installation costs	P Y T

Problem characteristics: **Y** multiple years; **P** multiple periods; **T** multiple technologies.

3. Problem Statement and Formulations

The ICLP is a multi-year planning problem centered on determining the locations of CSs and sizing their installed chargers. We consider minimizing the setup costs and installation costs, while guaranteeing that a target level of coverage is reached. Furthermore, we account for multiple demand periods (day and night) and multiple charging technologies (slow and fast).

We denote by P the set of periods, the set of available charging technologies as K, and the set of years as T. The objective of the problem involves the minimization of the total infrastructure investment cost while reaching a target level of coverage of Ψ^t percent of the total demand for each year $t \in T$. Let J denote the set of locations where CSs may be built, including locations already setup with existing CSs and chargers. Let I be the set of locations to be served (i.e., demand zones). We denote by d_{ij} the distance between location $i \in I$ and $j \in J$. Users are willing to walk for a maximum range of D to reach a CS. Thus, the charging demand of location $i \in I$ may only be allocated to locations within a D radius, i.e., $j \in J : d_{ij} \leq D$. We note that this potential coverage can be easily changed by the user. Indeed, what is needed in terms of input is a parameter indicating if location $i \in I$ could potentially be served by location $j \in J$. Thus, if the user possesses such information, it can easily be used instead of the maximum range.

A fixed cost of f_j^k is paid to set up location j for technology k. Each location $j \in J$ may have up to M_j^k chargers of technology k installed in total, and installing each charger entails a cost of c_j^k. We consider that certain charging infrastructures may exist before the start of the planning horizon. To capture this, parameter $\bar{z}_j^{0k}$ is one if location j is already setup for technology k at the start of the planning horizon. Similarly, parameter $\bar{x}_j^{0k}$ indicates the number of chargers with technology k that are already installed at location j at the start of the planning horizon. We denote by a_i^{tpk} the charging demand for location i for technology k, in period p, in year t. The demand for each node is expressed in kWh per period. We assume that demand is allocated to chargers in a system optimum approach. Thus, the optimization model establishes the assignment of demands to CSs, while respecting the range and capacity constraints. Furthermore, we consider that a single charger using technology k may provide up to q^k kWh per period.

3.1. Incremental Charger Location Problem Formulation

We formulate the ICLP as a MILP. We denote by x_j^{tk} the number of chargers installed at location j with technology k at the beginning of year t. Variable z_j^{tk} is one if location j is setup for technology k in year t, and zero otherwise. Variable y_{ij}^{tpk} is the percentage of allocated demand from location i for technology k in year t and period p to chargers at location j.

$$[ICLP] \min \sum_{t\in T}\sum_{k\in K}\sum_{j\in J}(f_j^k z_j^{tk} + c_j^k x_j^{tk}) \quad (1)$$

$$\sum_{j\in J:d_{ij}\leq D} y_{ij}^{tpk} \leq 1 \qquad \forall t\in T, p\in P, k\in K, i\in I \quad (2)$$

$$\sum_{p\in P}\sum_{k\in K}\sum_{i\in I}\sum_{j\in J:d_{ij}\leq D} a_i^{tpk} y_{ij}^{tpk} \geq \Psi^t \sum_{p\in P}\sum_{k\in K}\sum_{i\in I} a_i^{tpk} \qquad \forall t\in T \quad (3)$$

$$\sum_{i\in I:d_{ij}\leq D} a_i^{tpk} y_{ij}^{tpk} \leq q^k\Big(\sum_{t'\in T:t'\leq t} x_j^{t'k} + \bar{x}_j^{0k}\Big) \qquad \forall t\in T, p\in P, k\in K, j\in J \quad (4)$$

$$\Big(\sum_{t'\in T:t'\leq t} x_j^{t'k} + \bar{x}_j^{0k}\Big) \leq M_j^k\Big(\sum_{t'\in T:t'\leq t} z_j^{t'k} + \bar{z}_j^{0k}\Big) \qquad \forall t\in T, p\in P, k\in K, j\in J \quad (5)$$

$$\Big(\sum_{t\in T} z_j^{tk} + \bar{z}_j^{0k}\Big) \leq 1 \qquad \forall k\in K, j\in J \quad (6)$$

$$0 \leq y_{ij}^{tpk} \leq 1 \qquad \forall t\in T, p\in P, k\in K, i\in I, j\in J : d_{ij}\leq D \quad (7)$$

$$z_j^{tk}\in\{0,1\}, x_j^{tk}\geq 0 \qquad \forall t\in T, k\in K, j\in J \quad (8)$$

Objective function (1) minimizes the total setup costs and installation costs. Constraint (2) ensures that no more than the available charging demand is allocated to chargers. Constraint (3) enforces the target level of coverage for each year. Constraint (4) sets the available charging supply of the chargers. Constraint (5) links the setup activation variables with the charger installation variables. Constraint (6) ensures that the setup cost for each location is accounted for at most once.

Our proposed algorithm is based on iteratively solving subproblems of the ICLP. We present the relevant subproblems in the following subsections.

3.2. Single-Year Incremental Charger Location Problem

In this section, we present a rolling horizon approach to solving the ICLP. This entails optimizing the location of CSs and their installed chargers one year at a time. We use this approach to benchmark the performance of our proposed algorithm in the computational experiments.

When solving the ICLP in a rolling horizon fashion, all chargers installed in the previous years are considered fixed for the subsequent years. Thus, parameter $\bar{z}_j^{tk}$ is one if location j is setup for technology k before the beginning of year t. Note that parameter $\bar{z}_j^{0k}$ is a special case of this notation, as it denotes whether or not a location has been setup for technology k before the start of the planning horizon, i.e., year 0. Similarly, parameter $\bar{x}_j^{tk}$ represents the number of chargers with technology k installed at location j before the beginning of year t. We solve the ICLP for each year sequentially until the end of the planning horizon. We refer to the single-year version of the problem as the ICPL(t), where t is the year that is being optimized.

We formulate the ICLP(t) as a MILP. We denote by x_j^k the number of chargers installed at location j with technology k. Variable z_j^k is one if location j is setup for technology k, and zero otherwise. Variable y_{ij}^{pk} is the percentage of allocated demand from location i for technology k at period p to chargers at location j.

$$[\text{ICLP}(t)] \min \sum_{k\in K}\sum_{j\in J}(f_j^k z_j^k + c_j^k x_j^k) \tag{9}$$

$$\sum_{j\in J:d_{ij}\le D} y_{ij}^{pk} \le 1 \qquad \forall p\in P, k\in K, i\in I \tag{10}$$

$$\sum_{p\in P}\sum_{k\in K}\sum_{i\in I}\sum_{j\in J:d_{ij}\le D} a_i^{tpk} y_{ij}^{pk} \ge \Psi^t \sum_{p\in P}\sum_{k\in K}\sum_{i\in I} a_i^{tpk} \tag{11}$$

$$\sum_{i\in I:d_{ij}\le D} a_i^{tpk} y_{ij}^{pk} \le q^k(x_j^k + \bar{x}_j^{t-1,k}) \qquad \forall p\in P, k\in K, j\in J \tag{12}$$

$$(x_j^k + \bar{x}_j^{t-1,k}) \le M_j^k(z_j^k + \bar{z}_j^{t-1,k}) \qquad \forall p\in P, k\in K, j\in J \tag{13}$$

$$(z_j^k + \bar{z}_j^{t-1,k}) \le 1 \qquad \forall k\in K, j\in J \tag{14}$$

$$0 \le y_{ij}^{pk} \le 1 \qquad \forall p\in P, k\in K, i\in I, j\in J : d_{ij}\le D \tag{15}$$

$$z_j^k \in \{0,1\}, x_j^k \ge 0 \qquad \forall k\in K, j\in J \tag{16}$$

Objective function (9) minimizes the setup costs and installation costs for year t. Constraint (10) ensures that no more than the available charging demand is allocated to chargers for each location. Constraint (11) enforces the target level of coverage. Constraint (12) sets the available charging supply of the chargers. Constraint (13) links the setup activation variables with the charger installation variables. Constraint (14) ensures that the setup cost for each location is accounted for at most once.

3.3. ICLP(t) with Fixed Infrastructure

If all infrastructure variables are fixed (i.e., variables x and z are fixed), the ICLP(t) becomes a feasibility problem. Specifically, the problem pertains to whether or not the target level of demand coverage can be reached with a feasible allocation of the demand to chargers. Thus, it is possible to prove the feasibility of a given configuration of chargers by maximizing the total demand covered by the fixed infrastructure, and comparing it to $\Psi^t \sum_{p\in P}\sum_{k\in K}\sum_{i\in I} a_i^{tpk}$. We recall that $\bar{z}_j^{tk}$ is one if location j is setup for technology k before the beginning of year t and that parameter $\bar{x}_j^{tk}$ represents the number of chargers with technology k installed at location j before the beginning of year t. Thus, given the existing infrastructure at year t described by $\bar{x}$ and $\bar{z}$, the formulation of the problem of maximizing the covered demand is as follows:

$$[ICLP(t|\bar{x}\bar{z})] \max \sum_{p\in P}\sum_{k\in K}\sum_{i\in I}\sum_{j\in J:d_{ij}\le D} a_i^{tpk} y_{ij}^{pk} \tag{17}$$

$$\sum_{j\in J:d_{ij}\le D} y_{ij}^{pk} \le 1 \qquad \forall p\in P, k\in K, i\in I \tag{18}$$

$$\sum_{i\in I:d_{ij}\le D} a_i^{tpk} y_{ij}^{pk} \le q^k \bar{x}_j^{tk} \qquad \forall p\in P, k\in K, j\in J \tag{19}$$

$$0 \le y_{ij}^{pk} \le \bar{z}_j^{tk} \qquad \forall p\in P, k\in K, i\in I, j\in J : d_{ij}\le D \tag{20}$$

3.4. Maximum Flow Formulation of $[ICLP(t|\bar{x}\bar{z})]$

We now reformulate $[\text{ICLP}(t|\bar{x}\bar{z})]$ as a maximum flow problem on a directed graph $G(\bar{x}\bar{z}) = (N, A(\bar{x}\bar{z}))$. We note that maximum flow problems can be very efficiently solved by existing algorithms. Thus, the subsequent formulation is instrumental to our proposed algorithm in Section 4. Let s denote the source node, and t denote the sink node of the graph. Each potential charger location $j \in J$, for each technology $k \in K$ and each period $p \in P$, is associated with node β_j^{pk}. Similarly, each demand location $i \in I$ for each technology $k \in K$ and each period $p \in P$ is associated with node ξ_i^{pk}. Thus, the node set of the graph is $N = \{s,t\} \cup \{\beta_j^{pk} \forall j \in J, k \in K, p \in P\} \cup \{\xi_i^{pk} \forall i \in I, k \in K, p \in P\}$.

The arcs of $G(\bar{x}\bar{z})$ express how the demand is assigned to chargers. The source node s is connected through a directed arc to each charger node β_j^{pk} with capacity $q^k \bar{x}_j^{tk}$, $\forall j \in J, k \in K$. The flow on arc (s, β_j^{pk}) represents the total charging demand that is supplied by the chargers at location j with technology k in period p. Similarly, each demand node ξ_i^{pk} is connected to the sink node t with capacity a_i^{tpk}, $\forall i \in I, p \in P, k \in K$. The flow of those arcs represents the amount of charging demand of location i for technology k in period p, which is satisfied by the installed chargers. Lastly, each charger node β_j^{pk} is connected by an arc to all demand nodes in the range of j with technology k in period p, i.e., ξ_i^{pk}, $\forall i \in I : d_{ij} \leq D$. The flow on arc $(\beta_j^{pk}, \xi_i^{pk})$ represents the amount of charging demand at location i for technology k in period p, which is satisfied by chargers installed at location j. The capacity of an arc $(\beta_j^{pk}, \xi_i^{pk})$ is $\min(a_i^{tpk}, q^k \bar{x}_j^{tk})$, i.e., the minimum between the demand of location i for technology k in period p and the capacity of the chargers installed at j. Note, however, that the capacity of this last class of arcs can be relaxed without loss of generality. In Figure 1, we present a visual example of the construction of graph $G(\bar{x}\bar{z})$, for fixed infrastructure described by $\bar{x}\bar{z}$.

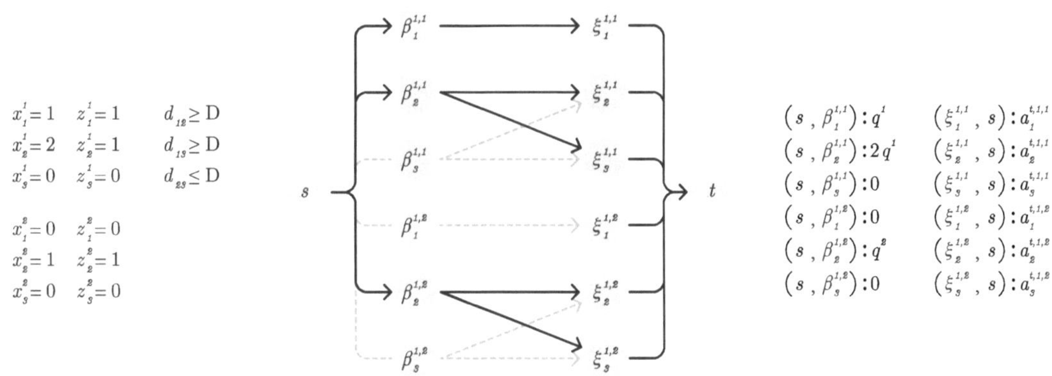

(**a**) Example charger configuration (**b**) Graph $G(xz)$ (**c**) Arc Capacities

Figure 1. Example of the construction of graph $G(\bar{x}\bar{z})$ from a charger configuration. For simplicity, $P = \{1\}$. Dashed arcs have a capacity of 0.

The maximum amount of demand covered by the charging infrastructure described by $\bar{x}\bar{z}$ equals the maximum flow on graph $G(\bar{x}\bar{z})$ from s to t. The charging infrastructure described by $\bar{x}\bar{z}$ is then feasible if the maximum flow is greater or equal to $\Psi^t \sum_{p \in P} \sum_{k \in K} \sum_{i \in I} a_i^{tpk}$.

4. Successive Incremental Algorithm

We propose an efficient successive incremental heuristic algorithm for the ICLP(t). We present our algorithm in two phases. We first present the main elements of the algorithm. Then, in Section 4.1, we introduce algorithmic enhancements to these elements.

Starting from the already existing infrastructure at year t, the heuristic iteratively installs chargers at promising locations, until the targeted level of demand coverage is reached. We use variable x_j^k, initialized to zero, to indicate the current number of chargers installed at location j with technology k in the solution. Thus, the value $x_j^k + \bar{x}_j^{t-1,k}$ represents the total number of chargers installed at location j with technology k at the current iteration of the heuristic. Similarly, we use variable z_j^k, initialized to zero, to indicate whether location j has been set up for technology k in the heuristic solution, and $z_j^k + \bar{z}_j^{t-1,k}$ to indicate whether location j is currently setup for technology k. At each iteration, we select a location, technology, and number of chargers, and update the corresponding x and z variables accordingly. Then, we measure the level of demand coverage achieved by the current solution and, if the targeted level of coverage has not been reached, the algorithm is iterated.

The location, technology, and number of installed chargers are selected at each iteration based on the two-step approach. In the first step, a restricted set of candidate locations is selected. This is done by identifying up to two locations for each technology: one representing the most promising location that has already been set up for that technology, and one for the most promising location that has not yet been set up for that technology.

The selection is based on the total uncovered demand that could be covered by installing any number of additional chargers at those locations. Formally, the computation of the total uncovered demand that could be covered is performed for each location i and technology k, by temporarily installing a single charger of infinite capacity at location i using technology k and measuring the increase in covered demand compared to the current solution. This results in a set of up to $2|K$ locations to be considered for the next step.

In the second step, for each location in the restricted set of candidate locations, we consider each possible number of installed chargers, and measure the increase in coverage, achieved by installing the number of chargers, divided by the required installation and setup cost (if necessary). Once all locations have been evaluated in this way, we select the best combination of location, technology, and number of installed chargers according to the measured coverage increase per unit investment. Then, those chargers are installed at the selected location. Afterward, if the targeted level of coverage is not achieved, the procedure is iterated. In the following, we will detail each step of the algorithm.

To perform the selection of the restricted set of candidate locations (performed in the first step of the algorithm), let us denote by δ_j^{pk} the maximum amount of additional charging demand that could be covered in period p by installing additional chargers with technology k at location j. To measure it, we consider a modified version of $G(\bar{x}\bar{z})$ where the capacity of arc (s, β_j^{pk}) is set to $+\infty$ and solves the associated maximum flow problem. The difference between the current covered demand and the maximum flow computed on the modified graph corresponds to δ_j^{pk}. For ease of notation, let $\delta_j^k = \sum_{p\in P} \delta_j^{pk}$. For each technology $k \in K$, we identify the most promising new location (π^{0k}) and the most promising already setup location (π^{1k}) as follows.

$$\pi^{0k} = \underset{j\in J\,:\, z_j^k+\bar{z}_j^{t-1,k}=0}{\arg\max} \quad \delta_j^k; \tag{21}$$

$$\pi^{1k} = \underset{j\in J\,:\, z_j^k+\bar{z}_j^{t-1,k}>0}{\arg\max} \quad \delta_j^k. \tag{22}$$

For the second step of the algorithm, each of the $2|K$ candidate locations is then evaluated by considering each possible number of chargers that could be installed at that location and dividing the achieved increase in covered demand reached by installing the number of chargers against the required installation and setup costs. Note that the setup cost is only accounted for when evaluating new locations. For each candidate location π with technology k, the possible number of chargers that could be installed is $\theta \in 1, \ldots, (M_\pi^k -$

$x_\pi^k - \bar{x}_\pi^{t-1,k}$). The amount of the charging demand that could be covered by installing θ chargers at location π for technology k is computed based on $\delta_\pi^{pk}, \forall p \in P$ as follows:

$$\sum_{p \in P} \min(q^k\theta, \delta_\pi^{pk}). \tag{23}$$

We denote by $\nu(k, \pi, \theta)$ the increase in covered demand divided by the potential setup and installation cost associated with installing θ chargers with technology k at location π:

$$\nu(k,\pi,\theta) = \begin{cases} \frac{\sum_{p \in P} \min(q^k\theta, \delta_\pi^{pk})}{f_j^k + \theta c_j^k} & \text{if } z_\pi^k + \bar{z}_\pi^{t-1,k} = 0; \\ \frac{\sum_{p \in P} \min(q^k\theta, \delta_\pi^{pk})}{\theta c_j^k} & \text{if } z_\pi^k + \bar{z}_\pi^{t-1,k} > 0. \end{cases} \tag{24}$$

Out of the candidate locations that we previously identified, we select the location, technology, and number of chargers according to the most promising value of ν. Thus, formally:

$$(k,\pi,\theta) = \underset{k \in K, \pi \in \{\pi^{0k}, \pi^{1k}\}, \theta \in 1,\ldots,(M_\pi^k - z_\pi^k - \bar{z}_\pi^{t-1,k})}{\arg\max} \nu(k,\pi,\theta). \tag{25}$$

Then, we install θ chargers of type k at location π, update the associated variables (i.e., x_j^k and z_j^k), and compute the total coverage achieved by adding these chargers via the maximum flow problem. If the achieved coverage reaches $\Psi^t \sum_{p \in P} \sum_{k \in K} \sum_{i \in I} a_i^{tpk}$, then we terminate; otherwise, the procedure is repeated until the targeted level of coverage is reached. The developed heuristic is described in Algorithm 1.

Algorithm 1: The successive incremental algorithm.

1 **Algorithm** `Constructive()`
2 $x_j^k \leftarrow 0 \quad \forall k \in K, j \in J$;
3 $z_j^k \leftarrow 0 \quad \forall k \in K, j \in J$;
4 $C \leftarrow [\text{ICLP}(\bar{x}\bar{z}|t)]$;
5 **while** $C \le \Psi \sum_{p \in P} \sum_{k \in K} \sum_{i \in I} a_i^{tpk}$ **do**
6 $\pi^{0k} \leftarrow \arg\max_{j \in J \,:\, z_j^k + \bar{z}_j^{t-1,k} = 0} \delta_j^k \quad \forall k \in K$;
7 $\pi^{1k} \leftarrow \arg\max_{j \in J \,:\, z_j^k + \bar{z}_j^{t-1,k} > 0} \delta_j^k \quad \forall k \in K$;
8 $(k,\pi,\theta) \leftarrow \arg\max_{k \in K, \pi \in \{\pi^{0k}, \pi^{1k}\}, \theta \in 1,\ldots,(M_\pi^k - z_\pi^k - \bar{z}_\pi^{t-1,k})} \nu(k,\pi,\theta)$;
9 $x_\pi^k \leftarrow x_\pi^k + \theta$;
10 **if** $z_\pi^k + \bar{z}_\pi^{t-1,k} = 0$ **then**
11 $z_\pi^k \leftarrow 1$;
12 **end**
13 $C \leftarrow [\text{ICLP}(\bar{x}\bar{z}|t)]$;
14 **end**
15 **return** x, z;

4.1. *Priority-Based Successive Incremental Algorithm*

The increase in the maximum demand that can be met by installing more chargers at a specific location, represented by δ_j^{pk}, is an important factor in the developed heuristic. We recall that the computation of δ_j^{pk} involves solving a maximum flow problem, which can be done very efficiently. However, lines 6 and 7 of Algorithm 1 require the computation of all values $\delta_j^{pk}, \forall p \in P, k \in K, j \in J$, which is still very computationally expensive. Therefore, we modify Algorithm 1 to take advantage of the fact that the value of δ_j^{pk} may only decrease at each iteration, $\forall p \in P, k \in K, j \in J$. For each technology $k \in K$, we introduce two priority queues: Q^{k0} and Q^{k1}. A priority queue is a queue data structure that is similar to a standard queue, with the additional feature that each element has a priority associated with it. When an element is added to the queue, it is added based on its priority, so that the element with the highest priority is at the front of the queue and the element with the lowest priority is at the back. The purpose of queues Q^{k0} is to facilitate the evaluation of expression (21), whereas queues Q^{k1} do so for expression (22). Within Q^{k0}, each location $j \in J \,:\, z_j^k + \bar{z}_j = 0$ is stored in the queue with priority $\bar{\delta}_j^k$, where $\bar{\delta}_j^k$ is an upper bound to δ_j^k. To compute π^{0k}, we examine the first element of the queue: $\bar{\pi}$. The priority associated with $\bar{\pi}$ is an upper bound to $\delta_{\bar{\pi}}^k$. Then, we compute $\delta_{\bar{\pi}}^k$ by solving the associated maximum flow problem and updating the priority of $\bar{\pi}$. If $\bar{\pi}$ is still the first element of the queue, $\pi^{0k} = \bar{\pi}$; otherwise, the process is repeated until the first element remains unchanged. Indeed, since the priority values of each element stored in Q^{k0} are upper bounds to their respective δ_j^k, we can conclude that $\bar{\pi} = \pi^{0k} = \arg\max_{j \in J \,:\, z_j^k + \bar{z}_j^{t-1,k} = 0} \delta_j^k$. This significantly reduces the number of times the δ_j^k needs to be computed and reduces the overall runtime of our algorithm. At the start of the algorithm, we populate Q^{k0} with each element $j \in J \,:\, z_j^k + \bar{z}_j^{t-1,k} = 0$ with priority $\sum_{p \in P} \sum_{i \in I \,:\, d_{ij} \leq D} a_i^{tpk}$. Similar considerations hold for Q^{k1}. We update Algorithm 1 to use these queues in Algorithm 2. We use Algorithm 2 in our experiments in a rolling horizon fashion, i.e., we optimize the location of chargers one year at a time. To solve the maximum flow problem in the heuristic, we use a push–relabel algorithm. This results in the worst-case complexity of $\mathcal{O}(M|K||P||T||I|^5)$ for the entire algorithm, where M corresponds to the maximum number of chargers of a single technology that may be installed at a single location. We remark however that, in practice, the effective computational complexity of the algorithm is much lower.

Algorithm 2: The priority-based successive incremental algorithm.

Algorithm `Constructive()`
  $x_j^k \leftarrow 0 \quad \forall k \in K,\ j \in J$;
  $z_j^k \leftarrow 0 \quad \forall k \in K,\ j \in J$;
  $C \leftarrow [\text{ICLP}(\bar{x}\bar{z}|t)]$;
  $Q^{k0} \leftarrow$ `PriorityQueue()` $\quad \forall k \in K$;
  $Q^{k1} \leftarrow$ `PriorityQueue()` $\quad \forall k \in K$;
  $Q^{k0}[j] \leftarrow \sum_{p \in P} \sum_{i \in I \,:\, d_{ij} \leq D} a_i^{tpk} \quad \forall k \in K, j \in J \,:\, z_j^k + \bar{z}_j^{t-1,k} = 0$;
  $Q^{k1}[j] \leftarrow \sum_{p \in P} \sum_{i \in I \,:\, d_{ij} \leq D} a_i^{tpk} \quad \forall k \in K, j \in J \,:\, z_j^k + \bar{z}_j^{t-1,k} > 0$;
  while $C \leq \Psi \sum_{p \in P} \sum_{k \in K} \sum_{i \in I} a_i^{tpk}$ **do**
    $\pi^{0k} \leftarrow$ `PopUpdate`$(Q^{k0}, k) \forall k \in K$;
    $\pi^{1k} \leftarrow$ `PopUpdate`$(Q^{k1}, k) \forall k \in K$;
    $(k, \pi, \theta) \leftarrow \arg\max_{k \in K, \pi \in \{\pi^{0k}, \pi^{1k}\}, \theta \in 1, \ldots, (M_\pi^k - z_\pi^k - \bar{z}_\pi^{t-1,k})} \nu(k, \pi, \theta)$;
    $x_\pi^k \leftarrow x_\pi^k + \theta$;
    if $z_\pi^k + \bar{z}_\pi^{t-1,k} = 0$ **then**
      $z_\pi^k \leftarrow 1$;
    end
    $C \leftarrow [\text{ICLP}(\bar{x}\bar{z}|t)]$;
    $Q^{k1}[\pi] \leftarrow \delta_\pi^k$;
  end
  return x, z;

Procedure `PopUpdate`(Q,k)
  $\bar{\pi} \leftarrow$ `Peek`(Q);
  while *true* **do**
    $Q[\bar{\pi}] \leftarrow \delta_{\bar{\pi}}^k$;
    if $\bar{\pi} =$ `Peek`(Q) **then**
      break;
    end
    $\bar{\pi} \leftarrow$ `Peek`(Q)
  end
  `Pop`(Q);
  return $\bar{\pi}$;

5. Computational Results

The successive incremental algorithm was implemented in Python version 3.9 using python-igraph version 0.10.2 [21] to solve the maximum flow problem used to evaluate the $ICLP(t|\bar{x}\bar{z})$. To validate the performance of our proposed algorithm, we compared its solutions to the exact solutions obtained by solving the $ICLP$ and the $ICLP(t)$ in a rolling horizon fashion using the CPLEX solver version 12.10. We considered a planning horizon of three years. Both formulations were implemented in Julia version 1.8, using the JuMP modeling interface version 1.7. The $ICLP$ formulation was run with a three-hour time limit, whereas for each year, the $ICLP(t)$ was solved with a one-hour time limit (totaling a maximum run time of three hours for three years). All experiments were run on 3.20 Ghz AMD Ryzen 5 1600, with 16 GB of RAM and 12 cores, running Windows 10, using a single thread.

We validated our algorithm on three test cases considered in the eCharge4Drivers; namely, the village of Gardone Val Trompia (Italy), the city of Barcelona (Spain), and the country of Luxembourg. Each of those instances is solved across a three-year planning horizon (2024–2026), considering two demand periods (day and night) per year and two charging technologies (fast and slow). The area of each instance is partitioned into the set of locations to be served I. After discussions with the project partners, the costs of setting up a location to accommodate chargers were set at EUR 20,000 for slow chargers and EUR 100,000 for fast chargers. Additionally, the cost of a single slow charger was set to EUR 7500 and EUR 80,000 for a single fast charger. Lastly, after discussions with the project partners, the charging capacities of the slow charger and fast charger were estimated (from charging session data) to be 28 kWh/day and 300 kWh/day, respectively. We note that these values correspond to about 25% of the theoretical capacity of the assumed chargers. This reduction is meant to capture the fact that chargers should not be assumed to be operational at full capacity, as users are not willing to wait for a CS to be free. Thus, this reduction implicitly acts as a buffer against demand uncertainty. An estimation of the demand (expressed in kWh/day) for each combination of location, period, and technology was provided to us by project partners. These data are not publicly available, as they were derived from confidential data.

The Gardone Val Trompia instance represents a small-sized instance (with $|I| = 113$ locations) in the context of a village area; the Barcelona instance represents a medium-sized instance (with $|I| = 656$ locations) within the context of a city, and lastly, the Luxembourg instance represents a country (with $|I| = 2674$ locations). The demand for each year was computed by assuming a linear growth of 5% per year over the estimated demand from the year 2023. For generality, the set of potential charger locations J was set equal to I, while accounting for all exiting charging infrastructure in J. For each area, we conducted tests with coverage target levels of 70%, 80%, and 90%.

Considering the three coverage target levels, Tables 2–4 show the results for Gardone Val Trompia, Barcelona, and Luxembourg, respectively. Under the title ICLP-[C1], the tables report the best-obtained cost values via CPLEX within the allotted time (i.e., three hours), where the costs include the setup and installation costs. The Gap column (%) shows the final optimality gap reported by CPLEX, while the t(s) column reports the runtime in seconds. The latter is lower than 10,800 only in cases where CPLEX found the optimal solution. Under the title ICLP(t)-[C2], we report CPLEX's results on the $ICLP(t)$ run in a rolling horizon of three years. Under the cost column, we report the best-obtained cost values via CPLEX for each year within the allotted time (i.e., one hour per year). The Gap(%) column shows the final optimality gap reported by CPLEX for each year. The Total cost column reports the cumulative yearly setup costs and installation costs, while the Total t(s) column shows the cumulative run time over the three years. The ICLP(t)-H -[C3] columns report the results of our proposed algorithm. The columns under this title have an analogous meaning to those presented by ICLP(t)-[C2]. The last three columns of Tables 2–4 compare the solutions obtained by the three methods. Specifically, we report the gaps computed as follows: Gap(A,B) $= \frac{(\text{cost}_A - \text{cost}_B)}{\text{cost}_A}$. For the sake of convenience, in the tables, we use **C1** to refer to the solutions computed using the ICPL (considering all three years), **C2** to refer to the solutions computed using the ICPL(t) (considering a single year in a rolling horizon fashion), and **C3** to refer to the solutions computed using the proposed algorithm.

Table 2. Computational results of the Gardone Val Trompia instance.

		ICLP-[C1]			ICLP(t)-[C2]				ICLP(t)-H-[C3]			Comparison		
Target Coverage Level (%)	Year	Cost	Gap(%)	t(s)	Cost	Gap(%)	Total Cost	Total t(s)	Cost	Total Cost	Total t(s)	Gap(C2,C1)(%)	Gap(C3,C1)(%)	Gap(C3,C2)(%)
70	1				280	0.0	280	204	280	280				0.0
	2				0	0.0	280	204	0	280				0.0
	3	295.0	3.1	10,800	15	0.0	295	204	15	295	1	0.0	0.0	0.0
80	1				330	0.0	330	464	350	350				5.7
	2				0	0.0	330	464	0	350				5.7
	3	345.0	2.2	10,800	15	0.0	345	464	0	350	1	0.0	1.4	1.4
90	1				400	0.0	400	2282	420	420				4.8
	2				0	0.0	400	2282	0	420				4.8
	3	415.0	3.0	10,800	15	0.0	415	2282	0	420	1	0.0	1.2	1.2

Tested approaches: **C1**: 3-year ICLP using CPLEX; **C2**: rolling horizon ICLP(t) using CPLEX; **C3**: rolling horizon ICLP(t) using a constructive heuristic.

We observe that our proposed algorithm produces solutions that are approximately 7% from optimality in a fraction of the computational time required by CPLEX to solve the same instances. In the large instances of the country of Luxembourg, the proposed heuristic produces solutions that are within 6.9% of the best solutions obtained using CPLEX in less than 90 s, whereas the same instances required well over one hour of computational time when using CPLEX. Additionally, we remark that when solving the *ICLP*, and considering all three years, CPLEX fails to converge to an optimal solution. In particular, for the large Luxembourg instances with target coverage levels of 80% and 90%, the solver performed very poorly. In those instances, CPLEX only managed to compute the naive solution obtained by installing the maximum number of chargers at each location, therefore trivially guaranteeing complete coverage of the demand. In the medium- and small-scale instances based in the city of Barcelona and the village of Gardone Val Trompia, we achieved solutions within 7.5% of the best solutions found by CPLEX in less than 5 s.

Table 3. Computational results of the Barcelona instance.

		ICLP-[C1]			ICLP(t)-[C2]				ICLP(t)-H-[C3]			Comparison		
Target Coverage Level (%)	Year	Cost	Gap(%)	t(s)	Cost	Gap(%)	Total Cost	Total t(s)	Cost	Total Cost	Total t(s)	Gap(C2,C1)(%)	Gap(C3,C1)(%)	Gap(C3,C2)(%)
70	1				6260	0.0	6260	16	6625	6625				5.5
	2				360	0.0	6620	53	310	6935				4.5
	3	6940.0	0.1	10,800	345	0.0	6965	64	505	7440	3	0.4	6.7	6.4
80	1				7490	0.0	7490	11	7885	7885				5.0
	2				415	0.0	7905	21	490	8375				5.6
	3	8285.0	0.0	515	420	0.0	8325	26	260	8635	3	0.5	4.1	3.6
90	1				8905	0.0	8905	388	9275	9275				4.0
	2				790	0.0	9695	399	890	10,165				4.6
	3	10,120.0	0.3	10,800	510	0.0	10,205	407	780	10,945	4	0.8	7.5	6.8

Tested approaches: **C1**: 3-year ICLP using CPLEX; **C2**: rolling horizon ICLP(t) using CPLEX; **C3**: rolling horizon ICLP(t) using a constructive heuristic.

Table 4. Computational results of the Luxembourg instance.

		ICLP-[C1]			ICLP(t)-[C2]				ICLP(t)-H-[C3]			Comparison		
Target Coverage Level (%)	Year	Cost	Gap(%)	t(s)	Cost	Gap(%)	Total cost	Total t(s)	Cost	Total cost	Total t(s)	Gap(C2,C1)(%)	Gap(C3,C1)(%)	Gap(C3,C2)(%)
70	1				18,690	0.2	18,690	3600	19905	19,905				6.1
	2				1000	0.0	19,690	3842	1135	21,040				6.4
	3	20,430.0	1.0	10,800	755	0.0	20,445	4216	915	21,955	67	0.1	6.9	6.9
80	1				23,245	0.2	23,245	3600	24,540	24,540				5.3
	2				1195	0.0	24,440	3902	1255	25,795				5.3
	3	672,545.0	96.3	10,800	1025	0.0	25465	3956	1055	26,850	77	−2541.1	−2404.8	5.2
90	1				28,260	0.5	28,260	3600	29,895	29,895				5.5
	2				1370	0.7	29,630	7200	1485	31,380				5.6
	3	672,545.0	95.6	10,801	1075	0.5	30,705	10,800	1130	32,510	89	−2090.3	−1968.7	5.6

Tested approaches: **C1**: 3-year ICLP using CPLEX; **C2**: rolling horizon ICLP(t) using CPLEX; **C3**: rolling horizon ICLP(t) using a constructive heuristic.

Thus, we conclude that our algorithm is effective at generating high-quality solutions within a short computational time.

Finally, the developed algorithm was also implemented in an online tool as part of the eCharge4Drivers project. In Figures 2 and 3, we show a visualization produced by the online tool of the solutions for the Barcelona and Luxembourg instances with 80% required coverage. Each panel displays the installed chargers by the end of a single year.

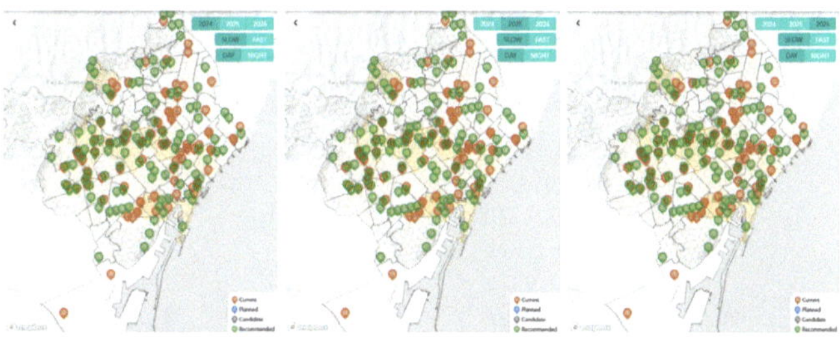

Figure 2. Visualization of the solution obtained from the online tool for the Barcelona instance, with 80% required coverage.

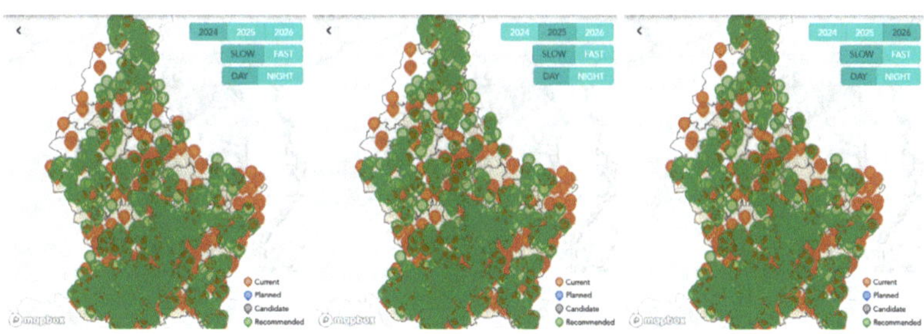

Figure 3. Visualization of the solution on the online tool on the Luxembourg instance with 80% required coverage.

6. Conclusions

We modeled a multi-year charger location and sizing problem, referred to as ICLP. This problem combines a number of realistic features, including multiple charging technologies and a distinction between the demand periods of the day. Furthermore, the setting up of new locations every year was considered along with increasing the number of chargers installed at locations set up in previous years. The objective of the ICLP is to minimize the total costs entailed by setting up new locations and installing chargers while guaranteeing a given service level. This is measured as the percentage of the total demand that is served by the charging infrastructure.

We developed MILP models for the ICLP and several of its subproblems. In particular, we showed that the single-year subproblem with a given charging infrastructure can be modeled as a maximum flow problem. We used this result to design a tailored successive incremental algorithm for the ICPL. This algorithm does not require the use of MILP solvers and, thus, may easily be applied in practice at no investment.

We validated the performance of our algorithm using a series of instances derived from realistic case studies. Specifically, we compared our algorithm against a commercial solver. We conclude that our algorithm achieves high-quality results in reasonable computational times.

Our algorithm could be used to provide guidelines for local authorities or private entities that are concerned with locating and sizing CSs. Such users would mainly need to possess demand estimates for the zones of a considered area and estimates of the effective capacity of the considered chargers.

Our model and proposed algorithm can be expanded to handle more realistic features of the problem. First, the demand of a zone can potentially be split and covered by two or more locations, which may not necessarily be within its vicinity. Furthermore, demand uncertainty could be incorporated by introducing appropriate uncertainty sets that model the operational behaviors of users. Moreover, various utilization rates can be further elaborated by assuming different input values for the effective capacity level for each location for each period. One of the assumptions made in our model entails that a charger installed at a given year will remain operational in the subsequent years of the considered planning horizon. Technological advances in terms of EVs, as well as their charging infrastructure, can make certain technologies obsolete. Therefore, a valid extension of our model may involve replacing (or removing) chargers during the planning period. Additionally, accounting for bidirectional charging in CS location problems would be an interesting extension. Finally, a highly relevant future research direction would be to include aspects related to the production of energy from renewable energy sources for the charging needs of electric vehicles.

Author Contributions: Conceptualization, T.S., M.d., O.J. and F.M.; methodology, T.S., M.d., O.J. and F.M.; software, T.S.; validation, T.S., M.d., F.F., O.J. and F.M.; formal analysis, T.S., M.d., O.J. and F.M.; writing—original draft preparation, T.S., F.F. and O.J.; writing—review and editing, T.S., F.F. and O.J.; All authors have read and agreed to the published version of the manuscript.

Funding: This work is part of the eCharge4Drivers project, which has received funding from the European Union Horizon 2020 research and innovation program under grant agreement no. 875131. This content reflects only the authors' view, and the European Commission is not responsible for any use that may be made of the information contained in this publication.

Data Availability Statement: The data presented in this study are not publicly available, as they were derived from confidential data.

Conflicts of Interest: The authors declare no conflict of interest.

References

1. Krishna, G. Understanding and identifying barriers to electric vehicle adoption through thematic analysis. *Transp. Res. Interdiscip. Perspect.* **2021**, *10*, 100364. [CrossRef]
2. Bryden, T.S.; Hilton, G.; Cruden, A.; Holton, T. Electric vehicle fast charging station usage and power requirements. *Energy* **2018**, *152*, 322–332. [CrossRef]
3. Kchaou-Boujelben, M. Charging station location problem: A comprehensive review on models and solution approaches. *Transp. Res. Part Emerg. Technol.* **2021**, *132*, 103376. [CrossRef]
4. Hemavathi, S.; Shinisha, A. A study on trends and developments in electric vehicle charging technologies. *J. Energy Storage* **2022**, *52*, 105013. [CrossRef]
5. Khalid, M.; Ahmad, F.; Panigrahi, B.K.; Al-Fagih, L. A comprehensive review on advanced charging topologies and methodologies for electric vehicle battery. *J. Energy Storage* **2022**, *53*, 105084. [CrossRef]
6. Brenna, M.; Foiadelli, F.; Leone, C.; Longo, M. Electric vehicles charging technology review and optimal size estimation. *J. Electr. Eng. Technol. J. Energy Storage* **2020**, *15*, 2539–2552. [CrossRef]

7. Alkawsi, G.; Baashar, Y.; Abbas U, D.; Alkahtani, A.A.; Tiong, S.K. Review of renewable energy-based charging infrastructure for electric vehicles. *Appl. Sci.* **2021**, *11*, 3847. [CrossRef]
8. Asamer, J.; Reinthaler, M.; Ruthmair, M.; Straub, M.; Puchinger, J. Optimizing charging station locations for urban taxi providers. *Transp. Res. Part Policy Pract.* **2016**, *85*, 233–246. [CrossRef]
9. Cavadas, J.; de Almeida Correia, G.H.; Gouveia, J. A MIP model for locating slow-charging stations for electric vehicles in urban areas accounting for driver tours. *Transp. Res. Part Logist. Transp. Rev.* **2015**, *75*, 188–201. [CrossRef]
10. Dong, G.; Ma, J.; Wei, R.; Haycox, J. Electric vehicle charging point placement optimisation by exploiting spatial statistics and maximal coverage location models. *Transp. Res. Part Transp. Environ.* **2019**, *67*, 77–88. [CrossRef]
11. Hu, D.; Zhang, J.; Liu, Z.W. Charging stations expansion planning under government policy driven based on Bayesian regularization backpropagation learning. *Neurocomputing* **2020**, *416*, 47–58. [CrossRef]
12. Vazifeh, M.M.; Zhang, H.; Santi, P.; Ratti, C. Optimizing the deployment of electric vehicle charging stations using pervasive mobility data. *Transp. Res. Part Policy Pract.* **2019**, *121*, 75–91. [CrossRef]
13. Baouche, F.; Billot, R.; Trigui, R.; El Faouzi, N.E. Efficient allocation of electric vehicles charging stations: Optimization model and application to a dense urban network. *IEEE Intell. Transp. Syst. Mag.* **2014**, *6*, 33–43. [CrossRef]
14. Ko, J.; Shim, J.S. Locating battery exchange stations for electric taxis: A case study of Seoul, South Korea. *Int. J. Sustain. Transp.* **2016**, *10*, 139–146. [CrossRef]
15. Xie, R.; Wei, W.; Khodayar, M.E.; Wang, J.; Mei, S. Planning fully renewable powered charging stations on highways: A data-driven robust optimization approach. *IEEE Trans. Transp. Electrif.* **2018**, *4*, 817–830. [CrossRef]
16. Bai, X.; Chin, K.S.; Zhou, Z. A bi-objective model for location planning of electric vehicle charging stations with GPS trajectory data. *Comput. Ind. Eng.* **2019**, *128*, 591–604. [CrossRef]
17. Zhu, Z.H.; Gao, Z.Y.; Zheng, J.F.; Du, H.M. Charging station location problem of plug-in electric vehicles. *J. Transp. Geogr.* **2016**, *52*, 11–22. [CrossRef]
18. Frade, I.; Ribeiro, A.; Gonçalves, G.; Antunes, A.P. Optimal location of charging stations for electric vehicles in a neighborhood in Lisbon, Portugal. *Transp. Res. Rec.* **2011**, *2252*, 91–98. [CrossRef]
19. Bouguerra, S.; Layeb, S.B. Determining optimal deployment of electric vehicles charging stations: Case of Tunis City, Tunisia. *Case Stud. Transp. Policy* **2019**, *7*, 628–642. [CrossRef]
20. Sadeghi-Barzani, P.; Rajabi-Ghahnavieh, A.; Kazemi-Karegar, H. Optimal fast charging station placing and sizing. *Appl. Energy* **2014**, *125*, 289–299. [CrossRef]
21. Csardi, G.; Nepusz, T. The igraph software package for complex network research. *InterJournal Complex Syst.* **2006**, *1695*, 1–9.

energies

MDPI

Article

Electrical Vehicle Smart Charging Using the Open Charge Point Interface (OCPI) Protocol

Sylvain Guillemin *, Romain Choulet, Gregory Guyot and Sothun Hing

Univ. Grenoble Alpes, CEA, Liten, Campus Ines, 73375 Le Bourget du Lac, France; romain.choulet@cea.fr (R.C.); gregory.guyot@cea.fr (G.G.); sothun-eric.hing@cea.fr (S.H.)
* Correspondence: sylvain.guillemin@cea.fr

Abstract: This paper proposes a new approach to the design of smart charging systems. It aims to separate the role of the Smart Charging Service Provider (SCSP) from the role of the Charge Point Operator (CPO) to provide real flexibility and efficiency of mass deployment. As interoperability is required for this purpose, the challenge is to use standard equipment and protocols in the design of the smart charging Energy Management System (EMS). The use of an Open Charge Point Interface (OCPI) is crucial for an interface between the EMS and the Charge Point Operator. The smart charging EMS developed has been implemented and successfully tested with two CPOs, with different use cases: (1) EV charging infrastructure at office buildings, and (2) EV charging infrastructure installed at a public car park facility.

Keywords: electric vehicle; smart charging; interoperability; digital twin; OCPP; OCPI; EMS

Citation: Guillemin, S.; Choulet, R.; Guyot, G.; Hing, S. Electrical Vehicle Smart Charging Using the Open Charge Point Interface (OCPI) Protocol. *Energies* **2024**, *17*, 2873. https://doi.org/10.3390/en17122873

Academic Editors: Michele Roccotelli, Evangelos Karfopoulos and Ioannis Karakitsios

Received: 25 March 2024
Revised: 2 May 2024
Accepted: 15 May 2024
Published: 12 June 2024

1. Introduction

Electric Vehicles (EVs) will continue to be deployed on a large scale around the world over the coming decades in order to reduce emissions caused by fossil-fuel-powered vehicles and to ensure the transition to a more renewable transport sector. The global EV market is expected to reach over one billion EVs by 2050 [1]. This projection poses three main challenges: the accessibility of charging for EV users, the availability of electric grid capacity to support the charging of electric vehicles, and the use of EV batteries as a source of flexibility for grid services [2]. To avoid a systematic reinforcement of the electricity grid due to the enormous power demand of EVs, several studies have been carried out in recent years on smart charging to manage the charging infrastructure for electric vehicles [1,3–9]. With smart charging management, EVs can be charged outside electricity market restrictions, so that local or global electricity markets and grids become more stable. The charge can be transferred as and when required, according to the user's preferences, namely, their departure time and loading level. In a nutshell, smart charging avoids high power peak demand at the local grid, enabling significant cost reductions for the electricity contract. It may prioritize the local Renewable Energy Source (RES) for EV charging, with the objective of maximizing the self-production ratio and reducing CO_2 emission, or may even be directly driven by the carbon emission market, as illustrated in [10].

The majority of studies present smart charging solutions integrated in the back-end of the Charging Point Operator (CPO) using the Open Charge Point Protocol (OCPP). In [5], the smart charging system is designed with four main functional components and directly incorporated in the CPO OCPP back-end. Smart charging is deployed in a real environment at SAP Labs France in Mougins to manage the enterprise EV fleets. References [10,11] propose an extension of the OCPP standard. This aims to include the user's advance reservation in the charging optimization process and enable the optimization of power grid management, considering the user requests and constraints. As the above studies show, the smart charging solution is directly incorporated into the CPO back-end. It appears to be less flexible in terms of deployment, since each CPO has its own specific development using its private and closed OCPP back-end.

Our work proposes an approach to the design and implementation of a smart charging solution using an interoperability protocol. This design considers a Smart Charging Service Provider (SCSP) as an independent provider. It can interface with any CPO through a standard protocol. In our case, the protocol is the Open Charge Point Interface (OCPI). Our approach clearly defines the CPO's responsibilities, the main purpose of which is to manage the charging infrastructure, and provides an implementation and testing flexibility by allowing the performances of smart charging solutions from different SCSP vendors to be evaluated before a large-scale deployment.

The remainder of the paper is organized as follows. Section 2 describes the new approach of smart charging solution architecture, including several factors such as e-Mobility Service Provider (eMSP), CPO and SCSP, and design. Section 3 presents the experimental setup of EV charging infrastructure located in office buildings at CEA-INES facilities, and the results that were obtained. Section 4 describes the potential benefits of our proposed solution. Finally, Section 5 presents the conclusions of this work.

This work has been conducted and founded by the eCharge4Drivers EU project.

2. Materials and Methods

2.1. Solution Architecture

The OCPI 2.2 specifications describe a communication architecture in which Smart Charging features and responsibilities can be delegated to a specific actor called the SCSP. In those specifications, the representation in Figure 1 is proposed.

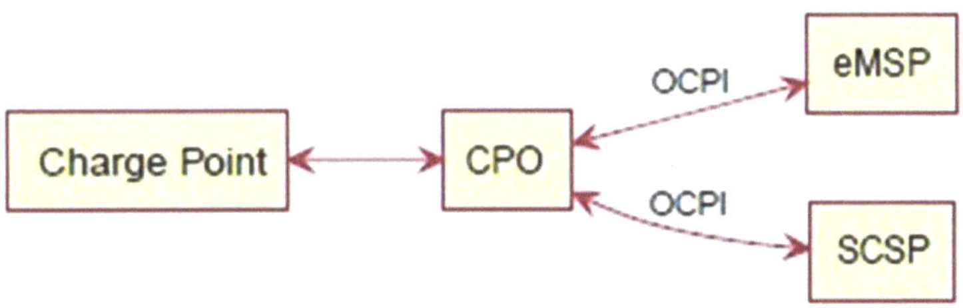

Figure 1. CPO, eMSP and SCSP architecture proposed by OCPI specifications.

It is this architecture that we chose when implementing our EV Smart Charging solution. Figure 2, below, gives some more details about that architecture, naming the various information systems that exchange data through OCPI and listing the endpoints involved in these communications.

The CPO information system is called the "ChargePoint Management System" or CMS, or often OCPP back-end. The eMSP information system is called the "User Management System" or UMS. The SCSP information system is called the "Energy Management System" or EMS.

The charging stations communicate with the CPO back-end using OCPP. In a typical session, when an EV arrives, the charging station checks that its user is allowed to charge (Authorize message), and then informs the CPO back-end that the session has started (StartTransaction message), regularly communicates about the energy consumption (MeterValues message) and finally notifies when the user has disconnected his vehicle (StopTransaction message). These data exchanges are shown in Figure 3.

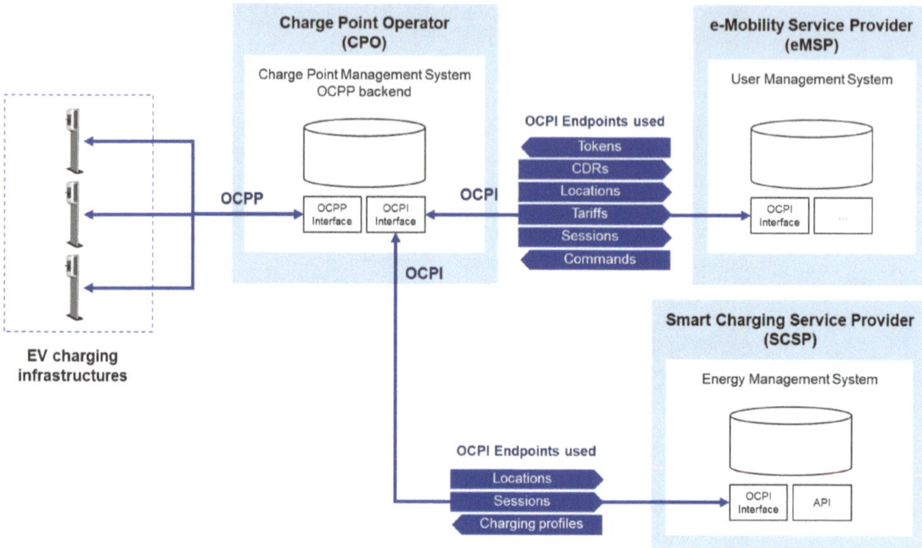

Figure 2. Global architecture including OCPI endpoints.

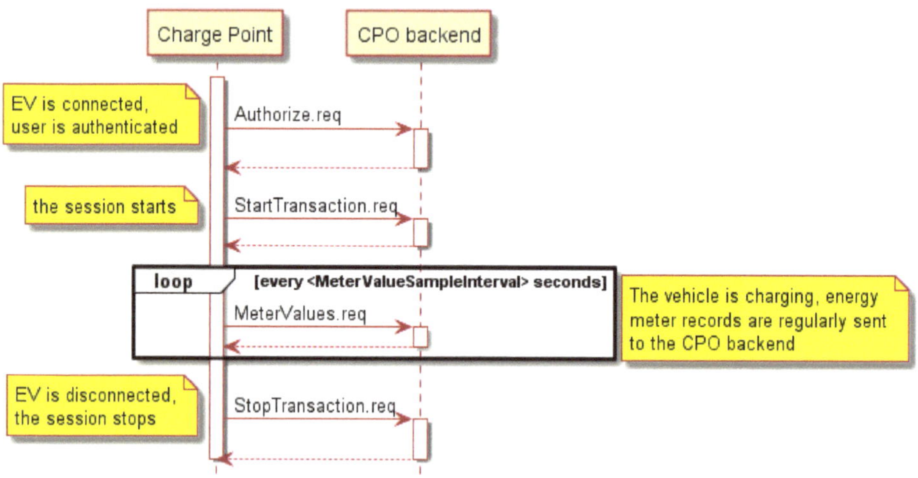

Figure 3. OCPP charge point/CPO back-end communication for a typical session.

If the CPO back-end offers an OCPI interface, these sessions' updates can be relayed to an eMSP information system. It is through such a link that an eMSP can allow its customers to access the CPO's charging stations and handle billing and payment for EV charging sessions. The eMSP can also offer other services, such as a user interface allowing its customer to follow up their charging sessions in real time, for instance. Figure 2 shows that the "sessions" endpoint is used, among other things. The CPO back-end sends a request to that endpoint each time it receives either a StartTransaction, a MeterValues or a StopTransaction OCPP message from a charging station. This mechanism, illustrated in Figure 4, allows the eMSP to learn about its customers' session statuses in real time.

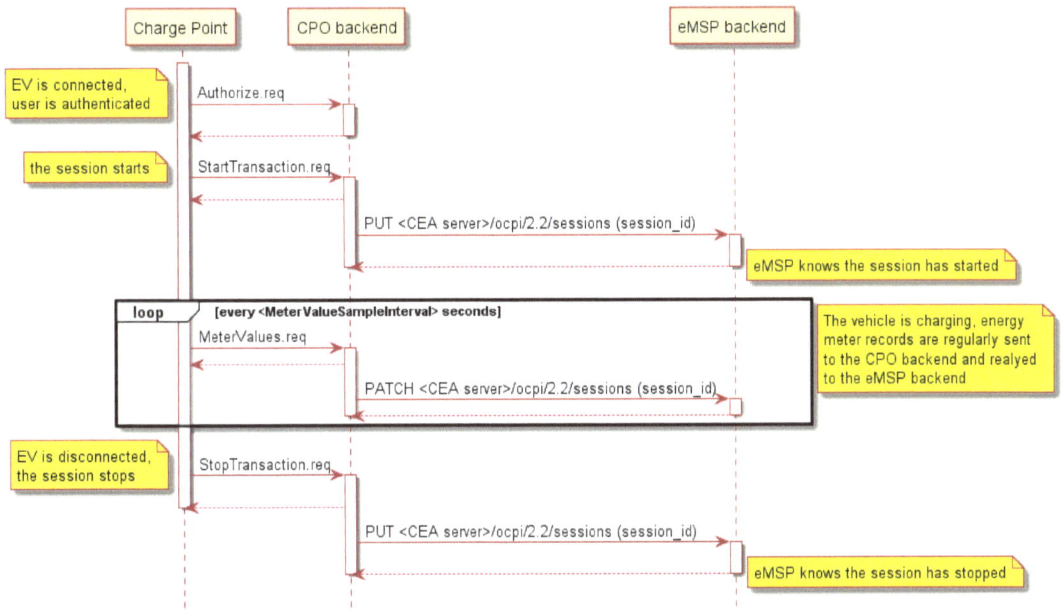

Figure 4. Charge point/CPO back-end/eMSP back-end communication.

The OCPI link between the CPO back-end and the EMS of the SCSP uses the same session updates principle that the eMSP uses. This means that the CPO back-end-relays the updates to both the eMSP and the SCSP. These session updates are key for the operations of the SCSP. They allow the EMS to (1) identify the user and then the type of vehicle being charged, and thus the vehicle battery capacity, and (2) know which amount of energy has already been transferred at what time during the session, allowing the processing of the actual average charging power between two session updates.

The EMS itself will use that OCPI link to send the optimized charging profiles it has calculated to the CPO back-end. The "chargingprofiles" endpoint implemented by the CPO back-end will then be used. These communications are illustrated in Figure 5.

After a call to its "chargingprofiles" OCPI endpoint, the CPO back-end acknowledges reception before transmitting the set points to the charging stations via SetChargingProfile OCPP messages. Then, the CPO back-end calls back the EMS to confirm that the profiles have been sent and accepted by the charging sessions.

The following paragraphs give details on how the EMS is architectured, as well as which data it manipulates.

2.2. EMS Architecture

The EMS is the information system of the SCSP. Its objective is to provide individual charging profiles for the charging sessions currently being executed by groups of charging stations in order to control the global power each group draws.

The grouping logic of charging stations can be of various forms depending on the optimization objectives: one might want to make sure that the charging stations at a given site keep their global power demand below the grid power available, or one might want to control the power of a whole territory, encompassing several charging sites. In our case, as our intent was to maximize the self-production ratio of the electricity produced by a local photovoltaic plant, we grouped together sessions executed at the same location.

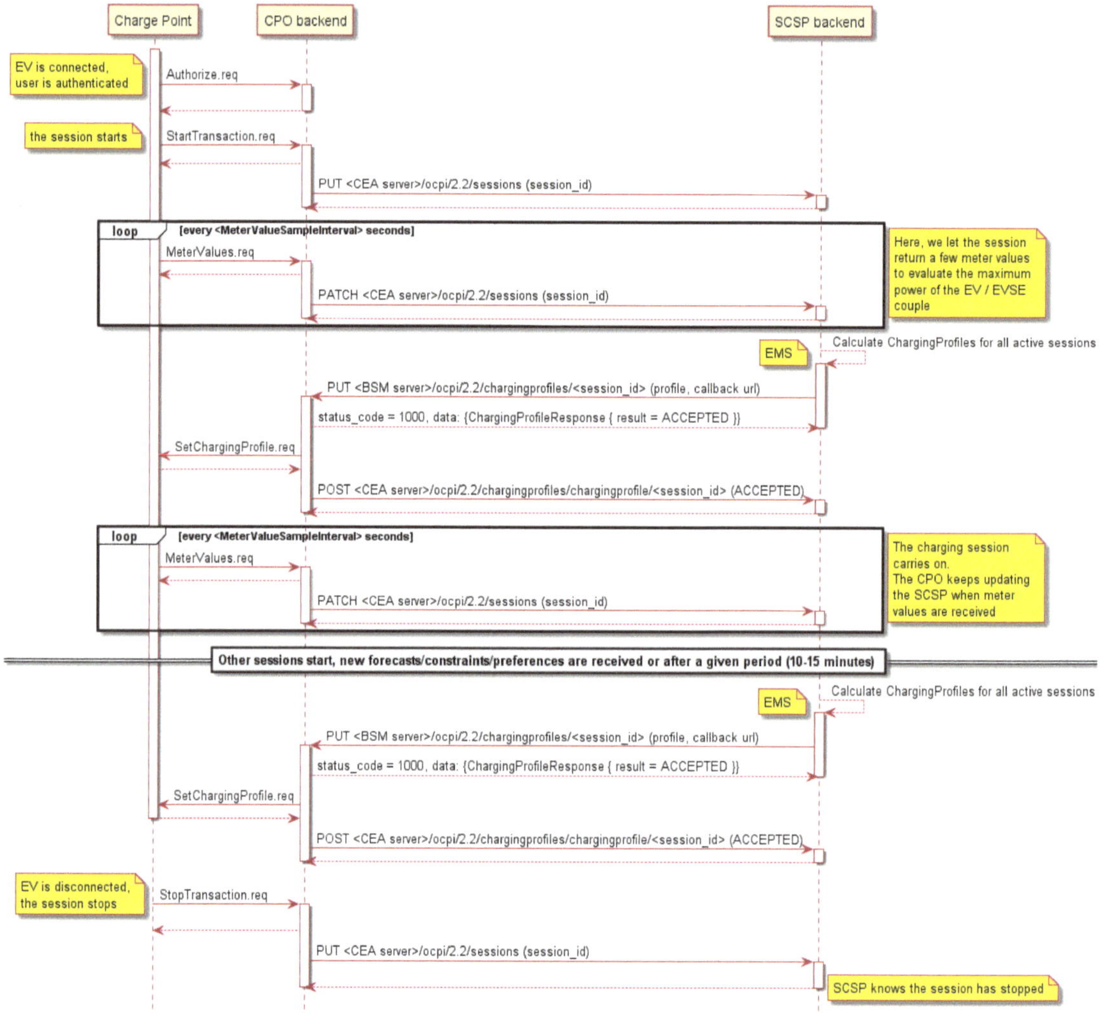

Figure 5. Charge point/CPO back-end/SCSP back-end communication.

Figure 6 reveals a representation of the internal architecture of the EMS.

Our EMS is made of several software components that we propose to detail hereafter to explain their utility. As a general rule, the idea is to hold the information that is needed by the optimization program, which is used to build the charging profiles.

2.2.1. OCPI Interface

First of all, as it is the main subject of this paper, it is important to mention that our EMS has an OCPI interface that it uses to communicate with the CPO back-end. It is through that interface that the EMS receives details about the infrastructure descriptions and the on-going charging sessions. That interface is (1) a set of web services which give access to the EMS database when the EMS acts as a server, and (2) a client program when it is necessary to call the CPO OCPI interface.

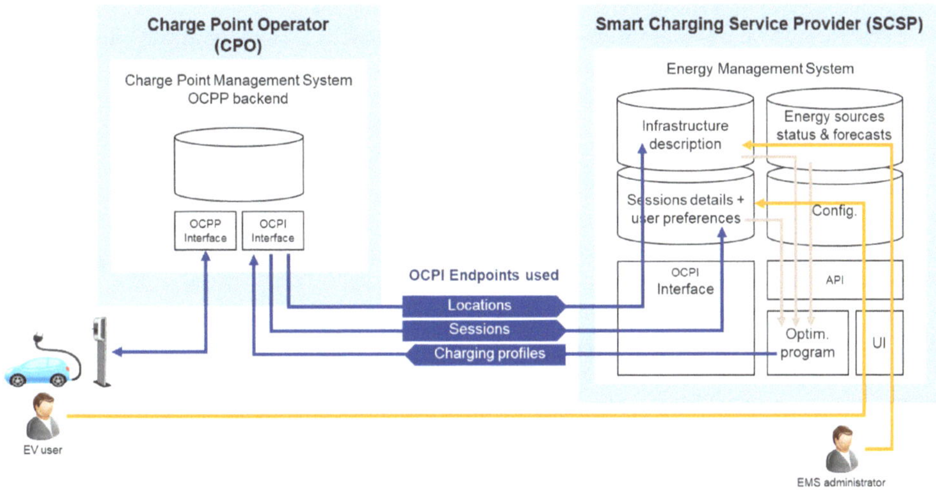

Figure 6. CPO-SCSP link and EMS architecture overview.

2.2.2. EMS Database/Infrastructure Description

Having a charging infrastructure description is a key feature for the EMS. The choice to power a given vehicle has an impact on the power flow of all upstream electrical nodes and protections. When powering several vehicles simultaneously, ensuring that these upstream nodes and protections are not overloaded is necessary and requires the availability of some models of the infrastructure electrical diagram.

For instance, let us consider the following partial diagram in Figure 7, showing three charge points (ChargePoint1–3) powered through the same circuit breaker (Protection1).

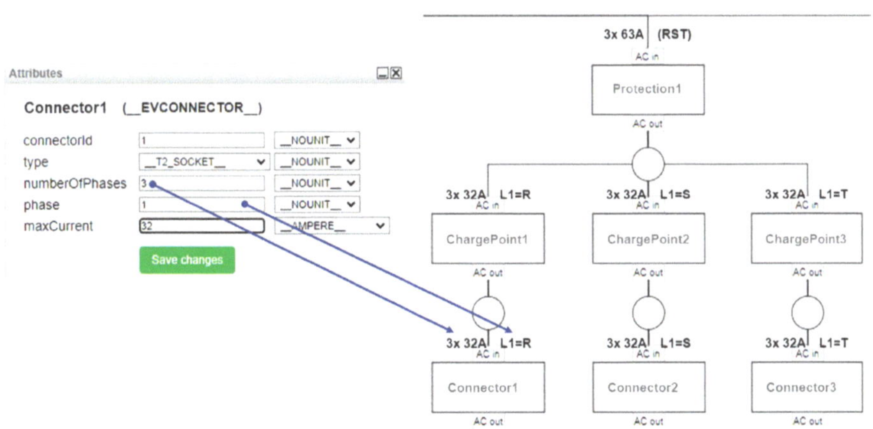

Figure 7. Partial infrastructure electrical diagram.

When vehicles are to be charged through the connectors (Connector1-3), the ability to anticipate how much electrical current is drawn for each phase of circuit breaker Protection1 means that we know whether each charge point and connector is tri-phase or not, and also which phase of the network is wired to the first phase of the connector (i.e., one current will be drawn from if the vehicle is mono-phase).

The description held in the EMS database has to contain these phase details, as well as the fact that the charge points ChargePoint1-3 are powered through circuit breaker Protection1.

In Figure 7, which is a screenshot of our EMS operator interface, we can see the technical attributes that are held for one of the connectors (Connector1) modeled in our energy system. Here, we see that the connector has its three phases wired and that its first phase corresponds to the first phase (R) of the grid. We also see that the maximum current it can let through is 32A per phase.

Through the "locations" of OCPI endpoint, the CPO back-end lets its partners (eMSPs and SCSPs) know about the technical details and status of the charging stations it manages. Unfortunately, the information we obtain is generally not sufficient to rebuild the electrical diagram and perform proper energy management. CPO back-ends often lack a detailed description of the electrical wiring of the charging infrastructure, the circuit breaker details and the phase rotation configuration. This means that the charging infrastructure description has to be carefully modeled and set at the EMS level, justifying the need for a specialized SCSP information system if the CPO back-end cannot hold that information.

The infrastructure description will then be set and updated by both the CPO back-end through the OCPI interface and also the EMS administrator through the EMS UI (User Interface). Our EMS UI makes it possible to edit CIM-style (Common Information Model) electrical diagrams like the one shown in Figure 8.

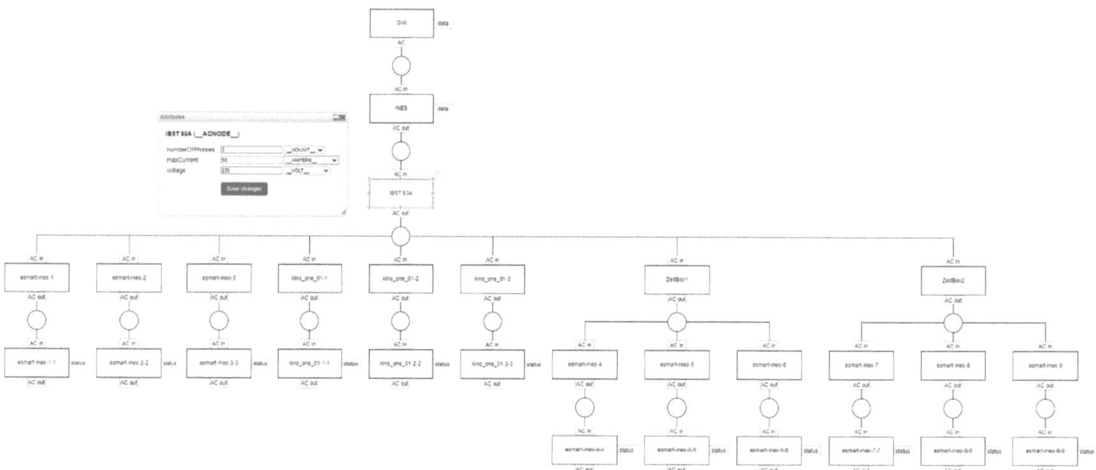

Figure 8. Electrical diagram of the charging infrastructure.

This kind of description makes it possible for us to draw a list of all technical constraints that need to be respected by the optimized charging profiles that will be calculated.

2.2.3. EMS Database/Configuration Data

Each charging session references a "cdr_token" that was used to start the session. It is out of the scope of this paper to discuss the various types of tokens that can be used, and we will just consider that an RFID tag was used. That token can be used to identify which EV user, and then which type of vehicle, is being charged. In the "configuration data" part of our EMS database, we have the ability to host, among other details, the energy capacity of the vehicle associated with that token. That piece of information is either set by the EMS administrator through the UI based on what the eMSP was able to communicate, or estimated using the previous charging sessions that were performed for that EV user. If 20 kWh were transferred in a session for which the user had declared a 50% state of charge on arrival, the vehicle capacity can be roughly estimated to be 40 kWh. Obviously, that estimation will grow in accuracy with a growing number of sessions.

Which algorithm the optimization program should use for which group of charging stations, and other specific details, are also held in the "configuration data" part of the EMS database.

2.2.4. EMS Database/Session Details and User Preferences

In the EMS database, for each session, we hold the details needed for the optimization program to make decision on which EV vehicle to charge in what priority order, when and at what charging power.

These details require answers to the three following questions for each session:

- How much energy will we still have to transfer to the EV?
- By what time?
- What are the minimum and maximum power set points we can use?

All this information is not necessarily directly available. To know how much energy is still to be transferred, we need to know:

- The state of charge of the vehicle on arrival (obtained through the user interface);
- The capacity of the vehicle's battery (either learned from history or configured by the EMS administrator as a characteristic associated with the user identifier);
- The amount of energy already transferred (directly accessible in the session OCPI update messages).

The UI interface used by the EV user is also the channel through which we will learn by what time the vehicle has to be charged.

By controlling the charge-point power limit, the SCSP can measure the maximum charging power which the charge point/vehicle couple can reach. That piece of information is key for the optimization program as it sets the boundaries of the charging profile it will be able to define. The maximum charging power value is also part of the data we host about each charging session.

2.2.5. EMS Database/Energy Source Forecasts

The optimization program objective is to distribute the energy, possibly available from many sources, in a smart manner. This implies that the amount of available energy for the time periods to come is known. These data might be obtained from the Distribution System Operator (DSO), the Transport System Operator (TSO), the electricity provider or from some other third-party data source. For the CEA-INES demonstrator, for which the objective is to maximize the self-production ratio of the infrastructure coupled with a photovoltaic plant, we can obtain photovoltaic production forecasts from a specialized company and automatically and regularly integrate them into the EMS database.

The prices at which this energy is available from the different sources are also a crucial indicator. They can be defined by contract and thus simply require a configuration by the EMS administrator, or can be much more dynamic and require that an automatic data flow is set up between the electricity providers and the SCSP.

The "Energy sources forecast" of the EMS database is then able to host time series representing the power that is supposed to be available for different time periods in the future and prices at which that energy is sold, as shown in Figure 9, below.

2.2.6. EMS User Interface

The EMS user interface addresses two types of users: the electrical vehicle users and the EMS administrators. The EV users need some way to let the EMS know how much energy they need and for how long their vehicle will stay connected to the charging infrastructure.

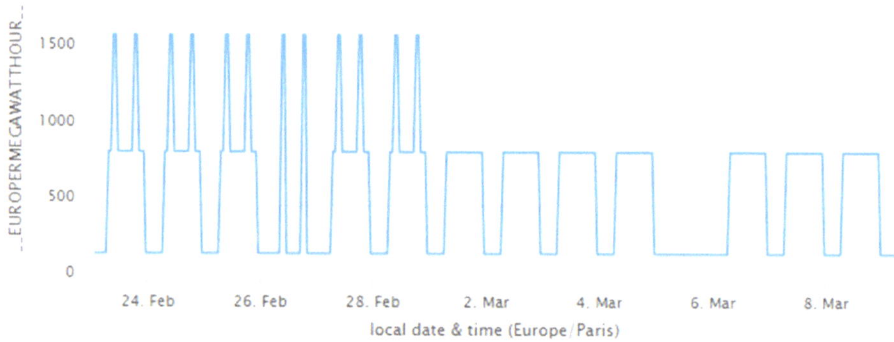

Figure 9. Grid electricity prices (day ahead).

For the e-charge4driver project, our EMS was used for two charging sites, (1) at CEA-INES and (2) in Barcelona. At CEA-INES, we are our own eMSP and we integrated the possibility for the users to communicate the state of charge of their vehicles and their expected departure time through our User Management System mobile website, which embeds a web widget served by our EMS UI; see Figure 10.

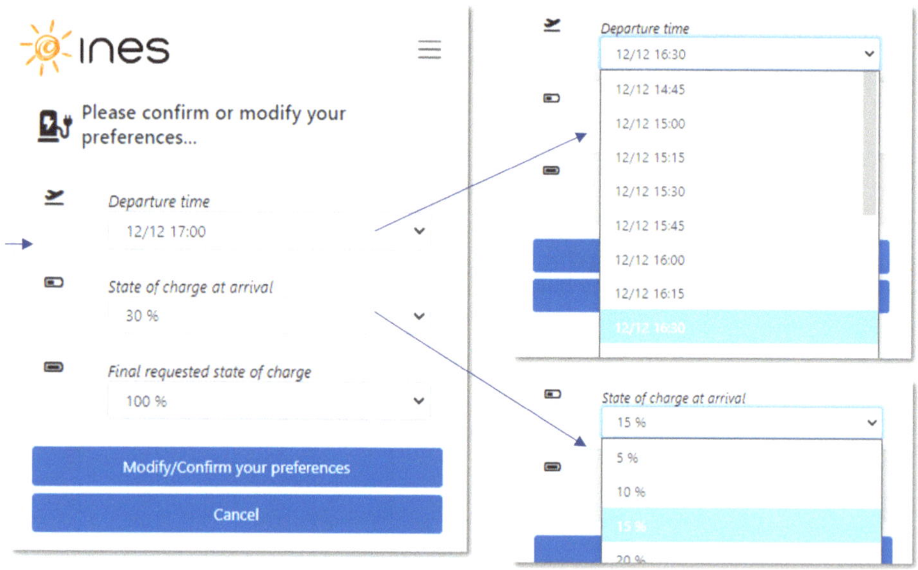

Figure 10. EV users' UI interface used at CEA-INES.

For the Barcelona demonstrator, the eMSP SMOU application used by EV users was amended to include those two questions, as shown in Figure 11. The requested amount of energy and the expected departure time were then communicated to the EMS by making use of an unused field (meter_id) of the OCPI session updates. This is a slight workaround for the fact that OCPI has not made provision for these details in its message structures.

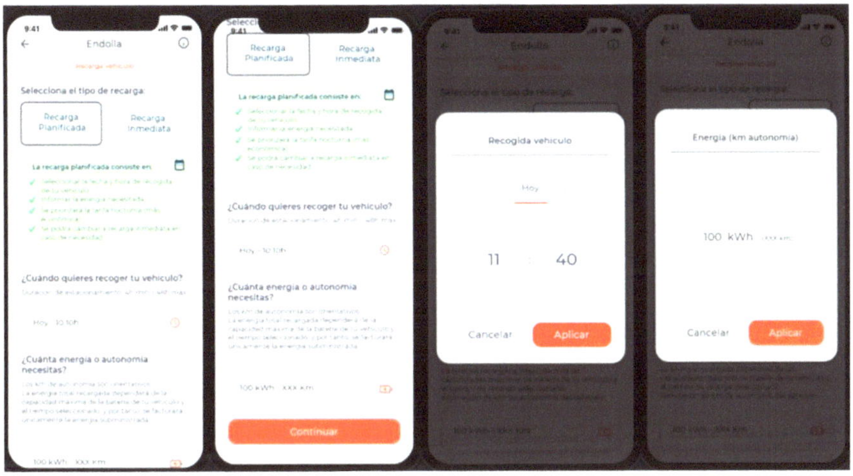

Figure 11. SMOU application used as the EV users' UI in Barcelona.

Our EMS user interface is also designed to be used by EMS administrators. These users will edit the charging infrastructure electrical diagrams (see Figure 12) and set other parameters used by the optimization program.

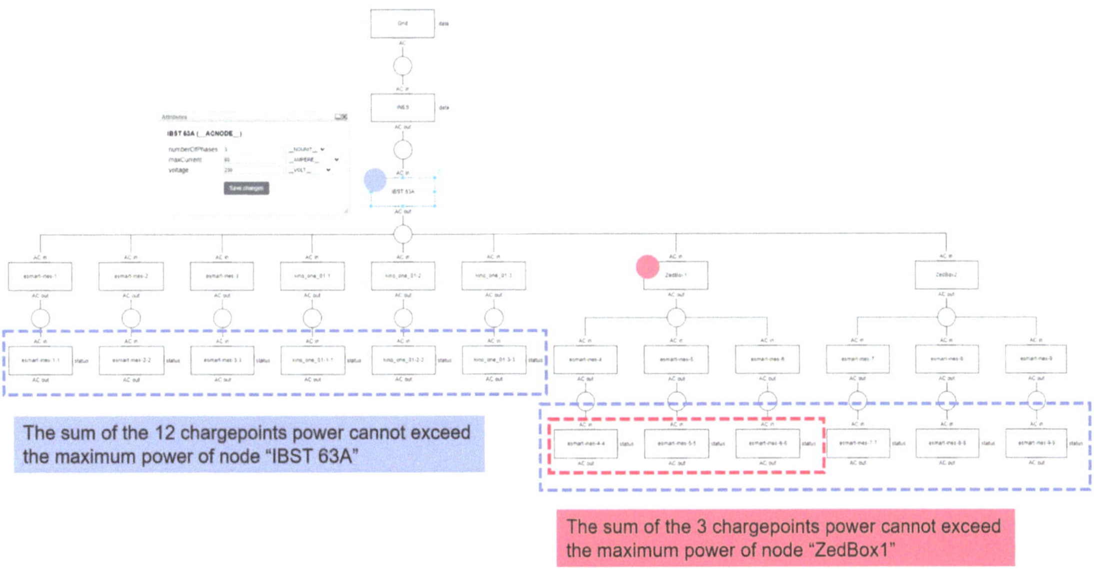

Figure 12. Power constraint identification.

2.2.7. Optimization Program

The optimization program calculates charging profiles for each and every charging session to be executed. Detailing how this program works internally is not in the scope of this paper. As described before, it makes use of the various data modeled and hosted in the EMS database and simulates the power flow within the infrastructure for the hours to come, aiming to minimize a cost function designed to translate the optimization objectives into numerical indicators. The data flow diagram of this program is shown in Figure 13.

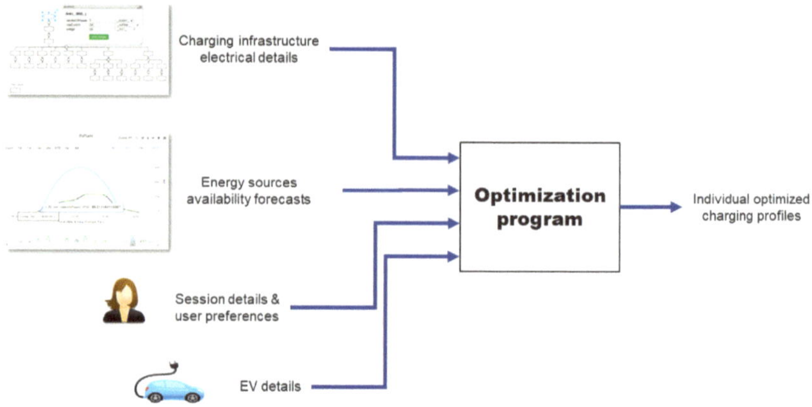

Figure 13. Optimization program synopsis.

It exploits the infrastructure description to identify constraints that need respecting. In Figure 12, we have illustrated, in lavender, the fact that the power drawn by the 12 EV connectors must not overload the node named "IBST 63A", and in red, the fact that connectors "esmart-ines-4-4", "esmart-ines-5-5" and "esmart-ines-6-6" must not overload the node named "ZedBox1". In the electrical diagram, the optimization program automatically identifies these constraints and makes sure that the charging profiles it generates respect them.

With forecasts for how much power will be available from the various energy sources and at what prices, and with estimations on the quantity or energy it has available to allocate to each charging session, by when and with what maximum power, the optimization program has the information it needs to build the set of charging profiles that best fits the optimization objectives.

Each time a new session starts or is stopped by an EV user, the global optimization problem changes and the optimized charging profiles need recalculating. This is when the optimization program is called by the EMS that detects those events.

When new charging profiles are ready, they are sent to the CPO back-end via OCPI for transmission to the charge points, as shown in Figure 5.

2.2.8. Digital Twin

Altogether, by modeling the equipment deployed on the charging infrastructure, by updating their status in real-time and by being able to simulate their physical behavior, these software components constitute a digital twin of the charging infrastructures managed by the EMS.

3. Results

3.1. Implementations

The smart-charging EMS and OCPI developed, capable of interfacing with the CPO back-end, was first tested in the laboratory on real-time simulators. It was then implemented on the CEA-INES research campus EV charging infrastructure in Chambéry, France. This infrastructure has a 43 kVA connection to the electricity grid and a local 35 kWp photovoltaic plant. It includes 12 AC charge points (3.7 kW, 7 kW and 22 kW) from different suppliers (Driveco and E-Totem). For this infrastructure, CEA-INES plays two roles: CPO and SCSP. We developed our own OCPP/CPO back-end, which offers an OCPI 2.2 interface able to link to our EMS. The complete system was set up so that the CEA-INES employees could charge their own vehicles at the office during the working day. More than 30 employees voluntarily took part in the evaluation test program, with vehicles from different car vendors (Renault, Nissan, Fiat, Volkswagen, Tesla, Hyundai, MG, Mazda, Peugeot, etc.).

After having been implemented and successfully tested at CEA's charging infrastructure (Figure 14) since December 2022, the EMS developed was linked to Barcelona Municipality's CPO back-end in order to manage one of their EV charging infrastructures. This infrastructure was equipped with five charge points of 3.7 kW and located at the Porta de Sarrià car park facility. The functional and operational tests were performed with success during the demonstration period of the eCharge4Drivers European project.

Figure 14. CEA-INES charging infrastructure.

3.2. Results

As mentioned above, the developed smart charging EMS is capable of optimizing and controlling the charging infrastructure of electric vehicles, with the aim of maximizing the photovoltaic self-production ratio and minimizing the charging cost, while respecting user preferences and taking into account local grid constraints.

Over a 12-month test period, with smart-charging EMS, the self-production ratio can be increased to 56% on average and to 95% in some cases (on days when photovoltaic production is higher and the number of electric vehicles to be charged is greater).

Figure 15 illustrates the effect that smart charging has on the global consumption of CEA-INES' charging infrastructure. In the right-hand-side figure, one can see the sums of the power consumption of 12 charging sessions carried out with smart charging. Each EV charging profile was calculated to use as much local renewable energy as possible for all vehicles charged before their planned departure time. The left-hand-side figure shows what would have happened without smart charging. This latter figure is obtained by simulating, for each session, the "uncontrolled" charging power. With each session's maximum power and transferred energy amount, this simulation is pretty straight forward. In this case, a high peak power demand is observed at the beginning of the day, while the photovoltaic plant produces less energy, leading to a consequent withdrawal of energy from the grid and a probable overshoot of the subscribed power.

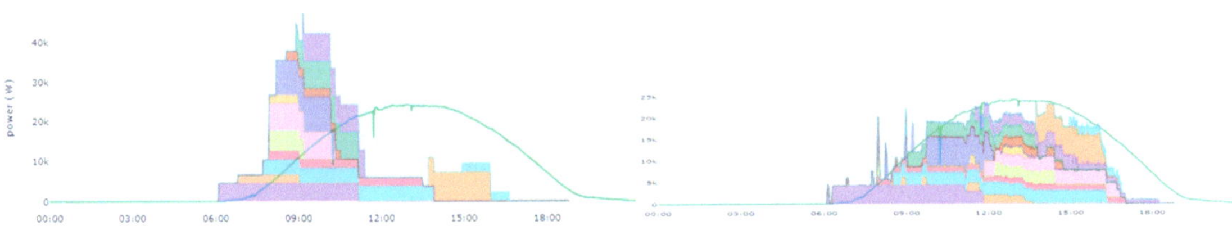

Figure 15. Photovoltaic self-production results, without and with smart charging.

4. Discussion

Using the architecture we have illustrated in the current paper, CPOs can concentrate on their core responsibilities: developing and maintaining charging infrastructure and ensuring that it is operational. The development of Energy Management Systems requires knowledge of the optimization opportunities that exist to allow CPO and EV users to benefit from variable electricity prices. Additionally, it requires mathematical skills to implement the optimization algorithms that will exploit these opportunities. Also, energy regulations change as more and more electricity consumers also become producers. In France, for instance, there is now the possibility of buying electricity from nearby photovoltaic plants through Power Purchase Agreements (PPAs), so SCSPs will leverage their expertise in setting up these contracts and adapting their algorithms to accommodate the new rules that will undoubtedly emerge.

Our proposed solution architecture is highly relevant and advantageous for CPOs, as it enables them to quickly implement and test the performance of smart charging solutions offered by different SCSP suppliers with a mind to selecting them. CPOs will be able to choose those with the best performance for their final large-scale deployments. It also allows them to change the SCSP supplier without further development, even during the operation phase, as our solution is based on standard protocols, making it flexible and interoperable.

Delegating the energy management to a specific operator through the OCPI protocol also has another advantage: as already stated, OCPI can work with two communication models, push and pull. In the push model, the CPO updates the SCSP each time a charging session evolves. That way, the SCSP can integrate these messages in real time and will be fully aware of the details of the current active sessions. With the pull model, the SCSP can ask the CPO to provide the full list of active charging sessions that need energy (by submitting a GET request on the session endpoint of the CPO). This particular feature, which has no equivalent in OCPP, for instance, makes it possible to easily benchmark different SCSP solutions. Some information systems offering, even partially, the OCPI interface of a CPO back-end could serve various lists of fake active sessions to the SCSPs being evaluated in order to understand the differences between the different sets of charging profiles that the SCSPs will send back. Obviously, the different SCSPs will also need to be fed with the charging infrastructure descriptions and energy source forecasts, and they might differ in the level of detail that they can cope with or exploit here, and this might show in their results. For instance, an SCSP performing single-line optimization rather than one that properly models the three-phase electric distribution might propose charging profiles that lead to an under-usage of the available grid power or, in the opposite case, to protection overloads.

5. Conclusions

This paper presented a new design approach for a smart charging system based on the separation between the SCSP and the CPO roles. It allows better flexibility for the CPO by testing and evaluating the performance of the smart charging solution before its final large-scale deployment. In the design of the SCSP (EMS), the standard protocol OCPI interface is developed to connect with any CPO back-end and to receive all necessary information, such as the infrastructure descriptions and the on-going charging sessions, for EMS computing. A mobile application has been developed to enable EV users to interact with the EMS system by providing their user preferences (the state of charge of the vehicle on arrival, the scheduled departure time and the need for energy if required) and also to be able to monitor their on-going charging session or change the scheduled departure at any time. The EMS is designed by taking into account, as input data, all forecast data (the local photovoltaic production, the local load consumption and electricity price), the charging infrastructure power constraint and the local grid constraint, in its Optimization Program, and then providing an optimal charging profile for each EV connected at the charge point through the OCPI interface linked to the CPO back-end. In addition, the energy management system prioritizes charging sessions according to user preferences, ensuring that their vehicle is fully charged before they leave. As a modular structure in the design, the EMS can be easily configured with one objective given for the optimization or combining of several objectives for optimization, for example maximizing the self-production ratio using the local RES and reducing the charging cost. The EMS has already been successfully deployed and operated in two use cases: (1) EV charging infrastructure at the CEA research center in Chambéry, France, and (2) EV charging infrastructure installed in a public car park facility in Barcelona, managed by Barcelona Municipality. In the first use case, the EMS was configured to maximize the self-generation ratio, whereas in the second use case it was configured to minimize the charging cost. More than 10,000 smart charging sessions have been carried out since the EMS system was set up.

Author Contributions: Conceptualization, S.G.; methodology, S.G.; software, S.G. and R.C.; assessment and operation of the systems, S.H. and R.C.; writing—review and editing, S.G., S.H., R.C. and G.G.; supervision and review, S.H.; project administration, S.H. All authors have read and agreed to the published version of the manuscript.

Funding: This work has been realized in association with INES.2S within the French national program Investments for the Future (ANR-10-IEED-0014-01) and received funding from the European Union's Horizon 2020 research and innovation program under the grant H2020-LC-GV-2019 program under Grant Agreement No. 875131.

Data Availability Statement: The original contributions presented in the study are included in the article, further inquiries can be directed to the corresponding author.

Conflicts of Interest: The authors declare no conflict of interest.

Abbreviations

Abbreviation	Meaning
CEA	Commissariat à l'Energie Atomique
CPO	Charge Point Operator
DSO	Distribution Service Operator
EMS	Energy Management System
eMSP	E-Mobility Service Provider
EV	Electrical Vehicle
INES	Institut National de l'Energie Solaire
kVA	Kilo Volt Ampere
kW	Kilo Watt
kWp	Kilo Watt Peak
kWh	Kilo Watt Hour
OCPI	Open ChargePoint Interface
OCPP	Open ChargePoint Protocol
PPA	Power Purchase Agreement
RES	Renewable Energy Source
SCSP	Smart Charging Service Provider
TSO	Transport Service Operator

References

1. IRENA. Electric-Vehicle Smart Charging. In *Innovation Landscape Brief*; IRENA: Masdar City, United Arab Emirates, 2019.
2. Grigorev, A.; Mao, T.; Berry, A.; Tan, J.; Purushothaman, L.; Mihaita, A.S. How will electric vehicles affect traffic congestion and energy consumption: An integrated modelling approach. In Proceedings of the 2021 IEEE Intelligent Transportation Systems Conference (ITSC), Indianapolis, IN, USA, 19–21 September 2021.
3. CEER. *CEER Report on Electric Vehicles: Network Management and Consumer Protection*; CEER: Ljubljana, Slovenia, 2023.
4. Amann, G.; Escobedo Bermúdez, V.R.; Boskov-Kovacs, E.; Gallego Amores, S.; Giannelos, S.; Iliceto, A.; Ilo, A.; Chavarro, J.R.; Samovich, N.; Schmitt, L.; et al. *E-Mobility Deployment and Impact on Grids, Impact of EV and Charging Infrastructure on European T&D Grids—Innovation Needs*; Publications Office of the European Union: Luxembourg, 2022.
5. Fleck, T.; Gohlke, S.; Nochta, Z. A Smart Charging System to Efficiently Run EV Fleets. In Proceedings of the 36th International Electric Vehicle Symposium and Exhibition (EVS36), Sacramento, CA, USA, 11–14 June 2023.

6. Frendo, O.; Graf, J.; Gaertner, N.; Stuckenschmidt, H. Data-driven smart charging for heterogeneous electric vehicle fleets. *Energy AI* **2020**, *1*, 100007. [CrossRef]
7. Marinelli, M.; Calearo, L.; Ried, S.; Pfab, X.; Cabrera, J.C.D.; Spalthoff, C.; Braun, M.; Sæle, H.; Torsæter, B.N.; Astero, P.; et al. Electric Vehicles Demonstration Projects—An Overview Across Europe. In Proceedings of the 2020 55th International Universities Power Engineering Conference (UPEC), Turin, Italy, 1–4 September 2020; IEEE: Piscataway, NJ, USA; pp. 1–6.
8. Robisson, B.; Guillemin, S.; Marchadier, L.; Vignal, G.; Mignonac, A. Solar Charging of Electric Vehicles: Experimental Results. *Appl. Sci.* **2022**, *12*, 4523. [CrossRef]
9. Sadeghian, O.; Oshnoei, A.; Mohammadi-Ivatloo, B.; Vahidinasab, V.; Anvari-Moghaddam, A. A Comprehensive Review on Eletric Vehicles Smart Charging: Solutions, strategies, technologies, and challenges. *J. Energy Storage* **2022**, *54*, 105241. [CrossRef]
10. Lei, X.; Yu, H.; Yu, B.; Shao, Z.; Jian, L. Bridging electricity market and carbon emission market through electric vehicles: Optimal bidding strategy for distribution system operators to explore economic feasibility in China's low-carbon transitions. *Sustain. Cities Soc.* **2023**, *94*, 104557. [CrossRef]
11. Flocea, R.; Hîncu, A.; Robu, A.; Senocico, S.; Traciu, A.; Remus, B.M.; Răboacă, M.S.; Filote, C. Electric Vehicle Smart Charging Reservation Algorithm. *Sensors* **2022**, *22*, 2834. [CrossRef] [PubMed]

energies

MDPI

Article

Development and Implementation of a Smart Charging System for Electric Vehicles Based on the ISO 15118 Standard

Jóni B. Santos [1,*], André M. B. Francisco [2], Cristiano Cabrita [2,3], Jânio Monteiro [2,3,*], André Pacheco [1] and Pedro J. S. Cardoso [3,4]

[1] Centre for Marine and Environmental Research—CIMA, Universidade do Algarve, 8005-139 Faro, Portugal; ampacheco@ualg.pt

[2] Centro de Investigação em Sistemas Ciberfísicos do Algarve—CISCA, Universidade do Algarve, 8005-139 Faro, Portugal; amfrancisco@ualg.pt (A.M.B.F.); ccabrita@ualg.pt (C.C.)

[3] Instituto Superior de Engenharia, Universidade do Algarve, 8005-139 Faro, Portugal; pcardoso@ualg.pt

[4] NOVA LINCS, Universidade do Algarve, 8005-139 Faro, Portugal

* Correspondence: jnsantos@ualg.pt (J.B.S.); jmmontei@ualg.pt (J.M.)

Abstract: There is currently exponential growth in the electric vehicle market, which will require an increase in the electrical grid capacity to meet the associated charging demand. If, on the one hand, the introduction of energy generation from renewable energy sources can be used to meet that requirement, the intermittent nature of some of these sources will challenge the mandatory real-time equilibrium between generation and consumption. In order to use most of the energy generated via these sources, mechanisms are required to manage the charging of batteries in electric vehicles, according to the levels of generation. An effective smart charging process requires communication and/or control mechanisms between the supply equipment and the electric vehicle, enabling the adjustment of the energy transfer according to the generation levels. At this level, the ISO 15118 standard supports high-level communication mechanisms, far beyond the basic control solutions offered through the IEC 61851-1 specification. It is, thus, relevant to evaluate it in smart charging scenarios. In this context, this paper presents the development of a charge emulation system using the ISO 15118 communication protocol, and it discusses its application for demand response purposes. The system comprises several modules developed at both ends, supply equipment and electric vehicles, and allows the exchange of data during an emulated charging process. The system also includes human interfaces to facilitate interactions with users at both ends. Tests performed using the implemented system have shown that it supports a demand response when integrated with a photovoltaic renewable energy source. The dynamic adjustment to charging parameters, based on real-time energy availability, ensures efficient and sustainable charging processes, reducing the reliance on the grid and promoting the use of renewable energy.

Keywords: electric vehicles; smart charging; demand response; ISO 15118; smart grids; renewable energy sources

Citation: Santos, J.B.; Francisco, A.M.B.; Cabrita, C.; Monteiro, J.; Pacheco, A.; Cardoso, P.J.S. Development and Implementation of a Smart Charging System for Electric Vehicles Based on the ISO 15118 Standard. *Energies* **2024**, *17*, 3045. https://doi.org/10.3390/en17123045

Academic Editors: Michele Roccotelli, Evangelos Karfopoulos and Ioannis Karakitsios

Received: 30 April 2024
Revised: 11 June 2024
Accepted: 17 June 2024
Published: 20 June 2024

1. Introduction

Introduced as a clean energy initiative due to their CO_2 emissions (relatively low or null) [1], electric vehicles (EVs) support a new paradigm of sustainable mobility. At a global level, taking into account the average carbon emissions used for electricity generation (518 g of carbon dioxide equivalent per kilowatt-hour [518 g CO_2-eq/kWh]) [2], an EV emits a smaller amount of greenhouse gases when compared with the average internal combustion vehicles. In 2023, the total number of EVs on the roads reached 40 million globally [3]; however, the uncontrolled charging requirements of this growing fleet represent a major challenge for the electrical grid, especially if the charging demand from EVs coincides with the peak consumption periods already existing in the grid, which can lead to its overload [4–6].

The two main solutions to this challenge could be, on the one hand, to introduce renewable energy sources (RESs), and particularly those that are generated at the distribution level, reducing the need for conventional energy sources, and, on the other hand, the introduction of demand response mechanisms that allow for adjusting the charging power of EVs according to the power output from these RESs. Through this mechanism, known as smart charging, EVs can minimize the requirements they place on electrical grids by shaping their demand pattern, varying their power intensity and charging periods [4–6]. At this level, EVs can be thought of as additional storage units either for the sole purpose of mobility or to support vehicle-to-grid (V2G) energy transfers [2]. In both cases, interoperability mechanisms between an EV and electric vehicle supply equipment (EVSE) are needed. One of the challenges associated with the implementation of grid-to-vehicle and V2G energy transfers is the limited capacity for exchanging information between the EV and the EVSE. Currently, most of this information exchange is performed via low-level control mechanisms (IEC 61851-1 [7]), through the control pilot and proximity pilot-signaling pins. Mostly used in the alternating current (AC) power transfer mode, the signaling carried out through these pins only indicates when the EV is connected to the EVSE and what the available electrical current that the supply equipment can supply to the vehicle is. That reduced set of control options limits the implementation of smart charging solutions [1,5,6].

In order to optimize charge management, improved interoperability between the EVSE and the EV is necessary. Information such as the state of charge (SoC) and its power limits are crucial to performing charge scheduling. Thus, there is a clear need for bidirectional communication between them, much beyond the control-level solutions offered through the IEC 61851-1 standard.

In this context, this work describes the development of a system that emulates a charging process between an EV and EVSE, and it uses the ISO 15118 [8] digital communication protocol. Its development required a comprehensive analysis of the ISO 15118 protocol, culminating in the design and implementation of the charging emulation system. Key steps included the following: (1) the characterization of system parameters; (2) architectural design; and (3) the development of human–machine interfaces (HMIs) for both EVs and EVSE (these interfaces play an important role in the process of demonstrating the system implemented). Two microcomputers were used to implement the communication between the Supply Equipment Communication Controller (SECC) and the Electric Vehicle Communication Controller (EVCC). Communication controllers were implemented to establish dynamic communication sessions, and they were developed using the open-source library RISE V2G [9]. Each controller (at the EVSE and EV) implemented a specific program/mechanism that managed a charge emulation process, while task-scheduling algorithms ensured a synchronized information exchange between EVs and the EVSE.

The system's performance was validated through the testing of the emulator when applying charging techniques. The results show that it realistically approximates the behavior of an actual direct-current (DC) charging process. Also, tests were performed to evaluate its adaptability to varying levels of power, such as those that occur in demand response scenarios. These tests, using data from a photovoltaic generation unit, show that the system reacts quickly to power changes from intermittent photovoltaic generation, resulting in a low percentual mismatch between generation power and charging power.

The rest of this work is structured as follows. Section 2 describes the state of the art, including the ISO 15118 standard and the RISE V2G library. In Section 3, a description is provided of the development of the charging emulator system, including its characterization, architecture, and task-scheduling algorithms. Section 4 presents the obtained system and an evaluation. Finally, Section 5 concludes this work, discussing future developments.

2. State of the Art

The increasing adoption of EVs presents a challenge to electrical grids due to the surge in demand that occurs during charging times. While RESs can offer a sustainable solution, their intermittent nature requires a method to balance grid load with generation. Smart charging, which benefits from communication mechanisms between EVs and charging stations, is a promising approach to optimizing energy consumption and integrating EVs seamlessly into the grid. The ISO 15118 standard provides a robust communication protocol for this purpose, enabling features beyond the basic control offered through previous standards.

This section explores the current literature on smart charging systems for EVs, focusing on their application in demand response and their alignment with the ISO 15118 standard. In Section 2.1, relevant studies in the area are analyzed and compared. Then, in Section 2.2, the ISO 15118 V2G protocol is described. Finally, in Section 2.3, the RISE V2G library is presented.

2.1. Literature Review

As EVs become more prevalent, their integration into the existing power grid becomes a critical issue. To create truly effective smart charging systems, a multifaceted approach is required that includes communication protocols, demand response functionalities, energy management strategies, and RES integration.

Recent works in this area delve into the critical role of smart charging in facilitating grid integration and promoting sustainable energy practices. The authors in [10] provide a comprehensive analysis of advances in smart grids, highlighting smart charging as a key enabler of the integration of intermittent RESs, such as solar and wind power. This emphasis on grid integration and sustainability is also highlighted in [11], where the potential of smart charging for managing peak demand periods and facilitating the integration of RES is explored. Smart charging provides benefits that go beyond individual EVs. In fact, as stated in [12], EVs can provide auxiliary services to distribution system operators (DSOs) through smart charging or smart charging and discharging (also known as Two-Way V2G).

The potential of the ISO 15118 standard for integrating electric vehicles (EVs) into the smart grid is described in [13]. The standard is considered crucial for enabling V2G applications, enabling a secure data exchange between EVs and charging stations.

At this level, several studies have investigated smart charging systems for EVs, exploring the ISO 15118 integration and/or system implementations. In [14], the authors propose a communication simulator, based on the ISO 15118 standard, applied in MATLAB and using Simulink toolboxes, that mimics the interactions between EVs, charging stations, and the grid, enabling demand response functionalities. In [15], a prototypical implementation of an ISO 15118 wireless V2G-based communication system is presented. The system enables EVs to act both as suppliers and consumers, contributing to demand response efforts. The work in [16] describes an experimental platform to evaluate and emulate V2G communication, the platform uses a Raspberry Pi for wireless communication and Texas Instruments development kits for processing analog signals (such as battery parameters). In [17], a simulator is proposed that leverages V2G communication (based on the ISO 15118 and OCPP 2.0 standards [18]). This approach aims to reduce stress on the grid during charging, especially when integrating variable RES. The study in [19] implements the ISO 15118 communication standard, combining it with an OBD-II interface. This integration allows EVs to respond to grid signals and optimize charging patterns. In [20], a simulation environment is developed that combines hardware and software components to simulate the interaction between EVs, charging infrastructure, and the grid. The study in [21] focuses on implementing the HomePlug Green PHY standard (based on ISO 15118) into EVSE. This integration improves communication performance and enables advanced features for smart charging systems. Expanding on the concept of V2G communication, the work in [22] proposes a novel approach to wireless authentication in ISO-15118-compliant systems. Their work allows for establishing a secure and efficient communication channel between EVs and charging stations, facilitating the two-way flow of energy and data.

In Table 1, a comparative analysis is made of the previously referenced studies, highlighting the different approaches employed, in terms of the following classifiers: (1) using the ISO 15118 standard; (2) considering demand response measures; (3) considering RES integration; (4) performing system emulation; and (5) performing system simulation.

Table 1. Comparative review of key references considering the primary topics of this work.

Ref.	ISO 15118	Demand Response	RES Integration	Emulation	Simulation
[14]	✔	✔			✔
[15]	✔	✔		✔	
[16]	✔	✔		✔	
[17]	✔	✔	✔		✔
[19]	✔	✔			✔
[20]				✔	✔
[21]	✔			✔	
[22]	✔	✔		✔	
This work	✔	✔	✔	✔	✔

By examining and comparing these relevant studies, we can gain valuable insights into areas where further research is needed.

2.2. ISO 15118 V2G Messages

The ISO 15118 standard plays a critical role in facilitating V2G communication between EVs and EVSE. It defines not only general information about the charging infrastructure but also communication protocols used during the charging process. The standard establishes a communication architecture with two endpoints: (1) the SECC located at the EVSE and (2) the EVCC within the EV. When connecting, these controllers establish a data link communication that supports the exchange of high-level messages. This communication facilitates the exchange of information for managing the charging process, including the initiation and ending of charging sessions, mutual authentication, and the utilization of security protocols [23].

Communication in the ISO 15118 standard follows a client–server model through request–response message pairs, as shown in Figure 1.

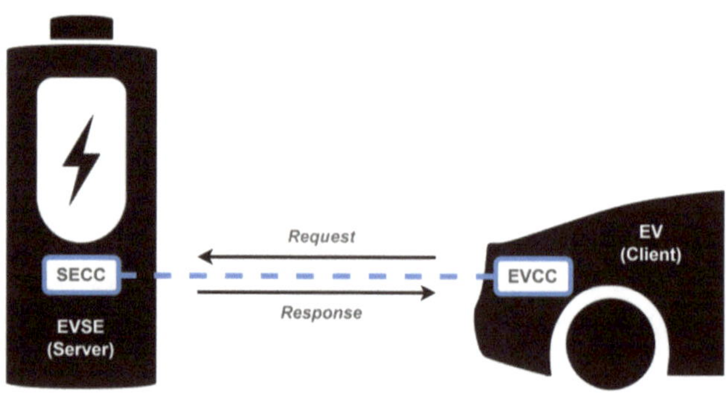

Figure 1. ISO 15118 client–server communication model between the SECC and the EVCC. For each type of V2G message, the client, on the EVCC side, sends a request message to which the server, on the SECC side, answers with a response message.

Each V2G message contains detailed information that depends on its type to satisfy its own purpose, ensuring efficient data exchange for various functionalities. Table 2 lists different V2G message types and its associated action, and the applicable power transfer modes (AC or DC) for each type. These message types are defined in the ISO 15118 standard for conductive charging, and can be exchanged by the EVCC and SECC during a V2G communication session, as illustrated in Figure 2 [24,25].

Table 2. ISO 15118 different V2G message types and their associated actions.

V2G Message Types	Associated Action
SessionSetup [AC & DC]	Establish the V2G communication session
ServiceDiscovery [AC & DC]	EVSE makes available all its services to an EV (e.g., charging services and payment options)
ServiceDetail [AC & DC]	An EV gets more information about an additional EVSE service
PaymentServiceSelection [AC & DC]	An EV chooses which services to use, as provided via EVSE previously
PaymentDetails [AC & DC]	Exchange details when certificates are chosen as a payment option (e.g., e-mobility account identifier)
Authorization [AC & DC]	EVSE allows, or not, the EV to have access to its energy, depending on the validation of the payment option
ChargeParameterDiscovery [AC & DC]	An EV and EVSE negotiate charging parameters (e.g., current, voltage, and power limits)
PowerDelivery [AC & DC]	EVSE supplies power to its outlet terminals so the EV can charge its battery
CertificateUpdate [AC & DC]	An EV requests a new certificate when it is about to expire
CertificateInstallation [AC & DC]	An EV requests a new certificate when it does not have a valid one; SECC may have to request this certificate from a secondary actor
SessionStop [AC & DC]	Finish the V2G communication session
MeteringReceipt [AC & DC]	EVSE digitally signs the charging energy metering information
ChargingStatus [AC]	Responsible for the charging loop in AC power transfer mode; EV verifies and validates the power consumed via EVSE
CableCheck [DC]	Checks whether the connector is locked and whether the EV is ready for charge
PreCharge [DC]	Adjust EVSE voltage to EV battery voltage
CurrentDemand [DC]	Responsible for charging loop in DC power transfer mode; control parameters are exchanged
WeldingDetection [DC]	Safety-checks the electrical contacts after the power transfer derived from charging

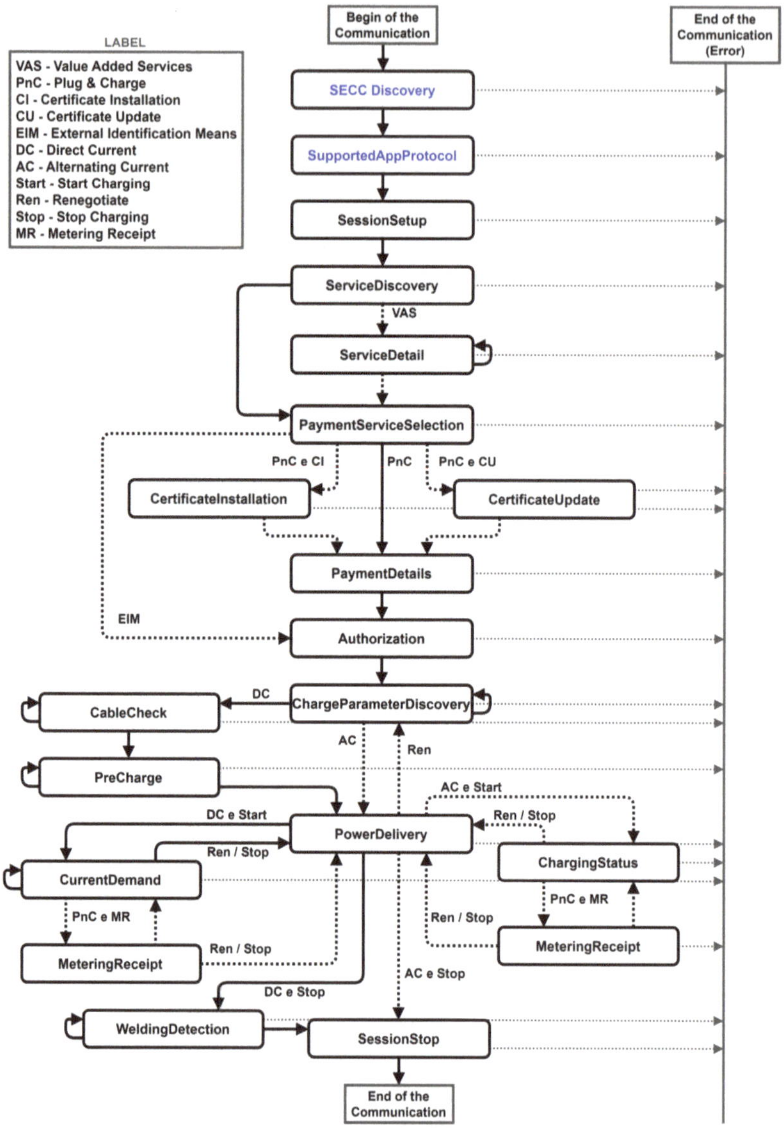

Figure 2. Flowchart of the types of V2G messages. In this flowchart, it is possible to see how the different types of V2G messages are used, depending on the power transfer modes (AC or DC), payment options (certificates and EIM), value-added services, metering receipt request, and charging progress ("Start", "Stop", or "Renegotiate"). The continuous black lines represent the V2G message flow implemented in the system that was developed in this work during a charging communication session.

2.3. RISE V2G Open-Source Library

The RISE V2G library [9] has emerged as a prominent, open-source tool for simulating communication systems between the EVCC and the SECC, implementing the ISO 15118-2 protocol. This Java-coded library enables the modeling of communication scenarios during an EV charging process, based on configuration files. These configuration files hold parameters such as the network interface through which the messages will be exchanged, the supported power transfer modes for conductive charging (AC and/or DC), and the payment options allowed, among a list of many others.

Using this library, the implemented communication controllers (SECC and EVCC) can exchange messages in the same machine or in separate ones. In this work, the communication controllers are implemented on separate machines.

In a practical charging situation where communication is set via V2G messages, while the EV is charging, the communication controllers must exchange the V2G messages referring to the charging loop during a certain period. A charging loop refers to the sequence of operations and communications that occur between an EV and a charging station during the charging process. This loop ensures that the vehicle and the charging infrastructure can negotiate and manage the charging session effectively and safely. The period corresponds to the time it takes for the EV battery to charge. Thus, the communication controllers must exchange the V2G messages referring to both the loop and the charging time.

Although the RISE V2G library implements communication through V2G messages according to the ISO 15118 standard, the parameters exchanged through the payload of these messages remain static during the communication session. In other words, this mechanism does not allow varying parameters like the SoC, charging current, or tension.

Before starting the communication session, the parameters must be configured, and the number of charge loops intended to be simulated must be defined a priori. Then, the communication session starts and ends a few seconds later. All messages exchanged via the communication controllers in these seconds are displayed in the console of a JAVA IDE. When observing these messages, it is possible to verify that the values of the parameter defined at the beginning of the communication session remain the same during the entire session. To overcome these limitations and create a platform capable of emulating the charging process through the ISO 15118 communication protocol, the implementation of changes to the source code of the RISE V2G library is required, as well as the development of additional code blocks. That process is described in the next section.

3. Charging Emulation System

After analyzing the ISO 15118 communication protocol and designing a solution for its implementation, the next step was to develop the charging emulation system between the EVSE and the EV. Figure 3 presents the general scheme of the implemented system, including the required interfaces for the EV user and the person responsible for the operation of the charging infrastructure (charge point operator, CPO).

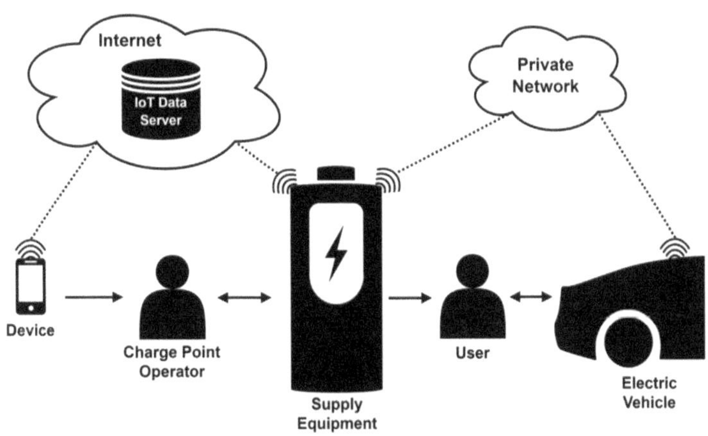

Figure 3. General structure of the implemented charging emulation system, comprising the vehicle endpoint, the supply equipment, and the required interfaces for the user of the EV and for the charging point operator, together with the communication infrastructure at both ends.

3.1. Characterization

The main objective of the charging emulation system is to monitor and manage the parameters exchanged during the charging process, according to the ISO 15118 digital communication protocol. This is done based on the open-source library RISE V2G.

According to ISO 15118 standard, the communication is performed through power-line communication over the control pilot line. However, the RISE V2G library does not implement or restrict the physical and data link layers that support the ISO 15118 standard, so any local communication network can be used as the interconnection medium between the SECC and the EVCC. In this work, the communication between both endpoints is done through a Wi-Fi wireless network.

In the implemented system, there is no real power transfer; i.e., the system only emulates the communication during the charging process between the EV and the EVSE. However, the selection of a power transfer mode is required to implement the communication through the corresponding V2G messages. Since this system was specified in the follow-up of a project, only the DC power transfer mode was chosen because the number of parameters exchanged between the EVSE and EV in this mode is wider, allowing better charge management. As a payment method, certificates were chosen since they were already implemented in the RISE V2G library and also because they offer greater security to the system. In this context, when the system starts, a communication session is established, followed by the V2G messages presented in a continuous black line in the flowchart of Figure 2.

In order to register the data exchanged during the charging process, a local Internet-of-Things (IoT) platform was integrated in the implemented system.

3.2. Architecture

The charging emulation system is divided into two main modules represented in Figure 4, the EVSEpi on the left side and the EVpi on the right side. Each one of these modules is divided into four sub-modules that support the implementation of similar functions at both ends, namely the communication controller, the parameter file, the management code, and the HMI. Each module was implemented on a dedicated Raspberry Pi (version 3—Model B).

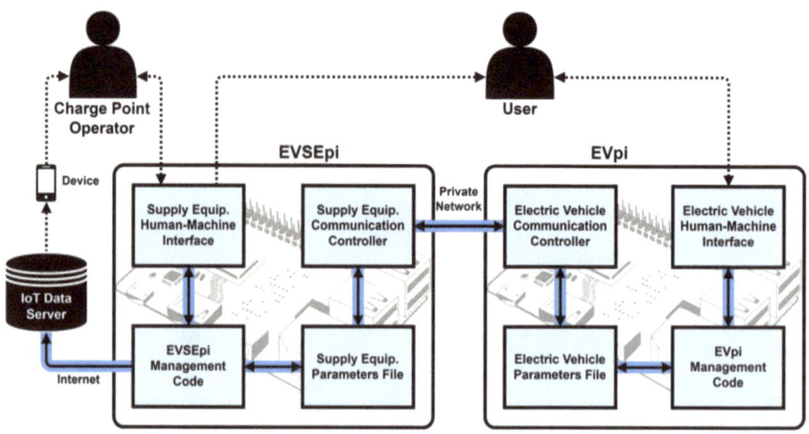

Figure 4. Architecture of the implemented system, comprising two modules. On the left side, the EVSEpi module is composed of the SECC, the EVSE parameter file, the EVSEpi management code, and the EVSE HMI. On the right side, the EVpi module is composed of the EVCC, the EV parameter file, the EVpi management code, and the EV HMI. Together, these eight sub-modules allow the implementation of the proposed charging emulation system according to the ISO 15118 standard, as well as the interaction with the charge point operator and the user and interfacing with the Internet-of-Things (IoT) platform.

3.2.1. Communication Controllers

The communication controllers in both modules aim to establish the communication session through an internet protocol connection, allowing the dynamic exchange of the charging parameters of the system. Without modification, the RISE V2G library only allows the exchange of static parameters. To overcome this characteristic of the RISE V2G library, firstly, the most relevant parameters of a communication session running on DC power transfer mode were selected. Then, these parameters were changed from static to a value that is read from the parameter file of the respective module. The values of the parameter file are updated every five seconds via the management code of the respective module. These communication controllers are handled via the management code of their respective modules and have an associated log file, which stores their communication records individually.

3.2.2. Parameter Files

The main purpose of the parameter files is to serve as an interface between the communication controller and the management code that exist in each module. Each file has its respective parameter list, with each parameter having its associated unit and type. The labeling is important because the communication controllers, implemented through the RISE V2G library, only allow the exchange of parameters if they are in accordance with a specific unit and, more importantly, with their specific type of variable. For successful communication, the names, units, and types of all parameters must be identical in both the management code and the communication controller.

In addition to the parameters of the ISO 15118 communication protocol, the parameter file also contains local interaction indicators (flags) so that the communication controller and the management code of the same module can synchronize (this will be further described in the next sections). Variables defined as flags are of a Boolean type with their initial value set to "False".

3.2.3. Management Codes

The management codes were developed in the JAVA programming language, and their main objective is to control the charging process, the communication controller, and the information processed on the HMI of the respective module. Each management code is implemented according to its respective finite state machine, in which each state is responsible for a certain set of tasks of the respective code.

One of the main objectives of the EVpi management code is to control the charging process of the EV battery. This approach allows for the definition of batteries with different characteristics in the EVpi management code. The charging process of an EV Li-ion battery applies both the constant-current (CC) and constant-voltage (CV) charging techniques, known as the CC-CV strategy. The CC-CV strategy consists of dividing the charging process into the two charging techniques [26,27]. Graph A of Figure 5 illustrates the voltage and current behavior during a full charge over time using the CC-CV charging strategy.

Since this work focuses on the communication between the EV and the EVSE, and to emulate the CC-CV strategy, a simplified model of the Li-ion battery was implemented. This model is fundamental to support the exchange of parameters between the controllers and emulate the typical functioning of the charging process of an EV, generating the corresponding curves, and bringing the developed system closer to a real charging process. Nonetheless, some influencing behaviors, such as temperature and cell balance, were set aside. To implement this model, the EVpi management code was adapted regarding the following approaches: (1) It was defined that the change from the CC to the CV charging technique would be performed when the EV battery was at 80% of its SoC. At that instant, the voltage reaches its maximum charging voltage, and the current begins to drop from its maximum value. Basically, the CC charging technique is used while the EV battery SoC remains in range [0, 79], and the CV charging technique is used from 80% upwards; (2) it is assumed that the voltage curve in the CC charging technique is linear, i.e., it can be approximated by a straight line; and (3) it was defined that the electrical current curve in the CV charging technique was also approximated, but in this case, another approach was used. The range [80, 100] is divided into 10 sub-ranges, where each sub-range represents 2% of the battery SoC. In these sub-ranges, the maximum current is multiplied by the corresponding coefficient, reducing the electrical current value, similar to a sequence of decreasing steps, whilst the battery SoC increases. When implementing these approximations, the current and voltage behavior during the charging process of the EV's virtual battery for the system follows the representation plotted in graph B of Figure 5. The characteristics of the Li-ion battery considered in Figure 5 and used in this work are represented in Table 3.

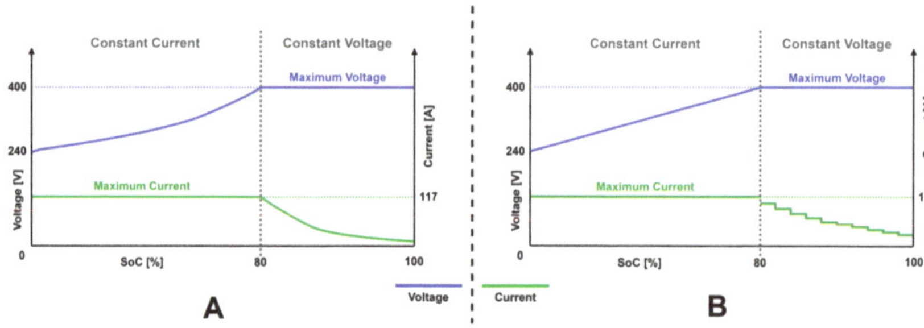

Figure 5. Graphs of the CC-CV charging strategies. (**A**) The left side represents the conventional behavior of the electrical current and voltage curves during the charging of a Li-ion battery using the constant-current (CC) and constant-voltage (CV) techniques. (**B**) The right side represents the approach made in this work to approximate the behavior of the virtual battery, implemented in the EVPi management code, to the behavior of a conventional Li-ion battery.

Table 3. Characteristics of the virtual battery implemented in the EVPi management code.

Total Capacity	Maximum Current	Maximum Voltage	Minimum Voltage	Nominal Voltage
235 Ah	117 A	400 V	240 V	350 V

The EVpi management code implements a finite state machine consisting of eight states, as shown in the diagram represented in Figure 6. These states are described in the following passage.

The EVpi State 1, named Initial Parameters Definition, translates the condition where the user of the vehicle starts defining, through the EV HMI, the limits of the current and voltage supported by the vehicle. The user can also define the initial state of charge of the EV for the subsequent charging emulation. The EVpi management code, in this state, is only responsible for accepting these values. To obtain a more realistic charge emulation, these values should be introduced according to the values of the virtual battery implemented in the management code. After having the initial parameters defined, the user allows the EVpi management code to change its state through interaction with the EV's HMI. If the vehicle battery is fully charged, the EVpi management code changes its state to number 6. Else if the vehicle battery needs to be charged, the EVpi management code changes to State 2.

At EVpi State 2, named Energy Request Calculation, the EVpi management code computes the amount of energy required to reach an SoC of 100%. It then gives the order to the EVCC to start the communication session. Through the EVCC start flag, the management code checks whether the EVCC has been able to establish the communication session with the SECC. In the case of success, the EVpi management code changes its state to State 3. After 25 s, if the EVpi management code is not able to establish a communication with the EVSEpi, the algorithm returns to State 1.

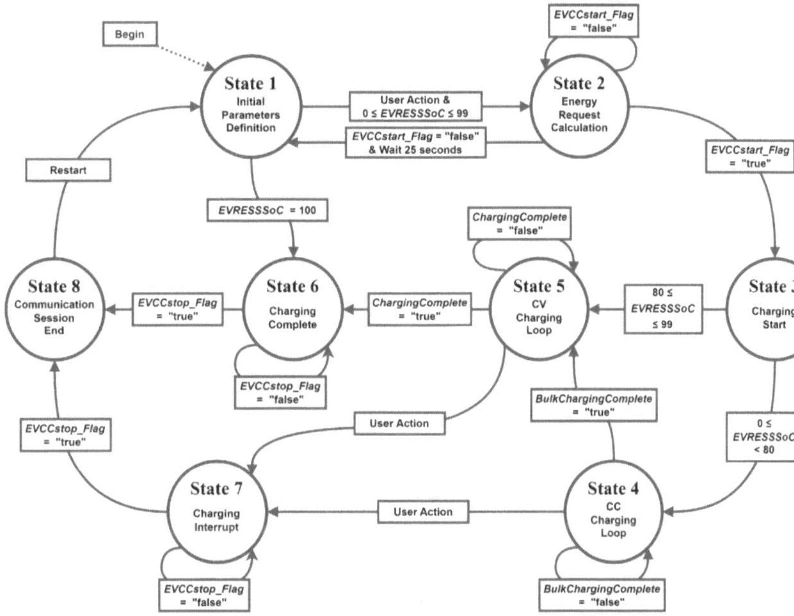

Figure 6. Finite state machine diagram of the EVpi management code, consisting of eight states.

In EVpi State 3, named Charging Start, the charging parameters are exchanged before the charging of an EV takes place. In other words, the EVpi management code sends the EV parameters to its own communication controller and receives the EVSE parameters from it. The parameters sent via the EVCC are as follows: (1) the EVCC identifier, (2) the values of the maximum current and voltage supported for the EV, and (3) the EV SoC. The parameters received at the EVCC are as follows: (1) the communication session identifier, (2) the EVSE identifier, (3) the values of the maximum current, voltage, and power supported for the EVSE, and (4) the minimum values of the current and voltage supported for the EVSE. After this exchange of information, the current, voltage, and power limits are defined for the charging process, respecting the limits of the EV and the supply equipment. Once these limits are established, the charging emulation process is initiated by sending the "Start" value through the charge progress parameter. Also, if the SoC of the vehicle is below 80%, the state changes from State 3 to State 4 or to State 5 if the SoC is already above or equal to 80%. This change in state that depends on the SoC results from the different charging techniques previously described and used in each of the next states.

EVpi States 4 and 5, respectively named CC Charging Loop and CV Charging Loop, translate the CC and CV charging techniques. In these states, the EVpi management code defines the target current and voltage values to charge the EV battery, according to the associated charging technique. These values are sent to the SECC via the EVCC so that the supply equipment can meet the vehicle's needs. In these states, the values of the EV SoC and the remaining time to reach a full charge level are also computed, according to the current and voltage that the EVSE supports. While the EVpi management code remains in one of these two states, the parameters mentioned above are constantly changing and updating, with a periodicity of 5 s. The transition between these states can be performed in two different ways, either as a result of the EV's user intervention through the vehicle's HMI or as a result of the EV's SoC. When the transition occurs due to the user's action, both states transition to State 7. This user action results from pressing the stop button on the EV's HMI. On the other hand, when a threshold value is reached for the EV's SoC, different situations occur in each of these two states. In State 4, when the vehicle's SoC reaches 80%, the value of the bulk (or fast)-charge-complete indicator is changed to "true", and the EVpi management code goes to State 5. This translates the change in the charging technique from CC to CV. In State 5, when the vehicle SoC reaches 100%, the value of the full-charge-complete indicator is changed to "true", and the EVpi management code goes to State 6.

States 6 and 7 of the EVpi respectively correspond to the Charging Complete and Charging Interrupt conditions, representing the last 5 s of the charging loop and, consequently, the stopping of charging. In State 6, the stopping results from the fully charged status of the EV, while in State 7, the stopping results from the interruption of the charging process via the vehicle's user. In these states, the EVpi management code is responsible for setting the EV target current and voltage values to zero, as well as the parameter charge progress value to "Stop". Through the charge progress parameter, the EVCC informs the SECC that a charging stoppage has been requested, thus ending the exchange of the V2G message related to the charging loop and starting the exchange of V2G messages responsible for terminating the communication session. Once the communication session is ended, the EVpi management code sets the EVCC end flag as "true" and transitions to State 8.

In EVpi State 8, named Communication Session End, the main task of the EVpi management code is to reset variables and terminate processes that were started during the previous communication session so that a new communication session can be started. After this update, the management code waits 10 s and transitions to State 1 (i.e., its initial state), when a new charge emulation can be started.

The main objective of the EVSEpi management code is to control the EVSE's charging process. To do so, it requires access to the voltage and current limits. Under real circumstances, these values are restricted due to the power limits of the electrical grid, or when there is no connection to the electrical grid, due to the power limits of the battery system. In a smart charging scenario, local generation from renewable energy sources should be used to modulate these values. In the charge emulator system, these limits are predefined by the CPO through the HMI of the EVSE. The EVSEpi management code is also responsible for calculating the energy transferred during each charging period and the associated cost, according to energy tariffs. It is also responsible for sending this information to an IoT platform.

The IoT platform used in this system is Emoncms [28]. Besides storing data, it also allows for generating dashboards and graphs of the charging parameters over time. These temporal graphs are extremely important for the CPO, as they can monitor all the information that was and is being processed during the charging processes.

The EVSEpi management code implements a finite state machine consisting of five states and is shown in the diagram in Figure 7. These states are presented in the following passages.

EVSEpi State 1, named Initial Parameters Definition, is where the CPO starts by defining, through the supply equipment's HMI, the maximum and minimum current and voltage values supported for the EVSE. After validating the entered parameters, the EVSEpi management code transitions to State 2.

In EVSEpi State 2, called Waiting Communication, the EVSEpi management code activates the SECC so that it waits for a connection from the EVCC. If a new communication session is detected, the SECC signals the EVSEpi management code through the "true" value of SECC start flag, which causes a transition to State 3. The EVSEpi management code can be kept in State 2 for an undefined time while waiting for communication establishment. However, the CPO can interrupt it through a stop button at the HMI of the EVSE, causing the management code to return to its initial state (State 1).

EVSEpi State 3, named Charging Start, functions similarly to the EVpi's State 3. The charging parameters are exchanged between the SECC and the SECC before the beginning of the power transfer. The parameters sent via the SECC are as follows: (1) the communication session identifier; (2) the EVSE identifier; (3) the maximum current, voltage, and power limits supported for the EVSE; and (4) the minimum current and voltage limits supported for the EVSE. The parameters received via the SECC are as follows: (1) the EVCC identifier; (2) the current and voltage limits supported for the EV; and (3) its SoC. This allows for the definition of the current, voltage, and power limits of the charging process, as well as its length and associated cost. This information is sent via the EVSEpi management code to the IoT platform. After completing this set of tasks, the EVSEpi management code checks the state of the progress parameter value. If its value is "start", then the charge emulation begins, as does the exchange of V2G messages related to the charging loop. This leads to the transition to State 4.

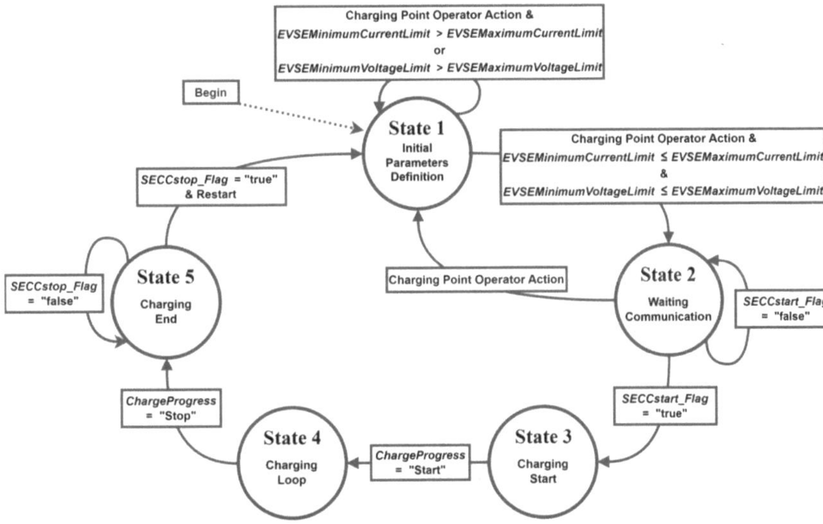

Figure 7. Finite state machine diagram of the EVSEpi management code, consisting of five states.

In EVSEpi State 4, called Charging Loop, the EVSEpi management code receives from the EVCC the values of the remaining time to reach a full charge, the target current and voltage to use in the charging process, and the EV's SoC. The management code then defines the values of the EVSE's electrical current and voltage, according to the power available from the electrical grid, renewable energy sources, and/or local batteries. In this state, the EVSEpi management code also sets up parameter values that are not part of the ISO 15118 communication protocol, such as the EVSE present power, the charge energy, and the charge energy cost. These parameters are updated at every charging loop. In State 4, the EVSEpi management code is also responsible for sending the charge information to the IoT platform. This information is sent with a periodicity of 1 min (i.e., every 12 charging loops). The EVSEpi management code transitions to State 5 only when the charging process ends, that is, when the charge progress parameter becomes equal to "Stop".

In EVSEpi State 5, named Charging End, the EVSEpi management code ensures that the EVSE's values of current, voltage, and power are zero. Thereafter, the connection between the communication controllers is finished, and the final information about the charge is sent for the last time to the IoT platform. Finally, the EVSEpi management code checks the value of the SECC's end flag to ascertain whether the communication session has ended. Once the communication session finishes, it resets the local variables, waits for 10 s, and returns to its initial state (State 1).

Figure 8 illustrates how the states of each module (EVpi and EVSEpi) are combined to make the implemented system work. During the system's operation, each EVSEpi state is associated with one or more states at the EVpi. However, at any given instant, only a one-by-one association might exist. For example, State 4 of the EVSEpi is related to States 4, 5, 6, and 7 of the EVpi, but no more than one of these four states can exist at the same time. So, if the EV is charging using the CC charging-loop technique, both the EVpi and the EVSEpi should be in State 4.

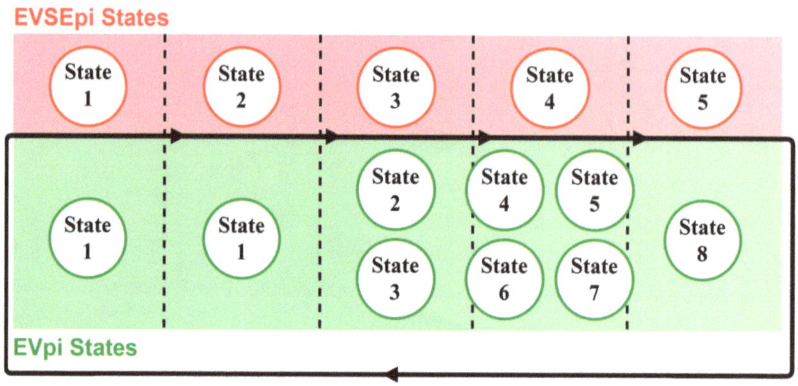

Figure 8. Articulation of states between the EVpi and EVSEpi. In order for the system to work, each state in the EVpi needs to have a corresponding state at the EVSEpi.

It should be noted that, for the system to operate correctly, the EVSEpi management code must be in State 2 before the EVpi management code transitions from State 1 to State 2, as shown in the diagram in Figure 8.

3.2.4. Human–Machine Interfaces

The HMIs of the EV and EVSE allow for defining the limits of the voltage and current supported for each device before the communication session starts. This way, the system allows different charging scenarios to be emulated. These interfaces also provide information about the EV's charging status that is updated every 5 s. The HMIs were developed in the JAVA programming language with the aid of the Swing graphical user interface tool of the Netbeans IDE, and they are controlled using the associated management codes. In the following passages, we describe each one of these interfaces in the EVpi and EVSEpi.

The HMI of the electric vehicle (EV HMI) allows the initial configuration (as shown in Figure 9), which corresponds to the behavior of State 1 of the EVpi management code.

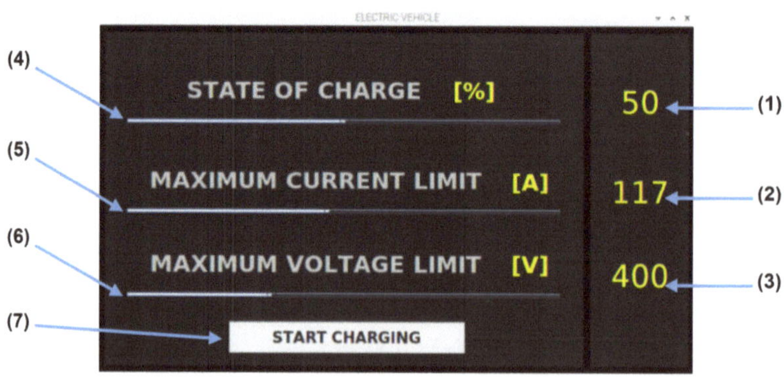

Figure 9. Initial configuration screen for the EV. In this interface, the values of the EV SoC (1) and the maximum limits of the current (2) and voltage (3) supported for the EV can be defined. These parameters are defined individually, respectively, through sliders (4), (5), and (6). The range of values of each slider can be changed in the source code. Once the initial parameters have been defined, button (7) initiates the transmission of the information to the EVpi management code, allowing the start of the communication session that will lead to the emulated charging of the EV.

When button (7) in Figure 9 is pressed, the EV's HMI presents a new interface with the EV's charging screen (shown in Figure 10). This interface allows for the monitoring of all the parameters of the EVpi management code states (except for the first state). In it, the user can observe the information associated with the status of both the charging process and the communication session.

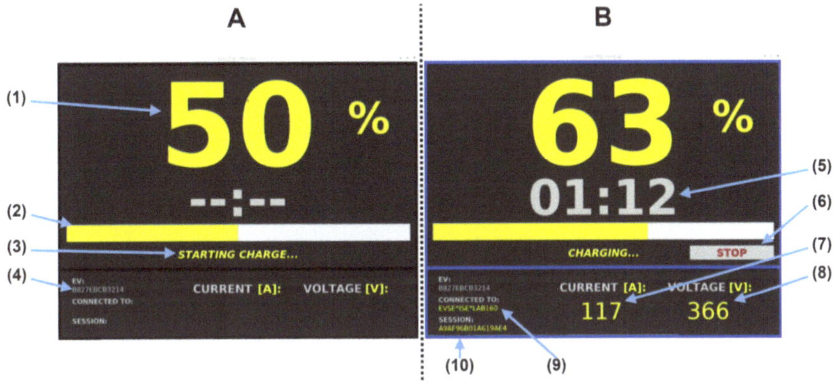

Figure 10. Charging screen of the EV. When the charging process has not yet started, only some information is made available at the charging screen (**A**), such as the numerical value of the EV's SoC (1), the level of the SoC (2), the information bar (3), and the EVCC identifier (4). When the charging process starts, the charging screen displays the remaining information (**B**), including the time left to reach the full charge (5) presented in an hour:minute format (hh:mm), the EV target current (7), the EV target voltage (8), the EVSE identifier (9), and the communication session identifier (10). While charging, the user can still request the charge to stop using the stop button (6) that is available in this interface.

The HMI at the supply equipment (EVSE HMI) has a structure that is similar with that of the EV's HMI but adapted to the charging station's requirements. In its initial screen, this interface displays the EVSE's initial parameterization screen, as shown in Figure 11.

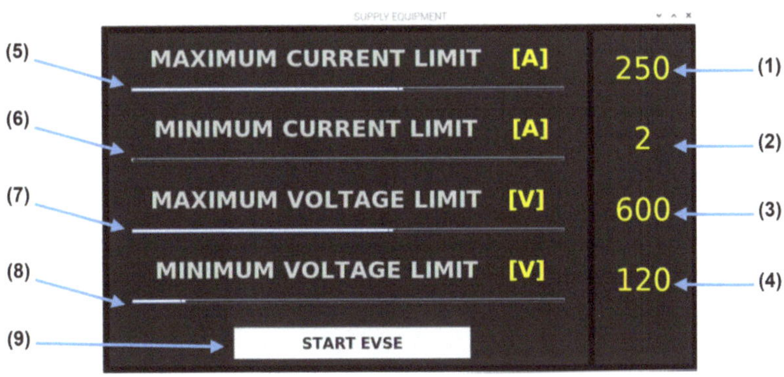

Figure 11. Initial parameterization screen of the EVSE. This screen allows for the definition of the maximum current (1), the minimum current (2), the maximum voltage (3), and the minimum voltage (4) supported for the EVSE. These parameters are individually adjustable, respectively, through sliders (5), (6), (7), and (8). Once the initial parameters have been defined, button (9) is responsible for initiating the transmission of the associated information to the EVSEpi management code.

After the definition of the maximum and minimum limits of both the current and tension, the EVSE's HMI displays the charging screen shown in Figure 12. The charging screen allows for the monitoring of all states of the EVSEpi management code except the first one. It also allows for monitoring information about the charging process, as well as the communication session. Different from the charging screen of the EV's HMI, the EVSE's HMI screen is made available before the communication session starts.

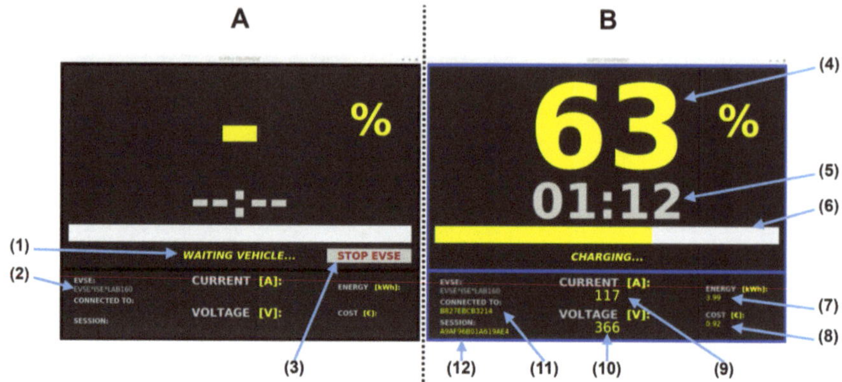

Figure 12. EVSE's charging screen. When the charging process has not yet started, only some information is made available at the EVSE charging screen (**A**), such as the information bar (1) and the EVSE identifier (2). It also includes a stop button (3) that can be used by the CPO only during this stage, allowing the EVSE HMI to get back to the initial parameterization screen, canceling the wait for the establishment of a new communication session. Once the charging process starts (**B**), the remaining information on the charging screen is made available, such as the EV's SoC (4), the time left to a full charge (5) displayed in an hour:minute format (hh:mm), the SoC level (6), the charged energy (7), the associated cost (8), the EVSE's charging current (9) and voltage (10), the EVCC identifier (11), and the communication session identifier (12).

3.3. *Task-Scheduling Algorithm*

The charging emulation system implements a task-scheduling algorithm to ensure a synchronized information flow, preventing failures or collisions. This algorithm enhances the system's performance by reducing the processing load and avoiding collisions when accessing the parameter file in each module. The set of tasks performed in the system is decomposed by the tasks performed via the EVpi and EVSEpi. The set of tasks performed via the EVpi result from the tasks performed via the EVpi management code and the EVCC. In turn, the set of tasks performed via the EVSEpi results from the tasks performed via the EVSEpi management code and the SECC. Figure 13 presents the task scheduling at both ends.

The scheduling algorithm operates in 5-s loops in which each second corresponds to a set of tasks that do not compromise the information flow, as can be seen in Figure 14. This cyclic approach maintains the integrity and efficiency of the emulation system.

The tasks are separated by 1-s intervals and have a running time of less than 1 s. However, a second is reserved for each set of tasks, reducing the system load. The scheduling of each of these tasks is repeated with a periodicity of 5 s. For instance, if a given task runs in the time interval between 0 and 1 s, the algorithm repeats this task in multiples of 5 s; i.e., the task can be started at seconds 0, 5, 10, 15, 20, 25, and so on, as exemplified in Figure 15.

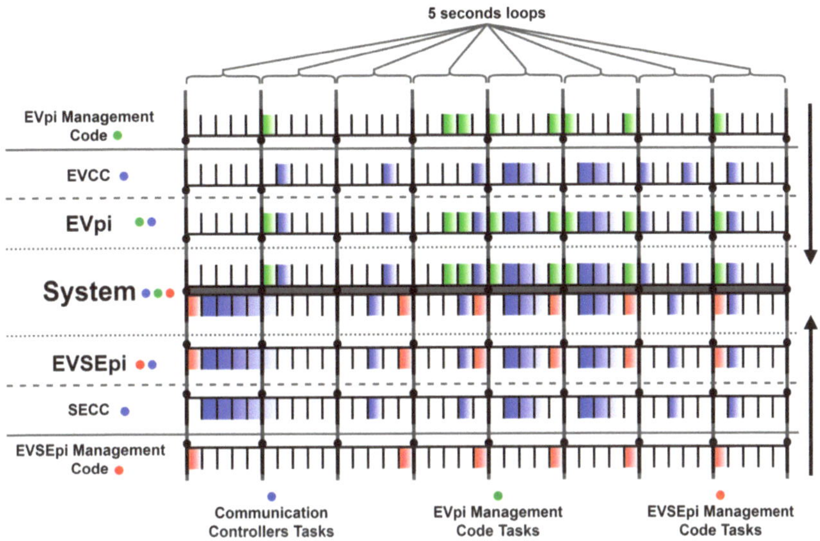

Figure 13. Task scheduling of the charge emulation system. In this figure, the tasks performed via the EVpi management code are represented in green, and the tasks performed via the EVSEpi are represented in red. The tasks performed via the communication controllers (EVCC and SECC), although implemented on separate machines, are represented with the same color (blue) because these tasks are performed together during the communication session. No task is overlapped between the SECC and EVSEpi management code or between the EVCC and the EVCC management code.

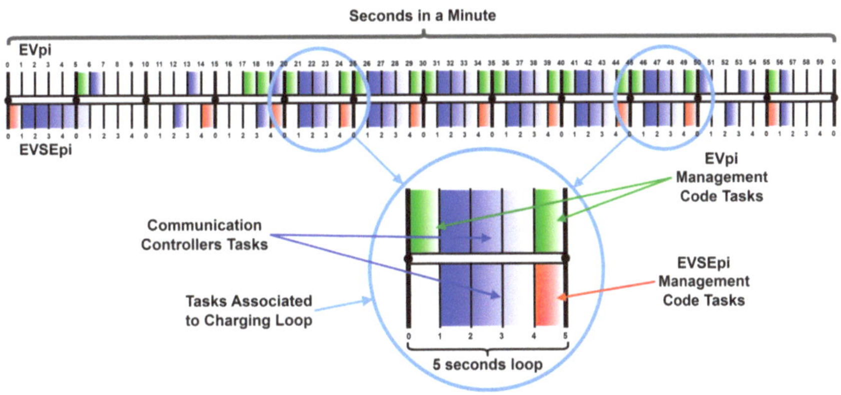

Figure 14. Loops of the task-scheduling algorithm. The 60 s of a minute are divided into 12 loops of 5 s. Each 5-s loop corresponds to a charging loop of the respective management code. Each interval of 1 s in a 5-s loop allows for the execution of a set of tasks associated with that charging loop.

The EVCC adopts the same temporization pattern as the SECC, making it possible to synchronize the processes during the communication session. The adoption of the 5-s loops resulted from the analysis of the number of tasks that the modules can perform without compromising the system's operability. Since each module is implemented in a Raspberry Pi 3—Model B microcomputer, and the controllers communicate through wireless networks, 3 s were reserved for the communication part. The remaining 2 s of the 5-s of the loop were reserved for the tasks of the management code at each module, including analyzing and updating charging parameters and updating their respective HMI.

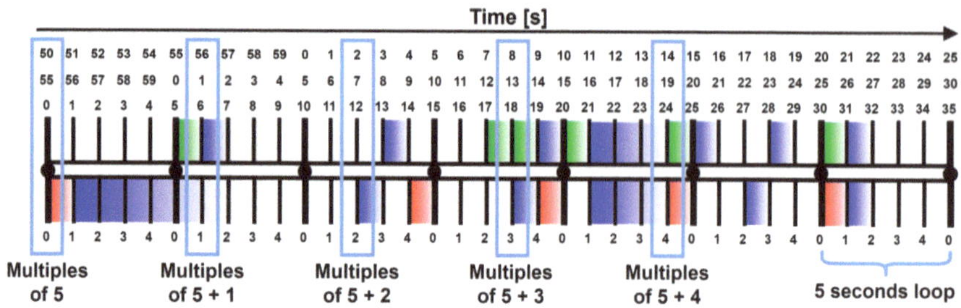

Figure 15. Task-scheduling algorithm. When a given task is executed, the algorithm repeats that same task with a periodicity of 5 s; e.g., if it is executed for the first time in the interval between 3 to 4, it will be repeated at 3 s, 8 s, 13 s, 18 s, 23 s, 28 s, etc.

One of the scenarios that led to the development of the charge-emulating system was its evaluation in contexts where renewable energy sources are integrated in, or close to, the charging infrastructure. The implementation of charge-responsive systems requires a continuous information exchange between both ends. To this end, in the following passages, we present how the charge parameters are exchanged between the EVPi and the EVSEPi.

3.4. Parameter Exchange during the Charging Process

The scheme represented in Figure 16 illustrates how the system parameters are exchanged during a charging emulation session. The *CurrentDemand* V2G message is used in the DC charging loop to illustrate the exchange of information between the EV (EVPi) and the EVSE (EVSEPi).

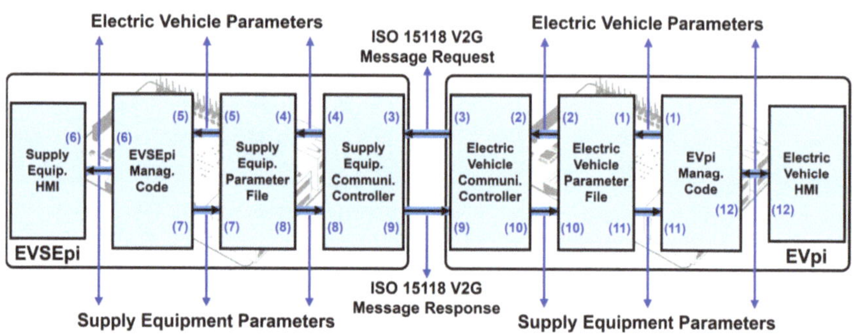

Figure 16. Parameter exchange in a charging emulation session between the EVSE (EVSEPi) and the EV (EVPi). In this figure, the flow of the CurrentDemand V2G message is depicted, showcasing the interaction between the EV and the EVSE during a DC charging session.

The numbering (1) to (12) in Figure 16 represents the key steps in the parameter exchange process:

1. The EVpi management code generates the updated values of vehicle parameters and sends them to the register list of the parameters file;
2. The EVCC, when preparing to send the request message (*CurrentDemandReq*), sends the updated values to the register list of the vehicle parameters file;
3. The EVCC sends the *CurrentDemandReq* to the SECC, containing in its body the vehicle's parameters;
4. The SECC, upon receiving the *CurrentDemandReq*, sends the vehicle parameters to the register list of the supply equipment parameters file;
5. The EVSEpi management code obtains the values of the parameters from the vehicle via the parameters file of the supply equipment. These parameters are used with the

EVSE for the processing of the charging level, tariff calculation, and data monitoring via the CPO. It also allows applies scheduling algorithms (not used in this system);

6. The EVSEpi management code, after receiving the vehicle's parameters and processing them, makes this information available to its HMI;
7. According to the available power, the EVSEpi management code generates the updated values of the parameters at the supply's equipment and sends them to the register list of its parameter file;
8. After receiving the request message (*CurrentDemandReq*) from the EVCC, the SECC creates the response message (*CurrentDemandRes*). It uses the updated values of the EVSE parameters through the list of registers in its parameter file;
9. The SECC sends the response message (*CurrentDemandRes*) to the EVCC, containing in its body the EVSE parameters;
10. The EVCC, upon receiving the *CurrentDemandRes*, sends the EVSE parameters to the register list of the parameters file on the EV side;
11. The EVpi management code obtains the values of the EVSE parameters via the EV parameter file. These parameters are used via the vehicle for charge processing;
12. The EVpi management code, after receiving the parameters from the EVSE and processing them, makes the resulting information available to its HMI.

Figure 16 highlights the continuous loop of communication and adjustment between the EV and EVSE during a charging session.

4. Resulting System and Demand Response Evaluation

This section starts by describing the implemented ISO 15118 EVSE and EV emulator system. It then evaluates the system's capability of supporting demand response when integrated with a photovoltaic renewable energy source.

4.1. Results of an Emulated Charging Process

Figure 17 shows the final product of the charging emulation system resulting from the architecture presented in the previous section.

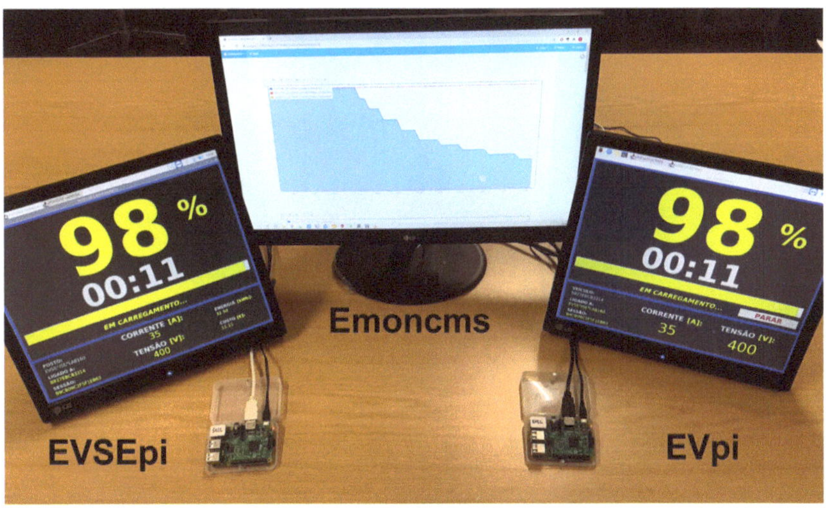

Figure 17. Implemented system. On the left side, the EVSE's HMI is shown, and close to it is the EVSEpi's microcomputer, responsible for the execution of the associated management code and the SECC source code. On the right side is the EV's HMI. A little further down is the microcontroller (EVpi) responsible for the execution of its management code and the EVCC source code. In the center of the figure, a chart from the IoT platform (Emoncms) is presented to illustrate the information logged during an emulated charging process.

The information shown in the charts of the IoT platform reflects the performed charge emulation. In each charge emulation, graphs are created in the IoT platform, with data from both the EV and EVSE. These data are sent to the IoT platform with a periodicity of 1 min via the EVSE. This allows for the monitoring of the parameters exchanged via the ISO 15118 communication protocol during the charging emulation process. Two of these graphs are presented in Figures 18 and 19, translating into an emulated charge with the initial conditions presented in Table 4.

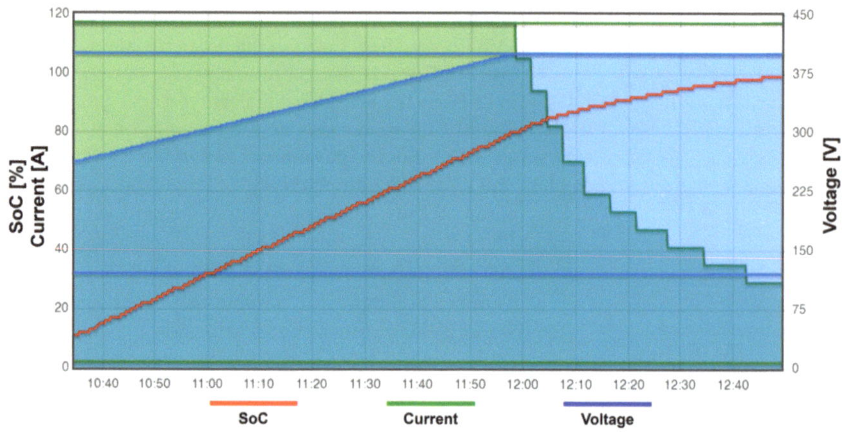

Figure 18. Chart of the electrical current, voltage, and SoC variables during an emulated charging process. The red line represents the SoC as a percentage, the green line represents the charging current in amperes, and the blue line represents the voltage of charging in volts. The top and bottom straight green lines represent the maximum and minimum currents that the charging process can reach. The top and bottom straight blue lines represent the maximum and minimum voltage that the charging process can reach.

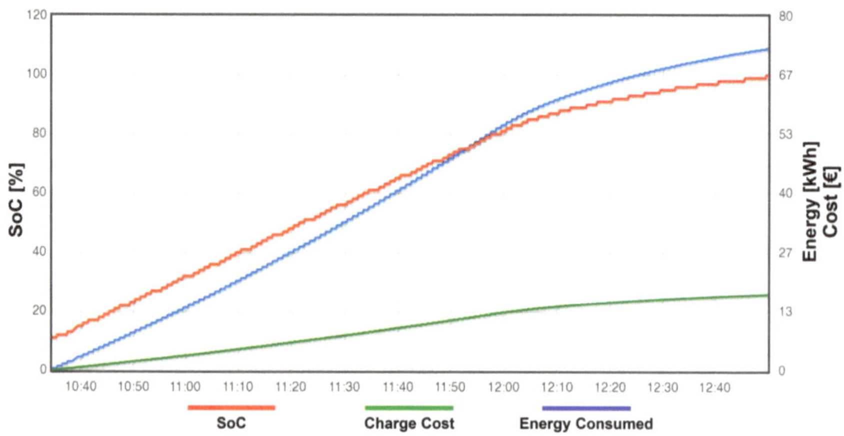

Figure 19. Graph of the energy and cost values during an emulated charging process. The red line represents the SoC in a percentage, the green line represents the charge cost in EUR, and the blue line represents the energy consumed during the charging in kilowatts hour.

Figures 18 and 19 illustrate the temporal evolution of the key parameters during a charging emulation session. The graph shown in Figure 18 illustrates the values of the current, voltage, and SoC variables during the charging emulation process. The graph in Figure 19 illustrates the values of the energy consumed and its associated cost during the charging emulation process. These graphs are generated based on the data sent to the IoT server (Emoncms) every minute. The graphs display the changing values of the current, voltage, state of charge, energy consumed, and cost over time, reflecting the dynamic nature of the charging process and the interaction between the EV and EVSE.

Table 4. Charging emulation initial conditions.

EVSE Maximum Current Limit	250 A	EV State of Charge	10%
EVSE Minimum Current Limit	2 A	EV Maximum Current Limit	117 A
EVSE Maximum Voltage Limit	600 V	EV Maximum Voltage Limit	400 V
EVSE Minimum Voltage Limit	120 V	Tariff Period	Peak

With the information in these graphs, it is possible to verify that the EV starts the charging emulation with a SoC of 10%, around 10:34 h and ends it, with a SoC of 100%, around 12:49 h. During these 2 h and 15 min, the EV charges 90% of its battery, with an energy transfer of approximately 72.7 kWh. This emulation was performed in a time period in which electricity had its highest cost (peak hours), so the 72.7 kWh was transferred from the electricity grid to the EV battery, resulting in a charging cost of 17.30 EUR, according to the tariff applied at the EVSE.

4.2. Evaluation of the Responsiveness of the System for Demand Response Purposes

Since the energy generated from some renewable energy sources can be highly intermittent, the system should be able to adapt swiftly to power variations. In the following, we evaluate its responsiveness to varying generation levels.

To evaluate the system's capability of supporting demand response, it was integrated with a photovoltaic RES. The system's performance was analyzed based on its ability to align the charging process with the availability of renewable energy, thereby minimizing reliance on the grid and optimizing the use of green energy. The responsiveness of the system in a demand response scenario was evaluated, focusing on the periodicity between SECC and EVCC transmissions. During the development of the system, a periodicity of 5 s was selected. However, it is important to assess whether this value is adequate in terms of: (1) the difference between charging power and generation power and (2) the number of packets per second it requires.

To perform this assessment, an energy generation dataset from a high-definition sampling of a photovoltaic unit was considered [29]. The dataset includes power output per square meter collected via a unit of photovoltaic panels over nearly 11 h of daylight on a cloudy day with high intermittency levels.

In order to evaluate the mismatch between generation levels and charging levels, the system's responsiveness was simulated with different communication intervals between the SECC and the EVCC. It was considered that the photovoltaic generation levels were transmitted to the SECC every 100 milliseconds. The transmission periodicity between controllers was then altered from 1 to 20 s. For each of these values, the percentual error between the generation power and the charging power was measured, as well as the number of packets per second required in each scenario. Figure 20 shows the associated results, highlighting that, as the communication interval increases, the percentual error between the generation and charging powers also increases, closely linearly. On the contrary, the number of packets per second decreases exponentially.

From the graph of Figure 20, it is evident that a 5-s communication interval used in the task management algorithm can be seen as a compromise, providing a relatively low percentual error between generation and charging levels (0.3798%) and maintaining a manageable number of packets per second (0.2 packets per second). This periodicity ensures that the system remains responsive to changes in the generation levels, thereby supporting an efficient demand response mechanism.

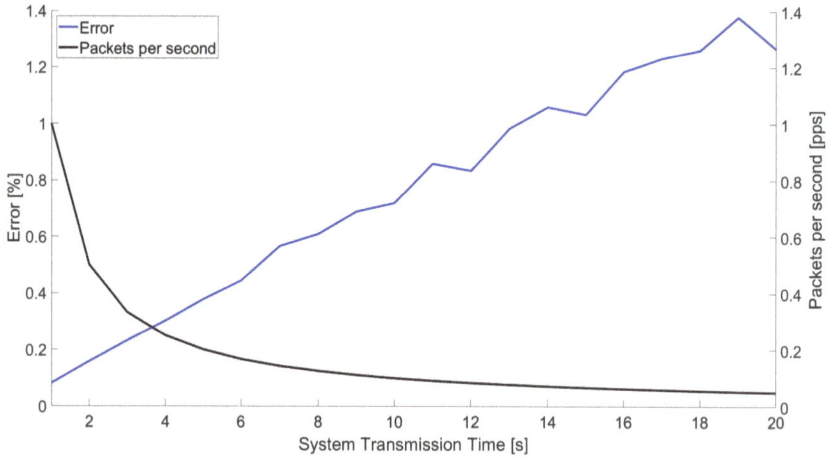

Figure 20. Power mismatch between charging and generation power. The blue line represents the percentual error between the generation power and the charging power, as the transmission periodicity is changed between 1 s and 20 s. The black line represents the rate of packets per second in the same conditions.

Considering a periodicity of communications between SECC and EVCC of 5 s, Figure 21 plots the difference between generation and charging levels. The values transmitted via the photovoltaic panel unit and the values transmitted between the communication controllers are very close throughout the daily sample. In light of these results, the utilization of the 5 s periodicity between transmissions in the setup for the task-scheduling algorithm is justified.

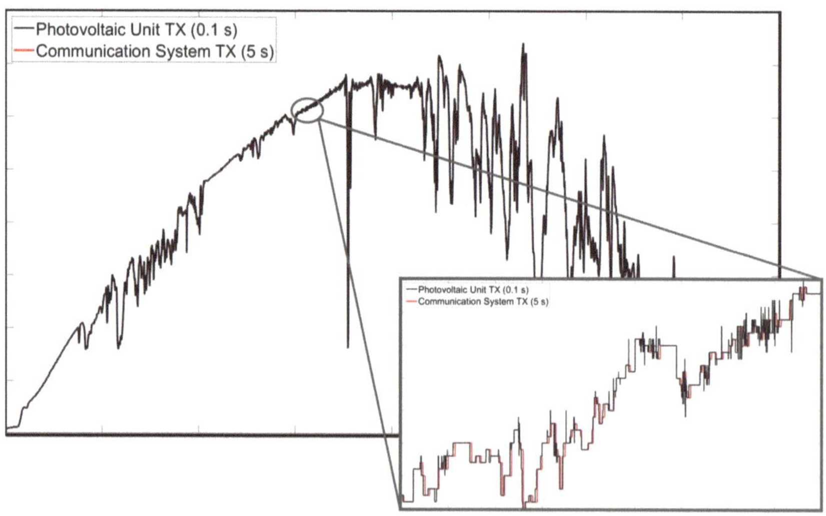

Figure 21. Power levels from the photovoltaic unit versus charging levels. The black line represents the photovoltaic unit transmission values, and the red lines represents the communication system transmission values. The excerpt, representing around 15 min, illustrates that the difference between the charging and generation levels is low.

5. Discussion, Conclusions, and Future Developments

The developed system successfully emulated a charging system using the ISO 15118 communication protocol between two microcomputers, providing a platform to observe the dynamic behavior of exchanged parameters throughout the communication session, mimicking real-world charging interactions. The implementation involved the development of management codes for both endpoints, i.e., the EV and EVSE. Within the EV module, a virtual Li-ion battery was implemented to emulate changes in charging parameters, employing CC and CV charging techniques to realistically approximate actual charging behavior. In the EVSE module, charging tariffs were implemented alongside a connection to an IoT platform, enabling the storage and future analysis of all emulation data. This enables researchers to observe how the system adjusts to real-time changes in energy availability, which is crucial for integrating RES into EV charging infrastructure. The emulation system also allows for the testing of different charging scenarios, providing a platform for optimizing algorithms and improving the overall efficiency of the charging process. Additionally, HMIs were developed, facilitating the interaction between an EV user and a CPO while providing real-time visualization of dynamic charging parameters.

A key finding of this work is the system's ability to adjust the charging parameters through the V2G communication interface in response to limitations on the electrical grid, a crucial feature for implementing charging stations that integrate RES. In scenarios where RES are utilized, the EVSE can dynamically communicate the available power limit to the EV, based on the generated power. The implemented system serves as an important tool for the testing and validating of the ISO 15118 protocol in a controlled environment. Ensuring a real-time data exchange between the EV and EVSE is crucial.

The use of Wi-Fi for communication, while convenient, may not fully replicate the power-line communication used in some EV charging systems. Future developments should consider implementing power-line communication over the control pilot line. Also, although the battery model implemented in this work effectively emulates real charging behavior, incorporating more complex models, such as the one used in [30], could enhance the system's realism even further.

Transitioning from the current emulation system to a real-world implementation poses several challenges and considerations. While the emulation system effectively mimics the behavior of an EV charging system using the ISO 15118 communication protocol, real-world deployment requires addressing factors such as scalability, interoperability with existing infrastructure, and user acceptance. Ensuring seamless integration with the diverse range of EV models and charging stations currently in use will be critical. Additionally, the infrastructure must support various communication protocols and adapt to the unique requirements of different regions and regulatory environments.

Future developments will focus on addressing the transition from controlled emulation to real-world implementation. Key areas for further research and development include the following: (1) scalability: extending the system to a more comprehensive framework that allows for testing V2G communication dynamics in larger and more complex scenarios, which includes scaling up the number of EVs and EVSEs to evaluate system performance under higher loads and more varied conditions; (2) physical system integration: integrating the EVpi and EVSEpi modules into a physical system that includes RES and batteries, building upon the digital twin introduced in previous research [30], which will involve testing the system's interoperability with actual hardware and ensuring that it meets real-world performance standards; and (3) renewable energy communities: applying the system within the context of renewable energy communities and extending it to also consider electric boats, as explored in previous research on Culatra Island [31,32], which involves assessing the system's effectiveness in supporting sustainable energy practices and community-based energy management.

While the developed emulation system provides a robust platform for testing and optimizing EV charging processes using the ISO 15118 protocol, several challenges remain before a real-world implementation can be achieved. Addressing scalability, interoperability, user acceptance, and regulatory constraints will be crucial for successful deployment. Additionally, overcoming technical hurdles and ensuring accurate real-time data management will enhance the system's reliability and effectiveness. Future research should focus on bridging the gap between controlled emulation and practical application, ensuring the system's readiness for real-world use and its contribution to sustainable energy solutions.

Author Contributions: Validation, P.J.S.C.; Investigation, J.B.S.; Writing—original draft, A.M.B.F., C.C. and A.P.; Project administration, J.M. All authors have read and agreed to the published version of the manuscript.

Funding: This work was supported by the Portuguese Foundation for Science and Technology (FCT) through the following individual research grants: 2021.08721.BD and 2021.08754.BD. The work was also supported by Project AGERAR+, Almacenamiento y Gestión de Energía Renovable para el fomento de la participación de pequeños y medianos prosumidores en redes eléctricas inteligentes (Project ID 0091_AGERAR_PLUS_6_E), funded by the European Union under the FEDER (Fundo Europeu de Desenvolvimento Regional) and INTERREG programs.

Data Availability Statement: The original contributions presented in the study are included in the article, further inquiries can be directed to the corresponding authors.

Conflicts of Interest: The authors declare no conflict of interest.

List of Abbreviations

AC	Alternate current
CC	Constant current
CPO	Charge point operator
CV	Constant voltage
DC	Direct current
EV	Electric vehicle
EVCC	Electric vehicle communication controller
EVSE	Electric vehicle supply equipment
HMI	Human–machine interface
IoT	Internet of Things
RES	Renewable energy sources
SECC	Supply equipment communication controller
SoC	State of charge
V2G	Vehicle-to-grid

References

1. Szumska, E.M. Electric Vehicle Charging Infrastructure along Highways in the EU. *Energies* **2023**, *16*, 895. [CrossRef]
2. International Energy Agency. Global EV Outlook 2019. Available online: https://www.iea.org/reports/global-ev-outlook-2019 (accessed on 25 April 2024).
3. International Energy Agency. Global EV Outlook 2024. Available online: https://www.iea.org/reports/global-ev-outlook-2024 (accessed on 25 April 2024).
4. Lopes, J.A.P.; Soares, F.J.; Almeida, P.M.R. Integration of Electric Vehicles in the Electric Power System. *Proc. IEEE* **2011**, *99*, 168–183. [CrossRef]
5. de Hoog, J.; Thomas, D.A.; Muenzel, V.; Jayasuriya, D.C.; Alpcan, T.; Brazil, M.; Mareels, I. Electric vehicle charging and grid constraints: Comparing distributed and centralized approaches. In Proceedings of the 2013 IEEE Power & Energy Society General Meeting, Vancouver, BC, Canada, 21–25 July 2013. [CrossRef]
6. Momoh, K.; Zulkifli, S.A.; Korba, P.; Sevilla, F.R.S.; Afandi, A.N.; Velazquez-Ibañez, A. State-of-the-Art Grid Stability Improvement Techniques for Electric Vehicle Fast-Charging Stations for Future Outlooks. *Energies* **2023**, *16*, 3956. [CrossRef]
7. International Electrotechnical Commission (IEC). IEC 61851-1: Electric vehicle conductive charging system—Part 1: General requirements. 2017. Available online: https://webstore.iec.ch/publication/33644 (accessed on 9 February 2024).
8. *ISO 15118*; Road Vehicles—Vehicle to Grid Communication Interface (Parts 1–20). International Organization for Standardization (ISO): Geneva, Switzerland, 2022.
9. SwitchEV. RISE-V2G. Available online: https://github.com/SwitchEV/RISE-V2G (accessed on 14 April 2024).
10. Alotaibi, I.; Abido, M.A.; Khalid, M.; Savkin, A.V. A Comprehensive Review of Recent Advances in Smart Grids: A Sustainable Future with Renewable Energy Resources. *Energies* **2020**, *13*, 6269. [CrossRef]
11. Sadeghian, O.; Oshnoei, A.; Mohammadi-ivatloo, B.; Vahidinasab, V.; Anvari-Moghaddam, A. A comprehensive review on electric vehicles smart charging: Solutions, strategies, technologies, and challenges. *J. Energy Storage* **2022**, *54*, 105241. [CrossRef]

12. Ohanu, C.P.; Rufai, S.A.; Oluchi, U.C. A comprehensive review of recent developments in smart grid through renewable energy resources integration. *Heliyon* **2024**, *10*, e25705. [CrossRef] [PubMed]
13. Mültin, M. ISO 15118 as the Enabler of Vehicle-to-Grid Application. In Proceedings of the 2018 International Conference of Electrical and Electronic Technologies for Automotive, Milan, Italy, 9–11 July 2018. [CrossRef]
14. Madhusudhanan, S.; Sivraj, P. Development of Communication Simulator for Electric Vehicle Charging following ISO 15118. In Proceedings of the 2022 IEEE North Karnataka Subsection Flagship International Conference, Vijaypur, India, 20–21 November 2022. [CrossRef]
15. Sayed, N.E. A Prototypical Implementation of an ISO-15118-Based Wireless Vehicle to Grid Communication for Authentication over Decoupled Technologies. In Proceedings of the 2019 AEIT International Conference of Electrical and Electronic Technologies for Automotive, Turin, Italy, 2–4 July 2019. [CrossRef]
16. Oulcaid, M.; El Fadil, H.; Njili, S.; Zytoune, O.; Bajit, A. Experimental Implementation of a Wireless Communication System for Electric Vehicle WPT Charger. In Proceedings of the 10th International Conference on Innovation, Modern Applied Science & Environmental Studies, Istanbul, Turkey, 12–14 May 2022. [CrossRef]
17. Wellisch, D.; Lenz, J.; Faschingbauer, A.; Pöschl, R.; Kunze, S. Vehicle-to-Grid AC Charging Station: An Approach for Smart Charging Development. *IFAC-PapersOnLine* **2015**, *48*, 55–60. [CrossRef]
18. *OCPP 2.0.1*; Open Charge Point Protocol (OCPP). Open Charge Alliance (OCA): Arnhem, The Netherlands, 2020.
19. Oumaima, B.; Hassan, E.F.; El Jeilani, S.; Halima, H.; Abdelaziz, K. Integration of Communication Between EV and EVSE Based on ISO 15118 with an OBD-II System. In *Lecture Notes in Electrical Engineering*; Springer: Berlin/Heidelberg, Germany, 2024. [CrossRef]
20. Stahleder, D. Hardware-in-the-Loop Co-Simulation Environment for Testing Grid Integration of Electric Mobility with Precise Charging Models. Master's Thesis, Technische Universität Wien, Vienna, Austria, 2018. [CrossRef]
21. Pallander, R. Implementation of HomePlug Green Phy standard (ISO15118) into Electric Vehicle Supply Equipment (Dissertation). Luleå University of Technology 2021. Available online: https://urn.kb.se/resolve?urn=urn:nbn:se:ltu:diva-83881 (accessed on 9 February 2024).
22. Jakó, Z.; Knapp, Á.; El Sayed, N. Wireless Authentication Solution and TTCN-3 based Test Framework for ISO-15118 Wireless V2G Communication. *Infocommunications J.* **2019**, *11*, 39–47. [CrossRef]
23. International Organization for Standardization (ISO). ISO 15118-1:2019. Available online: https://www.iso.org/standard/69113.html (accessed on 25 April 2024).
24. Ju, S.-H.; Lee, I.-H.; Song, S.-H.; Seo, H.-S. Communication Interoperability between EV Charging Infrastructure and Grid. *Int. J. Eng. Technol.* **2018**, *7*, 215–221.
25. International Organization for Standardization (ISO). ISO 15118-2:2014. Available online: https://www.iso.org/standard/55366.html (accessed on 25 April 2024).
26. Thomson, S.J.; Thomas, P.A.R.; Rajan, E. Design and Prototype Modelling of a CC/CV Electric Vehicle Battery Charging Circuit. In Proceedings of the 2018 International Conference on Circuits and Systems in Digital Enterprise Technology, Kottayam, India, 21–22 December 2018. [CrossRef]
27. Shen, W.; Vo, T.T.; Kapoor, A. Charging algorithms of lithium-ion batteries: An overview. In Proceedings of the 2012 7th IEEE Conference on Industrial Electronics and Applications, Singapore, 18–20 July 2012. [CrossRef]
28. Emoncms.org. Emoncms. Available online: https://emoncms.org/ (accessed on 14 April 2024).
29. Government of Canada. High-Resolution Solar Radiation Datasets. Available online: https://natural-resources.canada.ca/energy/renewable-electricity/solar-photovoltaic/18409 (accessed on 14 April 2024).
30. Francisco, A.M.B.; Monteiro, J.; Cardoso, P.J.S. A Digital Twin of Charging Stations for Fleets of Electric Vehicles. *IEEE Access* **2023**, *11*, 125664–125683. [CrossRef]
31. Santos, J.; Pacheco, A.; Monteiro, J. Implementation Process of a Local Energy Community in Portugal—The Case of Culatra Island. In *INCREaSE 2023—Advances in Sustainability Science and Technology*; Springer: Cham, Switzerland, 2023. [CrossRef]
32. Pacheco, A.; Monteiro, J.; Santos, J.; Sequeira, C.; Nunes, J. Energy transition process and community engagement on geographic islands: The case of Culatra Island (Ria Formosa, Portugal). *Renew. Energy* **2022**, *184*, 700–711. [CrossRef]

energies

MDPI

Review

Autonomous and Electric Vehicles in Urban Living Labs: Smart Mobility Strategies for the Future

Axel A. Mendoza-Armenta [1], Luis C. Félix-Herrán [2], Bartolomeo Silvestri [3], Andrea C. Valderrama-Solano [1], Juan C. Tudon-Martínez [1], Michele Roccotelli [3,*] and Jorge de J. Lozoya-Santos [1,*]

1 School of Engineering and Sciences, Tecnologico de Monterrey, Av. Eugenio Garza Sada 2501 Sur, Monterrey 64700, Mexico; axel.mendoza@tec.mx (A.A.M.-A.); andrea_valderrama@tec.mx (A.C.V.-S.); jc.tudon@tec.mx (J.C.T.-M.)

2 School of Engineering and Sciences, Tecnologico de Monterrey, Blvd. Enrique Mazón López 965, Hermosillo 83000, Mexico; lcfelix@tec.mx

3 Department of Electrical and Information Engineering, Polytechnic University of Bari, 70125 Bari, Italy; bartolomeo.silvestri@poliba.it

* Correspondence: michele.roccotelli@poliba.it (M.R.); jorge.lozoya@tec.mx (J.d.J.L.-S.)

Abstract: The deployment of electric, connected, and autonomous vehicles on public roads presents a significant challenge that can be addressed through previously established frameworks developed globally for implementing these technologies as part of an urban living lab (ULL). This systematic review, based on records from four distinct databases, focuses on projects that have conducted deployments of self-driving technologies in streets within urban environments. The review describes relevant information about various initiatives, including a classification of the stages of development reached according to the urban area covered, safety considerations, and lessons learned for optimal deployment. On-board sensing technology, digital infrastructure, and energy and communication systems emerge as the essential components of a ULL with autonomous vehicles (AVs). A crucial goal for smart cities is ensuring the scalability of large-scale deployments of such ULLs for safe, clean, and future mobility experimentation. This can only be achieved through effective coordination among academia, government, industry, and society to guarantee the successful integration of multiple projects in a unique environment.

Keywords: urban living lab; smart mobility; autonomous vehicles; electric vehicles; intelligent transportation systems; sustainability; service

Citation: Mendoza-Armenta, A.A.; Félix-Herrán, L.C.; Silvestri, B.; Valderrama-Solano, A.C.; Tudon-Martínez, J.C.; Roccotelli, M.; Lozoya-Santos, J.d.J. Autonomous and Electric Vehicles in Urban Living Labs: Smart Mobility Strategies for the Future. *Energies* **2024**, *17*, 5399. https://doi.org/10.3390/en17215399

Academic Editor: Tek Tjing Lie

Received: 25 September 2024
Revised: 23 October 2024
Accepted: 28 October 2024
Published: 30 October 2024

1. Introduction

To develop a smart city, a high degree of coordination among the government, academic institutions, and industry is required to have a high impact on society with sustainable solutions [1]. This is often referred to as a triple-helix model, where public institutions, research organizations, and the private sector cooperate to develop urban sustainability [2]. Thus, research needs to be integrated with real-world applications, which are often rife with systemic obstacles. A good strategy to introduce these kinds of sustainable solutions is the urban living lab (ULL) concept: a specific urban site or district supported by multiple stakeholders who want to promote economic and social development through innovations with sustainable goals, converting them in the ideal place to test and evaluate a product or service that offers an environmental solution for urban spaces [2]. Examples of this approach are demonstrated in ULLs around the world, where their objectives are aligned with the testing and innovation of mobility services, offering sustainable alternatives for the future of transportation systems. These efforts are developed in collaboration with government departments, universities, vehicle manufacturers, and other technological enterprises [3–6].

Along with the smart city concept is smart mobility, one of the main axes supporting the optimization of cities. It refers to initiatives promoting cleaner, safer, and more efficient mobility where transport systems are concerned [1]. Sustainable mobility is consistent with the principle of sustainability, with minimal environmental and cost impacts, while still meeting existing transport needs [7]. Smart mobility encompasses a wide range of initiatives and programs aimed at optimizing sustainable transport systems. The most important and promising applications are related to connected and autonomous vehicles (CAVs) [8] in terms of their ability to process real-time road information, together with revolutionary traffic system management. Autonomous vehicles (AVs) and electric vehicles (EVs) promise to be more efficient, cleaner, and safer options, but regulations and policies are still in progress [9], placing government departments as one of the key intermediaries to develop such ULLs.

An interesting example emerging in this area is the Automated Mobility District (AMD), developed by the National Renewable Laboratory (NREL), which describes it as "a geographically confined district or campus-sized implementation of connected and automated vehicle technology for the purpose of publicly accessible mobility" [10]. An important issue discussed by the NREL is the optimization of traffic operations, which is essential for increasing safety and sustainable energy use if connected and electric AV fleets operate in dense urban environments [11]. At this point, several aspects play a crucial role in the effective introduction of AVs in urban areas: automated SAE International's classification levels [12], the types of vehicles, the routes or services deployed, and intelligent infrastructure supporting smart vehicles. Regardless of the number of differences between AMDs, a consistent characteristic is that they are specific areas with well-defined boundaries where the optimization of traffic operations has been achieved [10,11].

The current review aims to evaluate the latest smart mobility initiatives globally, only considering real-world deployments of self-driving technologies in specific urban areas. AV experimentation depends on socio-technical progress. Trials occur not only in AMDs or ULLs but also in various urban contexts. AV trials can be classified as on-road, testbeds, or precinct experimentation [13]. On-road deployments are projects or services with AVs operating in public spaces within a city, such as Waymo [14], which is quite different from testbeds, where new mobility technologies are tested on closed tracks. The level of maturity for each type of urban deployment is discussed in the Section 3, along with their interrelationships. However, a growing tendency toward future and clean mobility using electric, connected, and autonomous vehicles has gained traction without providing a summary of its larger, system-level integration, referred to in this investigation as a ULL.

With the vast amount of information covering smart city initiatives and the formulation of concepts to identify specific areas with integrated smart mobility projects, a question arises: How have ULLs been implemented when their purpose is related to innovation and automation in mobility? The current systemic review aims to answer this question by analyzing specific areas or districts with a high level of technological development that actively demonstrates what the future of mobility looks like, illustrating how to provide citizens with cleaner, safer, and more efficient transport solutions.

By reviewing ULL initiatives around the world using self-driving technology, this study serves to describe the process of implementing a ULL with AVs, outlining the stages involved and the factors influencing the feasibility of such initiatives. To a certain extent, it can be considered a guideline for success, since lessons from previous strategies are discussed. Other studies have already highlighted the potential of smart cities in the transport domain, identifying the policies, standards, and technology needed to enable a new mobility era [15,16], but only from a theoretical perspective.

2. Materials and Methods

This research followed the guidelines of the Preferred Reporting Items for Systematic Reviews and Meta-Analyses (PRISMA) method. Only publications from Q1, Q2, and Q3 journals were used as the basis for further investigation. From the common topics and concepts addressed in the articles, reviews, and technical reports, further research was conducted to compile the most recognized global strategies, highlighting the future of mobility and contributing to scientific publications that communicate these advancements and findings.

To construct the database, three different combinations of keywords were applied using the Scopus database, considering common concepts and words aligned with the previously presented conceptual research question. The search schemes used to find the remaining studies are as follows:

- ALL (automation OR automated OR autonomous) AND ALL (electrification OR electric) AND ALL (transportation OR transport) AND ALL (mobility) AND ALL (smart OR future) AND ALL (vehicles) AND ALL (urban) AND ALL (laboratory OR district OR initiative OR center) AND ALL (test OR testing OR simulate OR innovate OR trialing) AND ALL (shared OR connected) AND ALL (sustainable OR sustainability OR clean) AND ALL (route OR circuit OR corridor) AND NOT TITLE-ABS-KEY (bike OR bicycle) AND NOT TITLE-ABS-KEY (metro OR subway OR train) AND NOT TITLE-ABS-KEY (airplane OR aerial OR airport).
- TITLE-ABS-KEY (automated AND mobility) AND TITLE-ABS-KEY (district OR city) AND TITLE-ABS-KEY (center OR lab OR corridor) AND TITLE-ABS-KEY (vehicle).
- TITLE-ABS-KEY (living AND lab) AND TITLE-ABS-KEY (smart AND mobility) AND TITLE-ABS-KEY (city OR district).

The first combination of keywords attempts to encompass a broad range of studies without exceeding the desired conceptual margin. This initial strategy provides us with a comprehensive collection of studies focusing on on-road smart mobility implementations. The search was conducted on 11 March 2024 and yielded a total of 459 studies. The second combination of keywords was restricted to the desired concepts found exclusively in the title, abstract, and keywords sections. This narrowed the search significantly to 25 documents, which were more representative of the purpose of this review. This refinement was performed on 8 April 2024. The last keyword combination was shorter than the others, maintaining the title, abstract, and keyword search strategy. However, the selected concepts are appropriate for describing initiatives or spaces taking action on smart mobility. This revised strategy was implemented on 22 April 2024, resulting in the discovery of 23 studies.

The search was delimited between 2019 and 2024; the Scopus database showed an increment in research studies published between those years when using the first combination of words, which considered a greater range of publications related to the desired concepts. However, it was necessary to apply exclusion criteria when performing a more in-depth analysis of the database.

Figure 1 shows the selection process in detail that led to the final number of studies included in this review. Database refinement was performed by limiting the search of publications to the English language (n = 392) and to the following document types: articles (n = 223), reviews (n = 128), book chapters (n = 26), and conference papers (n = 15). The reasons for exclusion are as follows:

1. Publications in languages that were not in English (n = 15).
2. Publications that were not an article, review, book chapter, or conference paper (n = 96).

Some of the records that were not analyzed due to language exclusion criteria were mostly written in Chinese. However, relevant initiatives from this country still appear in English publications. A careful analysis was performed on all the databases once the reports assessed for eligibility were identified. Due to the wide range of keywords considered in all the document sections in the first database, it was necessary to analyze the study to identify the purpose of the publication. If it was related to ULLs, any kind of self-driving technology, and cutting-edge technology for mobility being tested in real-world scenarios, it remained under consideration for the final studies included in the review. Taking into account these inclusion criteria, the final exclusion of records was performed as follows:

1. The publication was not related to automated land vehicles ($n = 37$).
2. The publication was more related to greenhouse emissions impact studies than new mobility strategies implementation ($n = 28$).
3. The publication was related to automated districts, but not implemented in a real-life scenario ($n = 68$).
4. The publication was related to social interests ($n = 42$).
5. The publication was related to public policies and regulatory frameworks ($n = 27$).

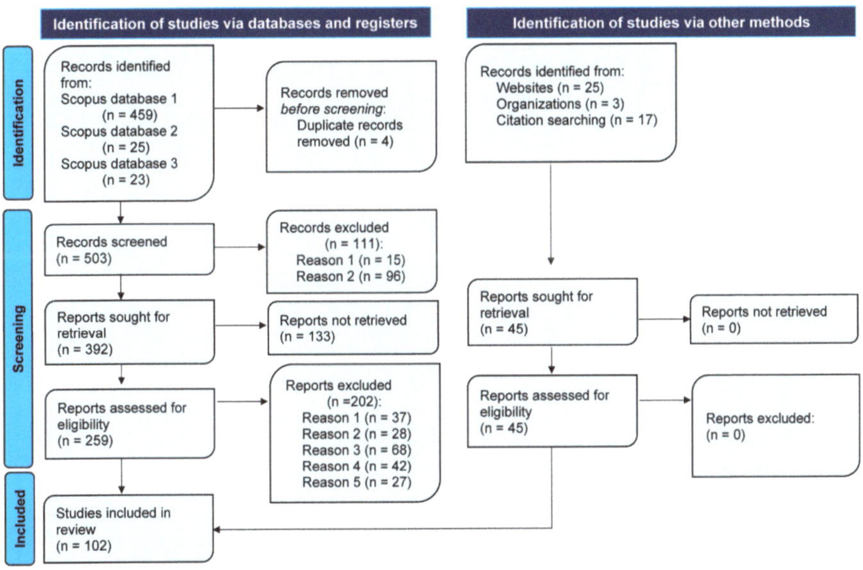

Figure 1. The methodology followed in a PRISMA flow diagram for reporting a systematic review [17].

This review includes 102 studies, of which 57 are from the Scopus database, comprising 49 articles and 8 conference papers. Additionally, 45 external studies were identified and included, 17 corresponding to paper publications, 3 to technical reports from public organizations, and 25 to references from official websites of ULL initiatives or government reports on technology growth and standardization.

3. Results

3.1. Summary

A total of 102 studies were included in this review, most of which are centered on real scenarios with some sort of technology or framework implemented to utilize AV technology. In this context, AV trials are strongly related to an urban ecosystem, as the system being tested interacts with urban infrastructure, socio-nature, and regulations [18]. Studies on smart cities have revealed a strong relationship between city development and cutting-edge information technology, regardless of the stage of experimentation [19]. City logistics face challenges at any stage of economic development. To address urban growth, traffic congestion, and environmental problems, developed countries constantly push initiatives involving AVs and EVs. However, developing countries often remain in the early stages due to internal constraints, leading to a lower distribution of technology and infrastructural adaptation [20]. Following these criteria, Table 1 presents the general characteristics of the studies applying some sort of implementation or initiative to promote the advancement of smart mobility.

Table 1. Records of AV deployments in real-life settings.

Study	Name of Initiative	Location	Classification	Purpose	Impact
[21]	Mcity	Ann Harbor, MI, USA	Testbed	To introduce AVs in community-based trip sharing.	The research center studies advanced mobility, showing that CAVs gain public acceptance. This has led to testing commute trip sharing and new mobility strategies, with large-scale implementation reducing vehicle usage by 92%.
[22]	Smart Nation	Singapore	Testbed	To transform transport services through technology.	The 2017 establishment of CETRAN made Singapore a leader in self-driving technology research and trials, contributing to international standards and large-scale deployments.
[23]	Florida-CAVI	Gainesville, FL, USA	Testbed	To field-deploy CAVs for testing and further development.	The Florida Department of Transportation developed the Florida-CAVI program to enhance traffic operations and document safety benefits for transportation automation. The I-STREET testbed facilitates data collection, analysis, and the distribution of CAVs in a safe demonstrative environment.
[24]	ADVI	Australia	Testbed	To support urban planning with AV technology.	Since 2015, Australia has stood out for AV trial announcements, contributing to international regulations and hosting conventions. AV deployments have sparked strong interest in policy development for Oceania's transition to shared mobility dynamics.
[25]	ITRL	Stockholm, Sweden	Testbed	To make a transition toward sustainable road transport.	The development of living labs and research platforms enables experimentation with automated vehicle solutions and smart mobility. The Automated Vehicle Traffic Control Tower serves as a testbed to explore connectivity, infrastructure, and sensing needs in AVs.
[26]	Innovation District	Las Vegas, NV, USA	Urban corridor	To demonstrate the first self-driving shuttle pilot project in the USA.	From November 2017 to October 2018, a self-driving shuttle bus operated in the Fremont Entertainment District, one of the first fully autonomous systems in real traffic conditions globally. The CAV gained significant public acceptance and project continuity.
[27]	CE-CERT	Riverside, CA, USA	Urban corridor	To evaluate CAVs' energy impacts.	An innovation corridor focused on sustainable mobility was developed. Using cooperative and automated driving technology can achieve a 4% reduction in total energy consumption compared to a scenario with 95.5% gasoline vehicles, 2% electric vehicles, and 2.5% plug-in hybrids.
[28]	Bay Street Innovation Corridor	Jacksonville, FL, USA	Urban corridor	To test and optimize CAVs.	Investigation for safer operations led to the creation of a testbed with CAVs that helped transform the transportation network, optimizing traffic operations and increasing safety within Downtown and South Downtown districts.
[29]	SURTRAC	Pittsburgh, PA, USA	Urban corridor	To create an adaptive traffic signal control system.	The project started as a pilot in June 2012 to manage traffic at fifty connected intersections. It achieved a 25% reduction in travel time and a 40% reduction in wait time compared to average traffic volumes.
[30,31]	Smart City Korea	Sejong, South Korea	Urban corridor	To develop a Cooperative Intelligent Transportation System.	The framework used in the autonomous bus field test proved to be a safe, comfortable, and efficient choice. It employed a cloud-based architecture for Sectionalized Speed Guidance, enhancing energy efficiency.
[32]	Babcock Ranch	Punta Gorda, FL, USA	Precinct	To gain insights into autonomous mobility on demand (AMoD).	The initiative developed an AMoD service to compare its benefits for city travel with conventional vehicles. The results revealed a high AV fleet efficiency, with 70% usage during operating hours. Although the service is no longer available, it inspired other self-driving projects in the state.

Table 1. *Cont.*

Study	Name of Initiative	Location	Classification	Purpose	Impact
[33]	University District AV Transit Circulator	Houston, TX, USA	Precinct	To study AV technology and its potential in public transportation.	The Phoenix Motorcar Zeus 400 battery-electric paratransit shuttle bus, featuring a level 4 ADS, aims to provide first- and last-mile connections to METRO services as part of a pilot project running through 2024 and 2025. In June 2019, an EasyMile EZ 10 Gen-1 vehicle completed around 7500 trips on Texas Southern University's one-mile Tiger Walk.
[34]	Regional Transportation District's	Denver, CO, USA	Precinct	To serve three distinct routes with an automated shuttle.	The AV deployment operated on a one-mile loop in 2019, offering 600 free rides near the 61st & Peña commuter rail station, passing Panasonic's and EasyMile's offices. Although no further deployments were supported by the Regional Transportation District, the pilot provided valuable insights.
[35]	Milo project	Arlington, TX, USA	Precinct	To improve public awareness of AV technology.	Launched in August 2017, one of the first self-driving shuttle deployments in the USA operated in a parking lot between AT&T Stadium and Globe Life Park. The pilot program concluded after serving 740 passengers, inspiring further AV initiatives in the city and nationwide.
[36]	City of Mobility	Hamburg, Germany	Precinct	To be considered the mobility capital with sustainable initiatives.	Eight projects were launched to improve public transport's flexibility, efficiency, and sustainability. One featured an autonomous shuttle in Hamburg's Bergedorf district, which transported up to 12 passengers between stops and homes for three months in 2021.
[35,37]	ELA	Alberta, Canada	Precinct	To test the feasibility of autonomous shuttles in transportation systems.	In Alberta, different projects were developed. One ran between Calgary Zoo LRT Station and Telus Spark Science Centre for 33 days in 2018, serving about 4500 passengers. Another operated in Beaumont for six months in 2019, running fully autonomously in mixed traffic. Both were part of ELA's Electric Autonomous Shuttle pilot program with EasyMile.
[38]	TOPS Laboratory	Madison, WI, USA	Precinct	To improve traffic operations and safety.	With an interdisciplinary focus, initiatives span from database creation to AV implementation. Four concept routes have been designed to test connected and self-driving technologies.
[39]	SOLUTIONSPlus	Nanjing, China	Precinct	To co-develop innovative and integrated e-mobility solutions through living labs around the world.	The development of mobility alternatives in Nanjing is driven by its high population and role as a transportation hub. The city has significantly adopted EVs and tested a self-driving minibus on regional roads in October 2021.
[40]	Rivium Business Park and Central District	Capelle aan den IJssel, Netherlands	Precinct	To provide frequent and flexible transport with AVs.	The Rivium ParkShuttle is unique globally as the only permanent autonomous service in Europe. This cost-effective system autonomously navigates and generates revenue, solving last-mile transport between Kralingse Zoom metro station and Rivium business park.
[41]	Darwin SatCom Lab	Oxford, UK	Precinct	To test self-driving car technology.	Multiple labs for CAV research and development have advanced 5G and satellite communications. A partnership between university research institutions and industry led to a level 4 autonomous shuttle deployment in September 2022 along Norham Gardens, Parks Road, and part of Broad Street, setting the stage for another unit in 2024.
[3]	Smart Columbus	Columbus, OH, USA	ULL	To enable technological solutions to improve mobility, sustainability, and digitization in the city.	The introduction of safe, innovative solutions aims to boost EV adoption. Self-driving shuttles and connected vehicles address transportation gaps, with the Smart Circuit Shuttle providing over 16,000 rides in a year.

Table 1. *Cont.*

Study	Name of Initiative	Location	Classification	Purpose	Impact
[4]	Smart Mobility Living Lab	London, UK	ULL	To accelerate the creation of clean, efficient, and safe mobility solutions.	TRL conceived service offerings through testing, simulation, and innovation to cover all aspects of connected, automated, shared, and electric mobility. This includes testbeds for automation, connectivity, energy, logistics, communication, and decarbonization. The initiative involves multiple projects and global partners, providing shared research, standards development, safety frameworks, and consultancy services for the industry.
[42,43]	OVIN	Ontario, Canada	ULL	To drive and shape the future of the automotive and mobility sector.	Since 2017, Ontario has been developing programs with smart mobility technologies to enhance transportation. In Toronto, the Technology Pilot Zone tests connected, autonomous, and electric vehicles, making it a leader in AV research and industrial policy application.
[44]	Aveiro Tech City Living Lab	Aveiro, Portugal	ULL	To provide new solutions with efficient traffic management and intelligent transportation systems.	Development includes connected infrastructure with a multi-protocol network that classifies vehicles, detects peak traffic times, exchanges collision-avoidance information, and supports comprehensive network data management with third-party integration for self-driving technologies.
[45]	Beijing E-Town	Beijing, China	ULL	To promote collaborative innovation and development in intelligent transportation.	The technological hub has enabled AV testing on public roads and closed sites. The Yizhuang Base has offered nearly 10,000 h of testing for 284 AVs, adapting to various urban traffic scenarios.
[46]	Modena Automotive Smart Area	Modena, Italy	ULL	To develop a regional ecosystem with connected and autonomous mobility.	The living lab in R-Nord, spanning 3 km^2, features smart cameras, sensors, 5G antennas, servers, and a data center, facilitating innovations in sustainability and urban regeneration.
[47]	AV Living Lab	Ljubljana, Slovenia	ULL	To transform the future of mobility with real-life and new mobility solutions.	Diverse pilot programs and deployments promote car sharing, smart mobility plans, and insights into real interactions with self-driving technologies.

3.2. Data Distribution

The included studies have a notable presence in the European Union (EU). In the case of Europe, it is worth mentioning that most of the projects with real-world applications are related to the funding program Horizon Europe, formerly known as Horizon 2020. One of its pillars is based on global challenges and European industrial competitiveness, allowing the intervention of a cluster on climate, energy, and mobility to promote clean, safe, and accessible transport, as well as smart mobility [48]. The EU has made an incredible effort to support harmonization in research and innovation for Cooperative, Connected, and Automated Mobility (CCAM), providing a repository called Connected and Automated Driving (CAD). This repository includes terminologies, regulations, procedures, and action plans for future research, encompassing projects and demonstration activities with AVs across Europe [49].

In North America, ULLs with AVs have a strong presence in the USA. Most of the cities involved in early deployments, as reported in "The Automated Mobility District Implementation Catalog" [11], remain active with new projects or are searching for alternatives. The use of self-driving technology was typically within periods no longer than a year to verify public acceptance and test transport system efficiency. Projects developed in Ohio, Texas, Nevada, and Michigan are frequently documented in automated shuttle deployment studies [10,34,35].

From all the research included in this review, Figure 2 highlights only the studies that mention the implementation of future mobility strategies involving connected, electric, and autonomous vehicles. This does not necessarily mean that the ULL ecosystem is fully developed, but the research provides a framework for urban planning, simulation, experimental setup, or urban testing that could support ULLs' integration in future studies. Figure 3 provides further insights into the records found via databases and registers by classifying each study geographically and by the level of maturity in urban deployment contexts. Records identified through other methods were used to enrich the bibliography of this study and to complete Table 1 presented earlier, classifying them according to the definitions and characteristic criteria from previous ULL studies [2,50].

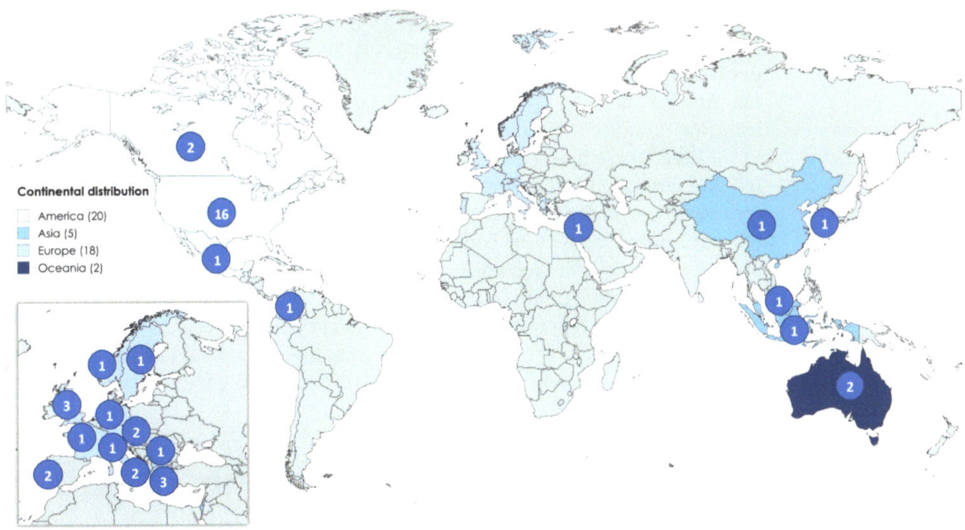

Figure 2. Worldwide records' distribution from the Scopus database with studies focusing on ULLs with the use of automated mobility in conceptual or experimental applications. Circled numbers indicate the total number of studies in each country, the sum of studies for each continent is also shown in the map's legend.

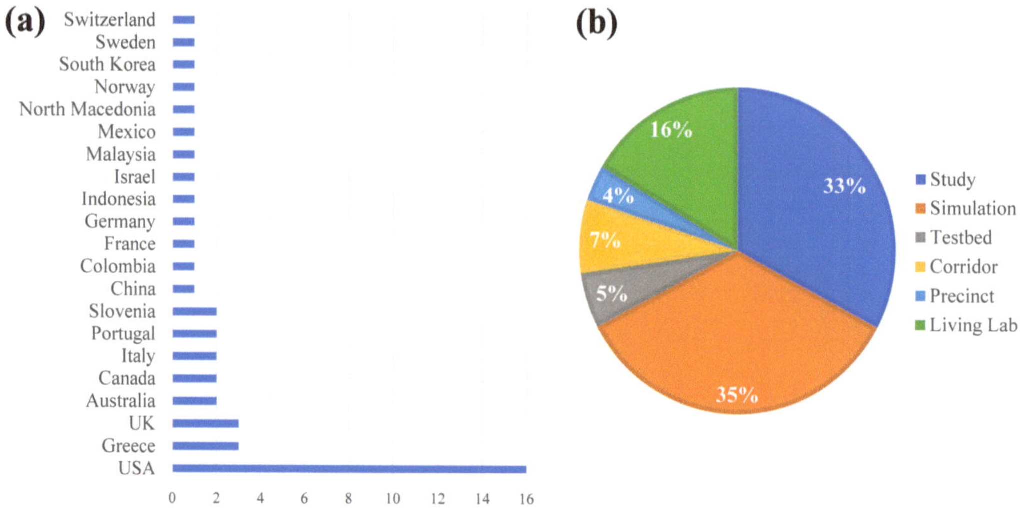

Figure 3. Records' classification for the Scopus database: the bar chart (**a**) shows the number of records found in each country, while the pie chart (**b**) displays the type of urban deployment for each record based on its own criteria.

Regardless of the continent where some cities are located, it is common to find studies on the concept of ULLs or actual deployments of technology to enhance mobility in communication, energy, and infrastructure without yet delving into the real-world implementation of AVs. This is the case for Latin American countries like Mexico and Colombia [1,51]. In Asia, a contrasting situation occurs, where some countries are actively implementing or studying the impact of urban deployments using self-driving technologies to improve mobility [30,52,53], while others use a living lab approach to address problems within the broader concept of a smart city [54]. This situation places China and Singapore as leaders on their continent, but special attention should be given to Japan, where the strong presence of Toyota Corporation and a dedicated institute for research in this area drive urban deployments related to battery-electric vehicles for autonomous mobility [55]. In Oceania, Australia and New Zealand are the representatives of such projects, promoting their adoption with the Australia & New Zealand Driverless Vehicle Initiative (ADVI) [13,24].

Considering that a ULL is an environment with multiple projects taking action in real-life settings, it is important to distinguish that proper implementation should include user involvement at any phase of development, as it aims to learn and experiment alongside the community [50]. In the context of EV and AV deployment, these characteristics rule out the majority of the records found in the database, but they help establish a reference framework for what can truly be considered a ULL with AVs. Describing the characteristics of a ULL can be extensive and vary depending on the projects and strategies of each ULL. Table 2 shows specific projects of the ULL initiatives described in Table 1 to understand the variety of action plans with AVs on public roads and how they serve specific purposes.

Table 2. Projects of AVs on public roads as part of ULL initiatives.

Flagship Project	Space	Service	Benefits for Community	Operating Years
'Smart Circuit and Leanden Leap from Smart Columbus https://smartcolumbus.com/projects/self-driving-shuttles (accessed on 20 August 2024)'	1.6-mile route around the Scioto Mile in downtown Columbus.	First-mile/last-mile mobility and food delivery service.	Service to 16,062 passengers.	2018–2021.

Table 2. *Cont.*

Flagship Project	Space	Service	Benefits for Community	Operating Years
'Area X.O from OVIN https://areaxo.com/smart-cities-and-infrastructure/ (accessed on 20 August 2024)'	1866-acre facility to develop, test, demonstrate, and implement smart city technologies.	Customized test cases and scenarios.	Open-source environment for data collection and Low-Speed Autonomous Shuttle deployment.	2019–present.
'GATEway from Smart Mobility Living Lab https://trl.co.uk/projects/gateway-project (accessed on 20 August 2024)'	3.4 km route around the Greenwich Peninsula.	Automated urban deliveries, remote teleoperation demonstrations, and automated vehicle systems for traveling.	Development and validation of new mobility solutions.	2016–2018.
'Smart Dynamic Area from Modena Automotive Smart Area http://www.automotiveacademy.unimore.it/site/home/masa.html (accessed on 20 August 2024)'	Smart Model Area through the city of Modena along a dedicated area for self-driving experimentation	ADAS testing.	Control and standardization in digital revolution for mobility.	2017–present.
'TRACE from AV Living Lab https://trace-horizon.eu/slovenian-pilot/ (accessed on 20 August 2024)'	11 km of roads.	Urban logistics and last-mile delivery.	Significantly reduce CO_2 emissions, cut down vehicle mileage, and improve the efficiency and reliability of delivery service.	2023–2026.
'Autonomous and Connected Mobility from Aveiro Tech City Living Lab https://www.navya.tech/en/first-deployement-in-portugal-of-a-navya-autonom-shuttle/ (accessed on 20 August 2024)'	Intra-urban route with three stops.	Autonomous shuttle testing using the V2X communication infrastructure of the zone.	Tech week to make citizens take part in technological and cultural activities.	2022.
'Yizhuang Base from Beijing E-Town http://www.beijingetown.com.cn/2023-05/15/c_886139.htm (accessed on 20 August 2024)'	Demonstration zone operated by Beijing Innovation Center for Mobility Intelligent (BICMI).	Testing for unmanned delivery vehicles.	290,000 km and nearly 10,000 h of closed test site.	2023–present.

3.3. Trends and Technological Advancements

Smart mobility implies the introduction of multiple technologies. In developing an autonomous and electric vehicle, several factors play crucial roles. However, to introduce it into an ecosystem where it can be connected and exchange information while interacting with real traffic, more considerations arise. Over the years, specific frameworks and standards have been proposed to indicate which types of technology are related to an automated driving system (ADS) and where the line of advancements is pointing. Nevertheless, it depends on the purpose of each city or district implementing these kinds of initiatives.

In Figure 4, the convergence of multiple technologies in a ULL is highlighted. Specific combinations result in the application of cutting-edge technologies across various sectors. The main idea is to demonstrate how ULLs implementing AV projects depend on the appropriate technological integration of three systems (transport, communication, and infrastructure) that continuously interact during the implementation of services or testing purposes with self-driving technologies. Physical components acting as sensors in the vehicle and in urban spaces have wireless communication with servers that create the roadway digital infrastructure [56].

Regarding transportation technology testing, hardware components are crucial for sustaining innovation in transportation. Their application, along with control algorithms, supports the development of ADSs and intelligent transportation systems (ITSs). The AMD Implementation Catalog incorporates a total of 10 sites, describing them as early deployment districts with AV technology. For instance, the University District AV Transit Circulator in Houston, Texas, and the Regional Transportation District's in Denver, Colorado, explore first-mile/last-mile solutions with an ADS. The Innovation District in Las Vegas, Nevada, explores some of the hardware components, including LiDAR sensors, cameras at the front and rear of the vehicle, inertial motor units (IMUs), and a global positioning system (GPS), providing fixed points of location in all operating sites [10,11]. The Center for Environmental Research and Technology (CE-CERT) has been developing on-board intelligent systems through a dedicated research group focused on innovation in transport, particularly through the application of sensors such as LiDARs, cameras, and radars, as previously mentioned. Additionally, if wireless communication devices are equipped to enable vehicle communication, the result is CAVs [27]. Similarly, the Integrated Transport Research Lab (ITRL) is considered a hub for innovation in intelligent transportation systems, developing large-scale automated platforms and testing the first 5G network in Sweden [25].

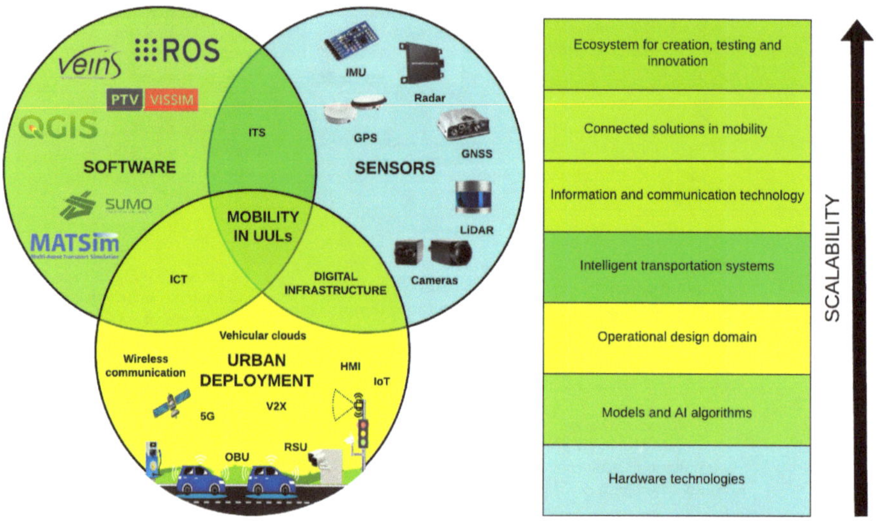

Figure 4. Interaction of technologies to support testing of connected, electric, and autonomous vehicles in a ULL environment.

Digital technologies play a crucial role in the innovation of products and services. Specifically, 5G networks enhance CAV functionality, extend coverage, and provide high-definition georeferencing [57]. A revision with evidence from the Smart Columbus projects identified high-tech and innovations as playing a crucial role in the growth of smart cities. An interesting initiative described as an emerging technology is the Connected Electric Autonomous Vehicle (CEAV), a testbed in specific areas of Ohio to evaluate the adoption of AV technology [58]. In 2019, Smart Columbus published a final report of the CEAV Test Plan, along with another called the CEAV Operational Concept for the Smart Columbus Demonstration Program, identifying real-time data transmission and software capabilities in the vehicles as crucial features to reach level 4 automation. The Smart Nation initiative in Singapore has a general vision of service digitization, but the government has ensured transformation in every possible area, evolving into a digital government, economy, and society [22]. As part of this, Nanyang Technological University developed a Mobility-as-a-Service (MaaS) testbed to provide an integrated mobility solution. This includes not only public transport digitization but also the deployment of an autonomous shuttle service between the university residence halls and the use of dynamic routing technology [59].

The promotion of autonomous EVs goes toward the development of smart cities, directly contributing to smart mobility strategies. Implementing this type of technology in

ULLs requires various procedures and extensive validation before the public can interact with an autonomous EV on public roads. At this point, simulation tools play a crucial role in gaining insights into traffic patterns, travel demand modes, and other mobility contexts of the study area [60]. The simulation of traffic models is essential, as it opens the investigation to any country trying to contribute to the field of AVs' impact. Using the MATSim software, a study was able to evaluate traffic congestion in the Jerusalem Metropolitan Area by introducing shared AVs, making them an attractive option for public transportation in the city [61].

Based on the high-quality requirements to test CAVs, simulation environments serve as a faster means of experimentation prior to a physical testbed implementation if appropriate control algorithms and vehicle dynamics are introduced in the simulation [62]. The Information and Decision Science Lab's Scaled Smart Digital City is a 1:25-scaled robotic testbed to investigate CAVs' impact on safety and transportation efficiency and is capable of replicating real-world traffic conditions [63,64]. It also implements digital simulations to complement the results of its model. Additionally, simulations allow for the interpretation of results related to energy consumption, optimal speed range, and pollutant emissions within the vehicle-to-everything (V2X) proof of concept for CAVs using eco-drive controllers [65,66].

When AVs are mentioned, EVs are often included in the concept, although they are not strictly related. Initiatives aiming for a positive impact on the environment often bring together e-mobility and self-driving technology on a large scale. In a living lab, innovations for a sustainable urban concept vary depending on the mobility approach taken, but ultimately, they serve the goal of emission reduction [67]. For example, initiatives in Nanjing focus on the development of new energy vehicles to reduce the impact of greenhouse gas emissions and improve traffic capacity for efficient road control using high technologies [39]. Europe has been considering commercial customers' requirements for CAV development. The Smart Mobility Living Lab in London offers assistance with research in a real-world testbed and its digital twin, allowing the validation of the performance of physical systems and components in virtual environments, simplifying the process before real-world testing is implemented [4]. Germany is another country characterized by strong projects related to automated mobility in living labs. The Hamburg Electric Autonomous Transportation (HEAT) project developed an autonomous shuttle to transport up to 10 passengers in mixed traffic conditions [36]. Additionally, the RealLabHH in Hamburg's Bergedorf district implemented an on-demand service with shuttle vehicles operating in automated mode within predefined areas. An on-board human–machine interface (HMI) displayed a street map, traffic surroundings, and a control panel [68].

3.4. Support Network

The network of a ULL is characterized by alliances that include universities, the government, and industry partners. For example, Transdev Group, a company operating public transport, has been involved in two different projects with AVs in Florida. May Mobility, a company focusing on autonomous technology, has participated in both the Smart Columbus and RAPID projects in Arlington, Texas, to deploy AVs in those districts [10]. Most of the initiatives in the USA are characterized by strong government support, sometimes with the help of federal departments, but also by universities procuring the latest advancements and conducting in-depth investigations. Notable universities include the University of Houston, the University of Michigan, and the University of Florida, to name a few [11].

The Smart Columbus initiative is supported by the U.S. Department of Transportation. Specifically, in the CEAV project, they relied on EasyMile's product system [11]. A large group of partners and stakeholders are involved in this initiative, with the City of Columbus, Michael Baker International (infrastructure USA-based company), and The Ohio State University playing crucial roles in testing responsibilities [58]. In Riverside, California, the investigation of shared, electric, connected, and automated operation levels, along with its data collection plan, is sustained through the collaboration of various state universities and the U.S. Department of Transportation, with matching funds from various sources [27].

Figure 5 denotes how the integration of multiple stakeholders plays a crucial role in these kinds of initiatives. Academia, industry, public bodies, and users make the execution of projects possible. This type of unique collaboration in Sweden's research labs has enabled the merging of inputs from all sectors [25]. Navya and EasyMile are two vehicle manufacturers that are deeply involved in projects with ULLs in Hamburg, Germany, providing them with low-speed autonomous shuttles [68]. To understand the necessary level of synergy in a ULL, the Smart Mobility Living Lab in London, one of the greatest initiatives around the world, needs government support via the Centre for Connected and Autonomous Vehicles and InnovateUK, along with a consortium including TRL, Cisco, DG Cities, and other partners [4]. Such strong partnering and networking have allowed it to become the most advanced center in CAV technology development.

Initiatives that are part of consolidated national growth plans can be directly managed by the government. Such is the case with the Smart Nation Group, a division of the Singapore government within the Prime Minister's Office and overseen by the Ministry of Communications and Information [22]. Specific projects developed within the Smart Nation vision can work independently and expand their network. For example, the MaaS testbed at Nanyang Technological University required an alliance with JTC and SMRT corporations, as well as mobility partners like Continental and 2getthere, to create a ULL [59]. This is what is called a multi-stakeholder ecosystem. There are other cases where initiatives are driven by international programs to achieve sustainable development goals. One example is SOLUTIONSplus, which works with the UN Environment and the International Energy Agency [39]. However, the ULLs that are part of this initiative can grow independently, depending on local support or additional projects involving self-driving technology; such is the case in Nanjing and Hamburg [68,69].

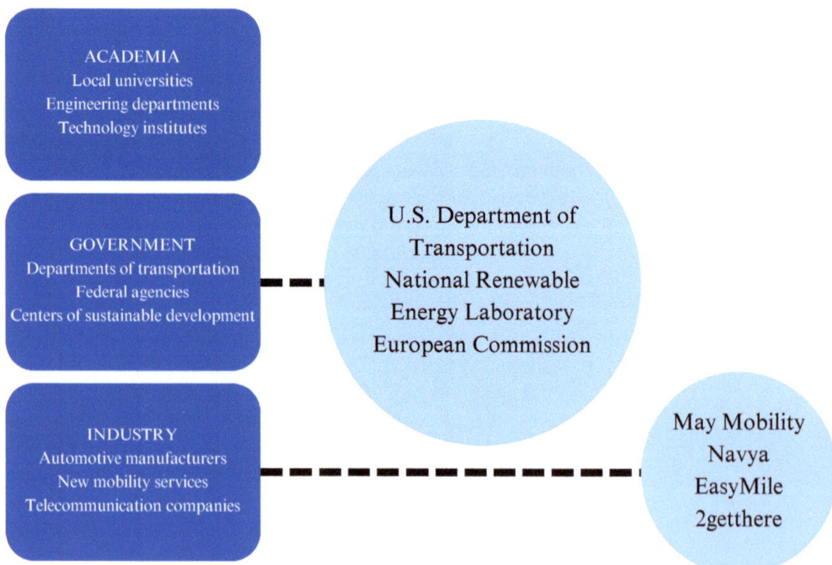

Figure 5. Principal stakeholders found in the literature.

3.5. Contributions

ULLs with AVs have a broad impact on sectors related to smart mobility, providing societal benefits, improvements in vehicle energy consumption, enhanced traffic control, and safer operations. These contributions are evident in any urban deployment of AVs, but the discussion about how the integration of automated mobility solutions within a unified ecosystem like a ULL creates a greater impact remains unexplored. On-road self-driving technologies have proven to be a viable solution for first-mile/last-mile mobility challenges. In general, the public describes the experience positively, and initiatives in Texas, Ohio, and Florida offer more inclusive transportation alternatives, allowing wheelchair users easy access to the vehicles. Additionally, zero-emission strategies using EVs have accumulated millions of miles while the service is active [11].

In Figure 6, it is stated how the level of development for each initiative can potentially impact technology, infrastructure, policies, and society for smart mobility strategies. Independent deployments are referenced as on-road trials of AVs, with Google's Waymo pioneering automated ride-hailing services since 2018 with great success [70]. Vehicle manufacturers, digital mobility companies, and other tech enterprises lead on-road tests to disrupt transport services and support ADS development, bringing cutting-edge technology with vehicle performance at any legally authorized location. However, these efforts are usually associated with motivations centered on constructing new business models in the private sector [13]. The initiatives presented in Table 1 have a positive effect on traffic operations, energy consumption, and safe mobility, with quantitative results demonstrating the significance of each type of urban deployment. Although all have a strong impact on specific issues, a qualitative analysis suggests that the potential of ULLs is greater, as they represent an ecosystem where such initiatives could be part of a supported network as they are continuously implemented, improved, and scaled up.

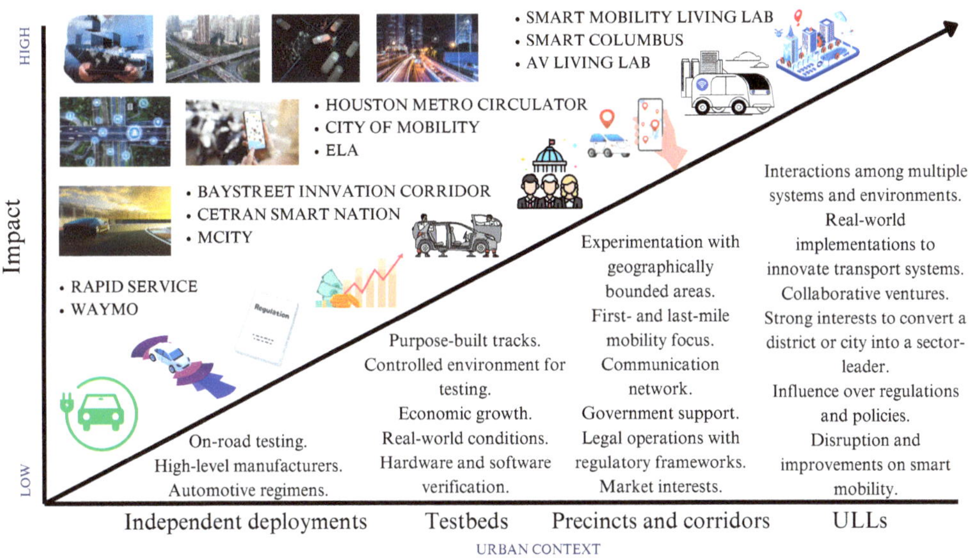

Figure 6. Overall impact on the deployment of electric and autonomous vehicles in different urban contexts.

Diverse cities in the USA are characterized by the presence of AVs in some form of implementation. Waymo One is the first service of autonomous trips providing a sustainable, efficient, and safe option for transportation in San Francisco, Phoenix, Los Angeles, and Austin [14]. San Francisco is a city with high-cost parking areas; a study with data on San Francisco's downtown central business district demonstrated more efficient use of parking spaces and a traffic congestion reduction when AVs performed drop-off and pick-up mobility services in the area [71]. The adoption of AVs for shared mobility services has been envisioned by companies with on-demand transportation services (Uber, Lyft, DiDi) to consider automation in vehicles as a viable option. Nevertheless, a study with data on preferences for shared AV services in New York revealed that ride-pooling options are usually avoided [72].

Testbeds are controlled environments where AVs can be deployed for testing purposes, featuring necessary urban elements to simulate real conditions within closed areas. Although testbeds represent a preliminary stage of implementation before deployment on public roads, research and development play a crucial role in preconfiguring technology for real conditions. These initiatives typically involve collaboration between academic institutions and the automotive industry [13]. The Mcity initiative provides a full-scale outdoor laboratory to test the performance and safety of CAVs, V2X communication, 5G networks, and state-of-the-art instrumentation and sensors for automation. This public–private partnership, led by the University of Michigan, brings innovation to mobility [73]. The Centre of Excellence for Testing & Research of Autonomous Vehicles—NTU (CETRAN) in Singapore has been open since 2017. While it does not develop new technologies, it applies research in collaboration with Nanyang Technological University to ensure the correct operation of ADSs according to international standards [74]. The NTU-JTC-SMRT MaaS Testbed of Smart Nation highlights specific points of success and valuable contributions. Firstly, the digitization of transport allows data to be shared in real time with the government. Secondly, it opens a new market with a proven business model and regulatory readiness. Finally, the ULL approach is centered on users, enabling continuous feedback and adaptations in subsequent iterations [59]. A study in China revealed that gender, age, educational background, income level, and frequency of use are factors in determining the level of acceptance of self-driving technologies [69].

Real-world data collection, simulation studies in shared mobility scenarios, experiments, and early deployments with CAV technologies conducted in Riverside estimated better energy efficiency in ride-hailing services using EVs. They also predicted a prevalence of shared AV fleets of 60% over the use of private cars in scenarios with dominant and high levels of penetration of these technologies [27]. Self-driving technology is not just limited to passenger mobility services: the freight transport system also represents a large scale of operations. A significant number of automated technology implementations in Sweden are related to driverless trucks. Some studies indicate a cost reduction of around 30–40% per ton-kilometer, amounting to SEK billions per year, with the introduction of driverless trucks in logistic hubs [75]. The introduction of CAVs impacts multiple areas, particularly in the emissions of greenhouse gases and other pollutants. Using the California Statewide Travel Demand Model framework, it was found that, in general, CAV penetration could increase vehicle miles traveled. For this reason, it is important to take action in implementing zero-emission vehicle (ZEV) policies. An analysis of dynamic shared-ride systems within an Internet of Things (IoT) network demonstrated that CO_2 emissions could be reduced by 15.16% to 19.13% if one-tenth of the riders adopted ride-sharing services with AVs in Sunway, Subang Jaya, Malaysia [53].

A more extensive implementation occurs in precincts or urban corridors, where similar synergies are sustained but with projects having a longer scope, predominantly focusing on first- and last-mile mobility solutions in public areas such as business parks, university campuses, or other geographically bounded urban zones [13]. Regulations and policies play crucial roles in the successful real-world implementation of AVs in precinct tests and urban corridors. Government agencies are involved in ensuring legal operations and providing resources for logistics, infrastructure, and funding. The ELA project has deployed up to 11 autonomous shuttle services across Canada, each varying in route, passenger capacity, and operational days. These deployments have significantly advanced autonomous shuttle technology, supported by various universities and government departments of transportation in each city [37]. The Innovation Corridor in Jacksonville has implemented projects to enhance IoT devices' value, expand mobility options with autonomous shuttles, and develop smart infrastructure, succeeding through partnerships among academia, industry, and government [76].

ULLs implementing AVs enable testing across various self-driving systems, offering comprehensive technical, legal, and societal solutions. This collaborative venture resembles precincts or urban corridors but operates at a higher level, transforming transportation systems and services to promote economic growth through disruptive innovations in mobility [13]. A ULL functioning with AVs integrates built environments and real-world trials of mobility services across diverse areas within the same district, representing the pinnacle of urban development and a precursor to city-wide implementation. This vision aligns with initiatives like Smart Columbus, whose diverse projects, applications, and programs shape future modes of transportation and public acceptance of smart mobility [3]. The Smart Mobility Lab in London has played a significant role in advancing CAV trials by establishing a safety case framework. Close collaboration with the government facilitates policy and legislation development, underscoring public and private sector interests in enhancing mobility [4].

The operation of Smart Columbus impacts the community by enhancing human services. Principal examples include enabling technologies for trip and mobility assistance with AVs and managing a food delivery service [58]. On the other hand, Europe has the Rivium Business Park and the Central District in Ijssel, Netherlands, where they are developing cutting-edge vehicle technology capable of moving up to 22 passengers using completely automated systems in mixed traffic conditions [10]. It is common to find that some applications and studies of AVs are about automation in delivery services and dispatching operations, supported by an increase in efficiency in last-mile operations with a reduction in waiting times [77].

The diversity of projects in London can be attributed to its Smart Mobility Living Lab, which is built for both the private and public sectors. This has resulted in collaborative projects like GATEway, ENDEAVOUR, ATLAS, and DRIVEN [4]. Considering urban development, intelligent transportation systems could meet the demands for transport and logistic services. However, stakeholders should understand the requirements in a simplified manner to grasp the global technological challenges in the urbanization, automation, and digitization of mobility services [78].

3.6. Challenges and Lessons Learned

AV adoption is complicated due to the numerous factors involved, but identifying socio-demographic variables and travel behaviors provides valuable information to delineate urban planning and policy adoption in real-world experimental settings. Long-distance trips are demonstrated to be less likely if AVs are adopted [79]. Figure 7 illustrates important factors in creating a ULL ecosystem for future smart mobility using shared, electric, connected, and automated vehicles. It shows that this process involves an integrated approach to specific issues that must be addressed to ensure cleaner, safer, and more efficient mobility, as well as gaining acceptance within the district where it is being developed.

The operational design domain (ODD) plays a crucial role in ADSs since it allows vehicles to operate autonomously in a designated area. In this context, multiple conditions can limit the maneuvers and features of the vehicles. Common causes of problems related to the ODD include mixed traffic flow, extreme weather conditions like heavy rainfall, and system constraints in detecting objects [10]. In the Netherlands, to achieve complete AV operation, Regional Transportation Districts rely on smart traffic signals [11].

During the testing of self-driving shuttles in the Smart Circuit of downtown Columbus, researchers encountered problems maintaining passenger safety conditions in the vehicle and other limitations in operational concept development [11]. However, the project found an alternative use for the AVs in the Linden community, providing an autonomous food delivery service for low-income families [58]. Studies from Riverside developed an innovative modeling suite framework to quantify the impact of new mobility technologies on traffic operations. While this framework is only useful for constrained system settings, it serves as an example of initiatives promoting CAV deployment in a sustainable energy direction, increasing confidence in this technology, and supporting further in-depth studies in distinct real-world applications [27].

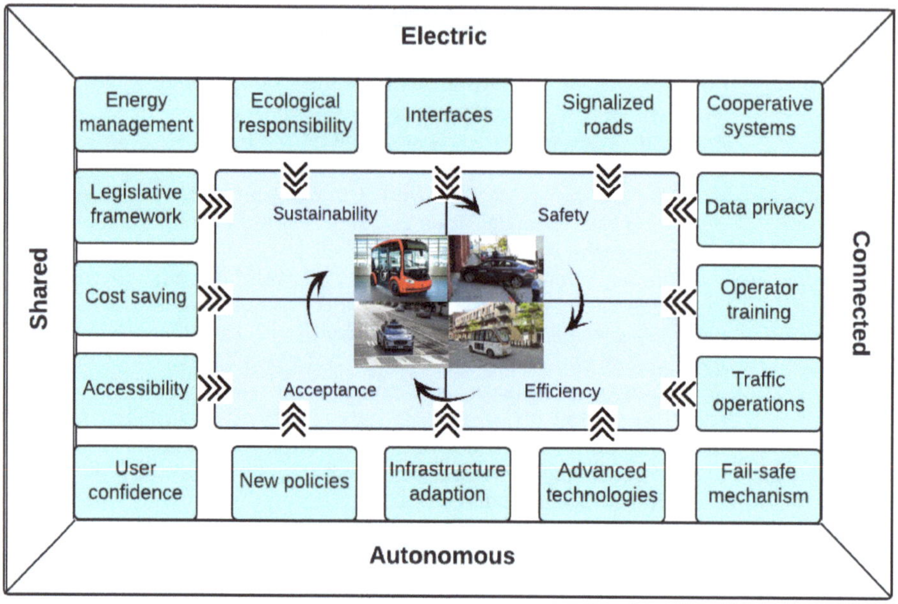

Figure 7. Important factors to address for the successful testing deployment of shared, electric, connected, and autonomous vehicles with positive effects on the pillars of the ULL ecosystem. All images from Wikimedia Commons, licensed under CC BY 4.0.

A project summary published in 2018 by the ITRL discussed the impact of AVs on transport systems, indicating that automation leads to an increase in vehicle kilometers traveled with reduced marginal costs. This can translate to greater energy consumption and higher emissions. However, to avoid this sustainability imbalance, policy instruments act as a regulating tool [75]. On the other hand, regulations should cover a wide range of considerations. In terms of safety, qualified operators and supervised interactions with AVs are essential to ensuring a safe experience. The operation of autonomous shuttles in Hamburg in 2021 provided information about the operators' perspectives on these technologies. The study concluded that a multimodal approach is necessary, incorporating auditory signals, visual cues, and accessible buttons with vehicle mechanism feedback [68].

Every initiative can gain different insights from CAV technology deployed in ULLs, but the roles and responsibilities in these kinds of services are among the most valuable aspects. AVs are an innovative concept that requires a controlled environment, continuous monitoring, a well-trained team to operate the vehicles, maintenance at charging or mobility stations, and safety marshals during operation; this has been shown to increase user confidence and comfort [59]. It is a fact that AVs require greater public acceptance, as society currently has a low level of trust in self-driving technologies. As secure and practical research on AVs continues to emerge, social acceptance should broaden [69]. Public acceptance of automation in transport may be low in some countries, even when opinions about AVs and their technological development are positive. This underscores the importance of a legislative framework and the adaptation of laws to support broad deployments and increase public awareness of the benefits of AVs [80].

Information and communications technology (ICT) is one of the most relevant aspects to implement in urban connected environments with an ITS, a key configuration in living labs using CAVs. In contrast to Europe, Asia, and North America, African and Latin American countries do not present active ULL initiatives using AVs. However, in Popayan, a medium-sized city in Colombia, a fleet management and control system was developed and tested to identify the appropriate requirements, technological components, and communication protocols in the ITS architecture for improvements in their transit services [51]. Connected vehicle algorithms and their evaluation have been proven to improve safety and mobility efficiency in corridors with signalized intersections [81]. Similarly, the Campus City project in Monterrey, Mexico, presents an analysis of an ecosystem of innovation within a pedagogic framework. It highlights the relevance of stakeholders to provide a first approximation of energy requirements and explains how a living lab relies on advanced data, information, and telecommunication systems to offer smart mobility solutions [1].

Consciousness about the increasing popularity of AVs is a reality in countries in the Middle East. In China, a case study of larger-size instances like districts to optimize vehicle routing algorithms has led to cost-saving solutions for the operation of shared autonomous electric vehicles (SAEVs) [52].

Numerous large-scale projects have been developed in the UK for CAV-related infrastructure development since 2014, emphasizing V2X systems and traffic control without an in-depth exploration of the concept of vehicular clouds. Oxford is characterized by the presence of CAVs, which led to a case study in this city to develop design principles in physical and virtual infrastructure [82]. Its application could expand ITS strategies and help manage parking facilities to successfully introduce CAVs into current urban environments. MERGE Greenwich and GATEWay projects are excellent examples of initiatives in the UK contributing to building this environment in the Smart Mobility Living Lab in London, where IoT constitutes the ecosystem communication to transform the infrastructures and materials of mobility [13].

4. Discussion

The key challenges in integrating various technologies to support the testing of connected, electric, and autonomous vehicles in ULL environments are related to the factors identified in this study. In particular, the interaction of CAVs with other actors in the urban environment is currently a crucial factor. Indeed, the various technologies developed, implemented, and adopted aim to control the interaction of all devices inside and outside CAVs. In most cases, especially in a ULL environment, where there is high interaction of CAVs with other road actors, it is necessary to use appropriate infrastructure to effectively develop an approach to future smart mobility. CAVs need seamless interaction among various technologies, such as vehicle-to-vehicle (V2V), vehicle-to-infrastructure (V2I), or vehicle-to-everything (V2X) communication. These systems need to work with existing road infrastructure, such as traffic lights, sensors, and public transportation systems. In addition, the use of dedicated lanes for AVs and EVs in key zones can enable more effective testing. ICT, IoT, and ITS are the most useful technologies, especially in view of developing multimodal mobility services, which, with the integration of CAVs, will enable levels of service for citizens that are unthinkable today. However, these technologies need further development to ensure higher levels of performance. Most of these technologies are developed autonomously, so it will be necessary to use a systemic and organic approach that can integrate and manage them. This is why it is important to introduce an unambiguous taxonomy and international standards for technologies and services to be adopted. Thus, the development of infrastructures able to support CAVs in a ULL environment will have to consider not only the technical aspect but also the need for the standardization and cost sustainability of these technologies and services. In particular, open standards for data exchange and communication protocols to ensure the compatibility of various technologies are required.

The scalability of initiatives using AVs in real-world conditions is a complex process. Investigations are required to evaluate the impact of self-driving technologies functioning in connected environments. It is common to find studies proposing agent-based simulations to evaluate the effect of AVs' introduction in urban environments [83]. These investigations enable the use of real case scenarios, and the results can vary depending on the variables selected for the agents, such as the mode of transportation, the length of trips, traffic conditions, and demand, among other characteristics of mobility [84]. The ability to control the dynamics of the simulation and apply a variety of scenarios with the desired characteristics for each city of interest allows for gaining insights into a wide range of implications, such as the operational efficiency of fleets, economic advantages, the impact on traffic congestion, and energy consumption [85,86]. This approach could also lead to real-world deployments that have already validated their proposals and impacts with existing urban data, facilitating the cooperation of public-sector transportation planners and policymakers. Nonetheless, it is more common to find these studies limited to simulation models.

The number of studies contributing to investigations regarding smart mobility is extensive, but focusing on autonomous, connected, and digital technologies reduces the number of initiatives to be included. It is important to highlight that developed countries in Europe are constantly supporting the adoption of future and clean mobility. Some studies may even focus on new street types to improve public transport efficiency, leading to low-carbon and low-speed multimodal solutions [87]. It is also evident that converting a study into a contribution with real-world testing is viable since the artificial intelligence (AI) algorithms and systems developed could be part of the chain to make a living laboratory platform with smart energy and mobility solutions function [88]. Nevertheless, even when investigations contribute to sustainable and intelligent mobility, some limitations prevent the scope from ever reaching an actual AV deployment on public roads. The main difference between precincts, urban corridors, and ULLs, all considered large-scale deployments of self-driving vehicle implementations, is that government departments are involved from the beginning of the project activation [89], or the initiatives are aligned with long-term funding programs that ensure their continuity [5]. In either case, it is evident that the government plays a crucial role in the policies and strategies that shape the direction of research and innovation in smart mobility interventions.

Even though the scope of some studies found in Scopus is limited to simulations and structural equation modeling for AV operation or impact research purposes [90,91], they are still included in the bibliography because their findings are related to the Horizon Europe program or previous editions, providing an assessment framework for future research and deployments. In other cases, it is interesting to see how specific initiatives supported by EU funding, such as the Ride-to-Autonomy project, are reported as scalable models for integrating autonomous shuttle solutions into public transport systems during designated periods, with pilot programs conducted across various European cities [92]. This enables an initial approach to future smart mobility strategies, serving as a basis for developing subsequent large-scale deployments such as ULLs [6,44,93] found in the literature.

For cases in USA cities, a situation similar to the European state of development can be observed; while industries and agencies sum up the AV tendency, urban areas present some type of on-road deployment with self-driving technologies. In this way, travel demand models and simulation scenarios with different levels of AV adoption serve as tools to evaluate the benefits of its use and predict how mobility could be reconstructed with more equity in metropolitan areas [94]. Data fusion models also contribute to investigations since they allow traffic signal control evaluation from a safety perspective, reducing collision risk for CAVs when implemented [95]. Even when research is limited to simulations, contributions based on modeling frameworks are oriented to corridors and street network implementations. Contributions in the design of a system architecture for V2X communication for connected corridors as a testbed allow the system's reliability and scalability so that it can also be applicable to similar urban deployment in other cities [96]. In addition, 5G networks are a crucial factor supporting the deployment of AVs, as they mediate the applications involved in service migration, as well as the security and privacy of data [97].

The urban deployment of self-driving technologies is often related to a CAV network, which refers to the operation of CAVs over a road network, corridor, or segment. The development of control algorithms in this context is what ensures safe operations in their movements [98]. The approach taken by institutions and centers conducting active research and development on AI algorithms typically involves simulations that consider real-world data and the characteristics of the network where they are hypothetically implemented.

Testbeds offer a viable option to validate the functionality of control systems, depending on the resources available for each investigation. Algorithms, communication systems, and any other technology related to the operation of CAVs can be tested in a reduced-scale environment [53,64] or within a specific spatial scope that replicates the urban conditions of a city [73,74]. It is important to recognize that successful urban deployment is only achieved when the CAV network operates within a multi-agent context, interacting seamlessly with other vehicles, users, and objects on the road [30,35]. Precincts and corridors that offer mobility services or provide other transportation solutions in urban areas are also included in this scope. A ULL does not necessarily refer to the next step after successfully implementing a CAV network but rather to an environment capable of encompassing all the previously discussed points to facilitate innovation and sustainable mobility [3–5].

Beyond the steps needed to achieve the urban deployment of AVs in public spaces, the core of ULL initiatives lies in addressing sustainability challenges to co-design the future of mobility [99]. Success will depend on the effective organization and prioritization of stakeholders' needs. Collaboration between cities and technology companies is another success factor in ensuring the implementation of these testing environments for future smart mobility solutions. A strong commitment by cities to the development of innovative mobility with the use of AVs is the main aspect that fosters collaboration with technology companies. Based on this aspect, data sharing (preserving data privacy and security), assisting companies in authorization and permit procedures, and the propensity for public investments in infrastructures and technologies useful for urban mobility, as well as forms of co-participation in public services, such as public–private procurement, are significant factors.

In this context, scalability can only occur when an ideal state is reached, offering tangible benefits for the involved parties and an appropriate design of strategies beyond testing purposes. Stakeholders will end up taking multiple roles in the management of the CAV supply chain as the adoption of these technologies increases in public transport. However, special attention should be given to preserving data privacy, along with all the adaptations in regulations and policies that this aspect involves [100]. The deployment of automated mobility strategies represents a challenge in a wide range of issues, but connectivity and data transmission still represent the principal obstacles to achieving the full development of smart infrastructure, where all the information coming from the mobility ecosystem is processed and interpreted successfully [101]. It has been demonstrated how the ULL plays a role in the introduction of innovations in mobility with AVs for the community, providing insights into transportation habits, public readiness, and acceptance to help policymakers create a safe and secure environment [102]. Considering that agenda-setting and policymakers need regulation-oriented measurements to ensure a safe and shared transition in automated mobility [103], ULLs offer a controlled urban space where this information can be obtained, adapted, and gradually expanded.

Indeed, regulatory bodies play a key role in ensuring the safe and effective use of AVs in test environments for future smart mobility solutions, as they clearly indicate which features and aspects need to be considered. They can influence different fields with specific indicator values that are necessary to achieve for future smart mobility solutions with AVs, such as defining the characteristics of AVs and the approval homologation process to be authorized for use on public roads; defining the infrastructural and environmental conditions for test implementations on public roads; defining harmonized authorization procedures for testing implementation; defining procedures in case of emergency events (passage of ambulances, police, etc.); defining administrative activities (insurance, etc.) required for test implementation; and defining priority rules with respect to other road users. In this vision, the European Commission adopted the first worldwide legislation (EU Regulation 2022/1426) concerning the conditions for the type-approval of Automated Driving Systems of fully automated vehicles to become the first market with a complete and unambiguous legislative framework.

5. Conclusions

The creation of a ULL is a complex process that can only be achieved through alliances and cooperation among stakeholders. Focusing on the development of ITSs and ADSs requires gradual steps to ensure scalability without issues. The strong presence of such projects and initiatives in the EU and USA highlights the need for grants and well-established programs to support sustainable solutions for the development of smart mobility in other parts of the world. In developing countries, the lack of regulations and adequate infrastructures must be overcome to achieve the necessary technological advancements for smart urban mobility.

In ULLs using AVs and EVs in public spaces, the necessary sensors, infrastructure, and ICT for safe deployments are revealed. Depending on the focus of each initiative, the simulations and studies presented may concentrate on assessing the impact of AV technology on mobility. However, it has been demonstrated that the use of software for microscopic simulations can lead to real-world deployments of electric, connected, and autonomous vehicles, supported by insights gained from these investigations into traffic operations, energy consumption, and travel demand preferences.

Universities lead research and develop strategies for AVs, while industrial partners enable the involvement of various private sectors to bring that technology to life. The government acts as the principal regulator, facilitating the legal use of this technology in urban areas. This collaboration is what distinguishes developing a model from actually having it functioning on the streets.

In summary, to implement a ULL for the innovation, testing, and scalability of AVs in mobility, it is necessary to have a diversified network of stakeholders covering the needs for regulatory frameworks, applied research, community engagement, and technological development under a sustainable vision. Once the models and control algorithms for ADSs have reached sufficient maturity through closed testing and simulations, the adaptation of urban infrastructure to create a connected environment where data transmission is successfully achieved could be considered. However, this is only possible when legal authorization for such operations is granted. Once again, the level of development achieved is restricted by the stakeholder network and how effectively they cooperate to enable experimentation

Author Contributions: Conceptualization, A.A.M.-A., L.C.F.-H., B.S., A.C.V.-S., M.R. and J.d.J.L.-S.; formal analysis, A.A.M.-A., L.C.F.-H., B.S. and M.R.; investigation, A.A.M.-A.; methodology, A.A.M.-A., J.C.T.-M. and J.d.J.L.-S.; project administration, M.R., A.C.V.-S. and J.d.J.L.-S.; resources, J.d.J.L.-S.; software, A.A.M.-A.; supervision, L.C.F.-H., B.S. and M.R.; validation, L.C.F.-H. and M.R.; visualization, A.C.V.-S., B.S. and J.d.J.L.-S.; data curation, A.A.M.-A.; writing—original draft, A.A.M.-A.; writing—review and editing, A.A.M.-A., L.C.F.-H., B.S., J.C.T.-M., M.R. and J.d.J.L.-S. All authors have read and agreed to the published version of the manuscript.

Funding: The research was funded by the CampusCity initiative from Tecnológico de Monterrey.

Acknowledgments: The authors would like to acknowledge the Conscious Technologies research group, the CampusCity initiative at Tecnológico de Monterrey, and the collaboration with Politecnico di Bari for their support in the development of this work.

Conflicts of Interest: The authors declare no conflicts of interest.

Abbreviations

The following abbreviations are used in this manuscript:

ULL	Urban living lab
AV	Autonomous vehicle
CAV	Connected and autonomous vehicle
AMD	Automated Mobility District
NREL	National Renewable Laboratory
AMoD	Autonomous mobility on demand
EU	European Union
CCAM	Cooperative, Connected, and Automated Mobility

ADS	Automated driving system
ITS	Intelligent transportation system
IMU	Inertial motor unit
GPS	Global positioning system
CEAV	Connected Electric Autonomous Vehicle
MaaS	Mobility-as-a-Service
AEV	Autonomous electric vehicles
SAV	Shared autonomous vehicles
V2X	Vehicle-to-everything
HMI	Human–machine interface
ZEV	Zero-emission vehicle
IoT	Internet of Things
ODD	Operational design domain
ICT	Information and communications technology
SAEV	Shared autonomous electric vehicles
AI	Artificial intelligence

References

1. Huertas, J.I.; Mahlknecht, J.; Lozoya-Santos, J.d.J.; Uribe, S.; López-Guajardo, E.A.; Ramirez-Mendoza, R.A. Campus city project: Challenge living lab for smart cities. *Appl. Sci.* **2021**, *11*, 11085.
2. Bulkeley, H.; Coenen, L.; Frantzeskaki, N.; Hartmann, C.; Kronsell, A.; Mai, L.; Marvin, S.; McCormick, K.; van Steenbergen, F.; Palgan, Y.V. Urban living labs: Governing urban sustainability transitions. *Curr. Opin. Environ. Sustain.* **2016**, *22*, 13–17.
3. Smart Columbus. 2024. Available online: https://smartcolumbus.com/ (accessed on 4 July 2024).
4. Smart Mobility Living Lab: London. 2020. Available online: https://smartmobility.london/ (accessed on 21 June 2024).
5. AV Living Lab. 2024. Available online: https://avlivinglab.com/ (accessed on 21 June 2024).
6. Aveiro Tech City. 2024. Available online: https://www.aveirotechcity.pt/en (accessed on 21 June 2024).
7. Huertas, J.I.; Stöffler, S.; Fernández, T.; García, X.; Castañeda, R.; Serrano-Guevara, O.; Mogro, A.E.; Alvarado, D.A. Methodology to assess sustainable mobility in LATAM cities. *Appl. Sci.* **2021**, *11*, 9592.
8. Wolniak, R. Smart mobility in a smart city concept. *Silesian Univ. Technol. Sci. Pap. Organ. Manag. Ser.* **2023**, *170*, 679–692.
9. Simonofski, A.; Handekyn, P.; Vandennieuwenborg, C.; Wautelet, Y.; Snoeck, M. Smart mobility projects: Towards the formalization of a policy-making lifecycle. *Land Use Policy* **2023**, *125*, 106474.
10. Young, S.; Lott, J.S. *The Automated Mobility District Implementation Catalog: Insights from Ten Early-Stage Deployments*; Technical Report; National Renewable Energy Lab. (NREL): Golden, CO, USA, 2020.
11. Young, S.; Lott, J.S. *The Automated Mobility District Implementation Catalog: Safe and Efficient Automated Vehicle Fleet Operations for Public Mobility*; Technical Report; National Renewable Energy Lab. (NREL): Golden, CO, USA, 2022.
12. SAE International. SAE J3016 Levels of Driving Automation: Updated for Clarity. 2024. Available online: https://www.sae.org/blog/sae-j3016-update (accessed on 15 July 2024).
13. Dowling, R.; McGuirk, P. Autonomous vehicle experiments and the city. *Urban Geogr.* **2022**, *43*, 409–426.
14. Waymo LLC. Waymo. 2024. Available online: https://waymo.com (accessed on 4 July 2024).
15. Pundir, A.; Singh, S.; Kumar, M.; Bafila, A.; Saxena, G.J. Cyber-physical systems enabled transport networks in smart cities: Challenges and enabling technologies of the new mobility era. *IEEE Access* **2022**, *10*, 16350–16364.
16. Cellina, F.; Castri, R.; Simão, J.V.; Granato, P. Co-creating app-based policy measures for mobility behavior change: A trigger for novel governance practices at the urban level. *Sustain. Cities Soc.* **2020**, *53*, 101911.
17. Page, M.J.; McKenzie, J.E.; Bossuyt, P.M.; Boutron, I.; Hoffmann, T.C.; Mulrow, C.D.; Shamseer, L.; Tetzlaff, J.M.; Akl, E.A.; Brennan, S.E.; et al. The PRISMA 2020 statement: An updated guideline for reporting systematic reviews. *BMJ* **2021**, *372*, n71.
18. Tamakloe, R.; Park, D. Discovering latent topics and trends in autonomous vehicle-related research: A structural topic modelling approach. *Transp. Policy* **2023**, *139*, 1–20.
19. Chocholáč, J.; Kučera, T.; Sommerauerová, D.; Hruška, R.; Machalík, S.; Křupka, J.; Hyršlová, J. Smart city and urban logistics-research trends and challenges: Systematic literature review. *Komunikácie* **2023**, *25*, 175–192.
20. Arvianto, A.; Sopha, B.M.; Asih, A.M.S.; Imron, M.A. City logistics challenges and innovative solutions in developed and developing economies: A systematic literature review. *Int. J. Eng. Bus. Manag.* **2021**, *13*, 18479790211039723.
21. Hasan, M.H.; Van Hentenryck, P. The benefits of autonomous vehicles for community-based trip sharing. *Transp. Res. Part C Emerg. Technol.* **2021**, *124*, 102929.
22. Smart Nation Singapore. 2024. Available online: https://www.smartnation.gov.sg/ (accessed on 21 June 2024).
23. Ponnaluri, R.; Heery, F.; Tillander, V.Y. The Florida connected and automated vehicle initiative: A focus on deployment. *ITE J.* **2017**, *87*, 33–41.
24. Legacy, C.; Ashmore, D.; Scheurer, J.; Stone, J.; Curtis, C. Planning the driverless city. *Transp. Rev.* **2019**, *39*, 84–102.
25. Integrated Transport Research Lab (ITRL). 2024. Available online: https://www.itrl.kth.se/ (accessed on 20 June 2024)

26. Dennis, S.; Paz, A.; Yigitcanlar, T. Perceptions and attitudes towards the deployment of autonomous and connected vehicles Insights from Las Vegas, Nevada. *J. Urban Technol.* **2021**, *28*, 75–95.
27. Barth, M.; Hao, P.; Wu, G.; Tanvir, S.; Wang, C.; Gonder, J.; Holden, J.; Devall, A.; Sun, B. *Evaluating Energy Efficiency Opportunities from Connected and Automated Vehicle Deployments Coupled with Shared Mobility in California*; Technical Report; University of California-Riverside: Riverside, CA, USA, 2020.
28. Omidvar, A.; Letter, C.; Elefteriadou, L. University of Florida Advanced Technologies Campus Testbed. 2017. Available online https://rosap.ntl.bts.gov/view/dot/35092 (accessed on 20 June 2024).
29. Smith, S. Smart infrastructure for future urban mobility. *AI Mag.* **2020**, *41*, 5–18.
30. Choi, S.; Lee, D.; Kim, S.; Tak, S. Framework for connected and automated bus rapid transit with sectionalized speed guidance based on deep reinforcement learning: Field test in Sejong city. *Transp. Res. Part C Emerg. Technol.* **2023**, *148*, 104049.
31. Smart City Korea. 2024. Available online: https://smartcity.go.kr/en/ (accessed on 15 July 2024).
32. Alotaibi, E.T.; Alhuzaymi, T.M.; Herrmann, J.M. Autonomous mobility on demand: From case studies to standardized evaluation *Front. Future Transp.* **2023**, *4*, 1224322.
33. METRO Houston. METRO Office of Innovation. Available online: https://www.ridemetro.org/about/who-we-are/innovation (accessed on 4 July 2024).
34. Haque, A.M.; Brakewood, C. A synthesis and comparison of American automated shuttle pilot projects. *Case Stud. Transp. Policy* **2020**, *8*, 928–937.
35. Nesheli, M.M.; Li, L.; Palm, M.; Shalaby, A. Driverless shuttle pilots: Lessons for automated transit technology deployment. *Case Stud. Transp. Policy* **2021**, *9*, 723–742.
36. Hamburg.com. Mobility in Hamburg. 2024. Available online: https://www.hamburg.com/mobility/ (accessed on 20 June 2024)
37. Pacific Western Transportation. ELA—The Future of Transportation. 2018. Available online: https://www.ridewithela.ca/ (accessed on 4 July 2024).
38. Traffic Operations and Safety Laboratory (TOPS Lab). 2024. Available online: https://topslab.wisc.edu/ (accessed on 21 June 2024).
39. SOLUTIONSplus. Demonstration City: Nanjing. 2024. Available online: https://www.solutionsplus.eu/nanjing (accessed on 21 June 2024).
40. Boersma, R.; Mica, D.; Arem, B.; Rieck, F. Driverless electric vehicles at Businesspark Rivium near Rotterdam (the Netherlands): From operation on dedicated track since 2005 to public roads in 2020. In Proceedings of the 31st International Electric Vehicle Symposium and Exhibition, EVS 2018 and International Electric Vehicle Technology Conference 2018, EVTeC 2018, Kobe, Japan, 1–3 October 2018.
41. Darwin Innovation Group. 2024. Available online: https://darwincav.com/ (accessed on 21 June 2024).
42. Mordue, G.; Karmally, D. Frontier technologies in Non-Core automotive regions: Autonomous vehicle R&D in Canada. *Can. Public Policy* **2020**, *46*, 73–93.
43. OVIN Hub. 2024. Available online: https://www.ovinhub.ca/ (accessed on 9 July 2024).
44. Rito, P.; Almeida, A.; Figueiredo, A.; Gomes, C.; Teixeira, P.; Rosmaninho, R.; Lopes, R.; Dias, D.; Vítor, G.; Perna, G.; et al. Aveiro Tech City Living Lab: A communication, sensing, and computing platform for city environments. *IEEE Internet Things J.* **2023**, *10*, 13489–13510.
45. Beijing E-Town. 2024. Available online: http://www.beijingetown.com.cn/ (accessed on 15 July 2024).
46. Leali, F.; Pasquale, F. The Living Lab for Autonomous Driving as Applied Research of MaaS Models in the Smart City: The Case Study of MASA—Modena Automotive Smart Area. In Proceedings of the International Conference on Technological Imagination in the Green and Digital Transition, Rome, Italy, 30 June–2 July 2022; Springer: Berlin/Heidelberg, Germany, 2022; pp. 273–284.
47. Zajc, I.; Sernec, R.; Lenart, G.; Pucihar, A. Autonomous mobility and user perception: A case of city as a lab in Slovenia. In Proceedings of the 2020 43rd International Convention on Information, Communication and Electronic Technology (MIPRO), Opatija, Croatia, 28 September–2 October 2020; IEEE: Piscataway, NJ, USA, 2020; pp. 1843–1847.
48. European Commission. Horizon Europe. 2024. Available online: https://research-and-innovation.ec.europa.eu/funding/funding-opportunities/funding-programmes-and-open-calls/horizon-europe_en (accessed on 21 June 2024).
49. Connected Automated Driving. 2024. Available online: https://www.connectedautomateddriving.eu/ (accessed on 21 June 2024).
50. Steen, K.; Van Bueren, E. The defining characteristics of urban living labs. *Technol. Innov. Manag. Rev.* **2017**, *7*, 21–33.
51. Rojas, B.; Bolanos, C.; Salazar-Cabrera, R.; Ramírez-González, G.; Pachon de la Cruz, A.; Madrid Molina, J.M. Fleet management and control system for medium-sized cities based in intelligent transportation systems: From review to proposal in a city. *Electronics* **2020**, *9*, 1383.
52. Ma, B.; Hu, D.; Wang, Y.; Sun, Q.; He, L.; Chen, X. Time-dependent vehicle routing problem with departure time and speed optimization for shared autonomous electric vehicle service. *Appl. Math. Model.* **2023**, *113*, 333–357.
53. Bakibillah, A.S.M.; Paw, Y.F.; Kamal, M.A.S.; Susilawati, S.; Tan, C.P. An Incentive Based Dynamic Ride-Sharing System for Smart Cities. *Smart Cities* **2021**, *4*, 532–547.
54. Firmansyah, H.S.; Supangkat, S.H. Smart world living lab: A living lab approach to improve smart city implementation (introducing the DDG areas as an integrated living lab). In Proceedings of the 2022 IEEE International Smart Cities Conference (ISC2), Paphos, Cyprus, 26–29 September 2022; IEEE: Piscataway, NJ, USA, 2022; pp. 1–6.

55. Nieoczym, A.; Caban, J.; Dudziak, A.; Stoma, M. Autonomous vans-the planning process of transport tasks. *Open Eng.* **2020**, *10*, 18–25.
56. Khan, S.M.; Chowdhury, M.; Morris, E.A.; Deka, L. Synergizing roadway infrastructure investment with digital infrastructure for infrastructure-based connected vehicle applications: Review of current status and future directions. *J. Infrastruct. Syst.* **2019**, *25*, 03119001.
57. Rizopoulos, D.; Laskari, M.; Kouloumbis, G.; Fergadiotou, I.; Durkin, P.; Kaare, K.K.; Alam, M.M. 5G as an Enabler of Connected-and-Automated Mobility in European Cross-Border Corridors—A Market Assessment. *Sustainability* **2022**, *14*, 14411.
58. Chen, Z.; Cheng, J. Economic Impact of Smart City Investment: Evidence from the Smart Columbus Projects. *J. Plan. Educ. Res.* **2024**, *44*, 1881–1897.
59. Jin, Z.R.; Qiu, A.Z. Mobility-as-a-Service (MaaS) testbed as an integrated approach for new mobility-A living lab case study in Singapore. In Proceedings of the HCI in Mobility, Transport, and Automotive Systems: First International Conference, MobiTAS 2019, Held as Part of the 21st HCI International Conference, HCII 2019, Orlando, FL, USA, 26–31 July 2019; Proceedings 21; Springer: Berlin/Heidelberg, Germany, 2019; pp. 441–458.
60. Saha, B.; Fatmi, M.R. Simulating the impacts of hybrid campus and autonomous electric vehicles as GHG mitigation strategies: A case study for a mid-size Canadian post-secondary school. *Sustainability* **2021**, *13*, 12501.
61. Ben-Dor, G.; Ogulenko, A.; Klein, I.; Benenson, I. Modal shift and shared automated demand-responsive transport: A case study of Jerusalem. *Procedia Comput. Sci.* **2022**, *201*, 581–586.
62. Zayas, R.M.; Beaver, L.E.; Chalaki, B.; Bang, H.; Malikopoulos, A.A. A digital smart city for emerging mobility systems. In Proceedings of the 2022 IEEE 2nd International Conference on Digital Twins and Parallel Intelligence (DTPI), Boston, MA, USA, 26 October–13 November 2022; IEEE: Piscataway, NJ, USA, 2022; pp. 1–6.
63. Chalaki, B.; Beaver, L.E.; Mahbub, A.I.; Bang, H.; Malikopoulos, A.A. A research and educational robotic testbed for real-time control of emerging mobility systems: From theory to scaled experiments [applications of control]. *IEEE Control Syst. Mag.* **2022**, *42*, 20–34.
64. Beaver, L.E.; Chalaki, B.; Mahbub, A.I.; Zhao, L.; Zayas, R.; Malikopoulos, A.A. Demonstration of a time-efficient mobility system using a scaled smart city. *Veh. Syst. Dyn.* **2020**, *58*, 787–804.
65. Wang, Z.; Wu, G.; Barth, M.J. Cooperative eco-driving at signalized intersections in a partially connected and automated vehicle environment. *IEEE Trans. Intell. Transp. Syst.* **2019**, *21*, 2029–2038.
66. Ma, J.; Hu, J.; Leslie, E.; Zhou, F.; Huang, P.; Bared, J. An eco-drive experiment on rolling terrains for fuel consumption optimization with connected automated vehicles. *Transp. Res. Part C Emerg. Technol.* **2019**, *100*, 125–141.
67. Sofeska, E.; Sofeski, E. Developing Projects for Realizing of the Program "Skopje 2020 Smart Strategy" by Enhancing Citizen Approach, Engineering, Digitalization, and Sensing of the City District Toward Smarter Sustainability Urban Potential in the Small Ring of Skopje. In *Resilient and Responsible Smart Cities*; Abdalla, H., Rodrigues, H., Gahlot, V., Salah Uddin, M., Fukuda, T., Eds.; Springer: Cham, Switzerland, 2022; pp. 69–84.
68. Schrank, A.; Kettwich, C.; Oehl, M. Aiding Automated Shuttles with Their Driving Tasks as an On-Board Operator: A Case Study on Different Automated Driving Systems in Three Living Labs. *Appl. Sci.* **2024**, *14*, 3336.
69. Li, Z.; Niu, J.; Li, Z.; Chen, Y.; Wang, Y.; Jiang, B. The impact of individual differences on the acceptance of self-driving buses: A case study of Nanjing, china. *Sustainability* **2022**, *14*, 11425.
70. Ryan, M. The future of transportation: Ethical, legal, social and economic impacts of self-driving vehicles in the year 2025. *Sci. Eng. Ethics* **2020**, *26*, 1185–1208.
71. Chai, H.; Rodier, C.J.; Song, J.W.; Zhang, M.H.; Jaller, M. The impacts of automated vehicles on Center city parking. *Transp. Res. Part A Policy Pract.* **2023**, *175*, 103764.
72. Krueger, R.; Rashidi, T.H.; Vij, A. A Dirichlet process mixture model of discrete choice: Comparisons and a case study on preferences for shared automated vehicles. *J. Choice Model.* **2020**, *36*, 100229.
73. Mcity. 2024. Available online: https://mcity.umich.edu/ (accessed on 21 June 2024).
74. Centre of Excellence for Testing & Research of Autonomous Vehicles—NTU (CETRAN). 2024. Available online: https://cetran.sg/ (accessed on 21 June 2024).
75. Engholm, A.; Pernestål, A.; Kristoffersson, I. System-Level Impacts of Self-Driving Vehicles: Terminology, Impact Frameworks and Existing Literature Syntheses. 2018. Available online: https://www.diva-portal.org/smash/get/diva2:1268871/FULLTEXT02.pdf (accessed on 15 July 2024).
76. Introducing The BayJax Innovation Corridor. Innovation Corridor. 2024. Available online: http://thebayjax.com/ (accessed on 21 June 2024).
77. Li, L.; Pantelidis, T.; Chow, J.Y.; Jabari, S.E. A real-time dispatching strategy for shared automated electric vehicles with performance guarantees. *Transp. Res. Part E Logist. Transp. Rev.* **2021**, *152*, 102392.
78. Richter, A.; Löwner, M.O.; Ebendt, R.; Scholz, M. Towards an integrated urban development considering novel intelligent transportation systems: Urban Development Considering Novel Transport. *Technol. Forecast. Soc. Change* **2020**, *155*, 119970.
79. Faisal, A.; Yigitcanlar, T.; Paz, A. Understanding driverless car adoption: Random parameters ordered probit model for Brisbane, Melbourne and Sydney. *J. Transp. Geogr.* **2023**, *110*, 103633.
80. Gaitanidou, E.; Bekiaris, E. Consumer Acceptance in Measuring Greece's Readiness for Transport Automation. *Future Transp.* **2022**, *2*, 644–658.

81. Cvijovic, Z.; Zlatkovic, M.; Stevanovic, A.; Song, Y. Conditional transit signal priority for connected transit vehicles. *Transp. Res. Rec.* **2022**, *2676*, 490–503.
82. Liu, H.; Yang, M.; Guan, C.; Chen, Y.S.; Keith, M.; You, M.; Menendez, M. Urban infrastructure design principles for connected and autonomous vehicles: A case study of Oxford, UK. *Comput. Urban Sci.* **2023**, *3*, 34.
83. Bucchiarone, A.; De Sanctis, M.; Bencomo, N. Agent-based framework for self-organization of collective and autonomous shuttle fleets. *IEEE Trans. Intell. Transp. Syst.* **2020**, *22*, 3631–3643.
84. Poliziani, C.; Hsueh, G.; Czerwinski, D.; Wenzel, T.; Needell, Z.; Laarabi, H.; Schweizer, J.; Rupi, F. Micro Transit Simulation of On-Demand Shuttles Based on Transit Data for First-and Last-Mile Connection. *ISPRS Int. J. Geo-Inf.* **2023**, *12*, 177.
85. Hyland, M.; Mahmassani, H.S. Operational benefits and challenges of shared-ride automated mobility-on-demand services *Transp. Res. Part A Policy Pract.* **2020**, *134*, 251–270.
86. Ahn, K.; Du, J.; Farag, M.; Rakha, H.A. Evaluating an eco-cooperative automated control system. *Transp. Res. Rec.* **2023** *2677*, 1562–1578.
87. Tsigdinos, S.; Karolemeas, C.; Bakogiannis, E.; Nikitas, A. Introducing autonomous buses into street functional classification systems: An exploratory spatial approach. *Case Stud. Transp. Policy* **2021**, *9*, 813–822.
88. Frey, C.; Hertweck, P.; Richter, L.; Warweg, O. Bauhaus. MobilityLab: A living lab for the development and evaluation of AI-Assisted services. *Smart Cities* **2022**, *5*, 133–145.
89. Modena Automotive Smart Area. 2024. Available online: https://www.automotivesmartarea.it/ (accessed on 21 June 2024).
90. Ziakopoulos, A.; Oikonomou, M.G.; Vlahogianni, E.I.; Yannis, G. Quantifying the implementation impacts of a point to point automated urban shuttle service in a large-scale network. *Transp. Policy* **2021**, *114*, 233–244.
91. Nguyen, P.H.; Hugo, Å.; Svantorp, K.; Elnes, B.M. Towards a simulation framework for edge-to-cloud orchestration in c-its. In Proceedings of the 2020 21st IEEE International Conference on Mobile Data Management (MDM), Versailles, France, 30 June–3 July 2020; IEEE: Piscataway, NJ, USA, 2020; pp. 354–358.
92. SUMMALab. Ride2Autonomy EU Project. 2024. Available online: https://summalab.nl/r2a/ (accessed on 21 June 2024).
93. Amaral, J.; Viegas, J.; Lemos, B.; Almeida, P.; Rosmaninho, R.; Perna, G.; Rito, P.; Sargento, S. Autonomous Shuttle Integrated in a Communication and Sensing City Infrastructure. In Proceedings of the 2023 IEEE International Conference on Mobility, Operations, Services and Technologies (MOST), Detroit, MI, USA, 17–19 May 2023; IEEE: Piscataway, NJ, USA, 2023; pp. 96–104
94. Cohn, J.; Ezike, R.; Martin, J.; Donkor, K.; Ridgway, M.; Balding, M. Examining the equity impacts of autonomous vehicles: A travel demand model approach. *Transp. Res. Rec.* **2019**, *2673*, 23–35.
95. Lin, W.; Wei, H. CAV-enabled data analytics for enhancing adaptive signal control safety environment. *Accid. Anal. Prev.* **2023**, *192*, 107290.
96. Wu, K.; Cheng, Y.; Parker, S.T.; Ran, B.; Noyce, D.A. Development of the Data Pipeline for a Connected Vehicle Corridor. In Proceedings of the International Conference on Transportation and Development 2023, Austin, TX, USA, 14–17 June 2023; pp. 218–230.
97. Hakak, S.; Gadekallu, T.R.; Maddikunta, P.K.R.; Ramu, S.P.; Parimala, M.; De Alwis, C.; Liyanage, M. Autonomous Vehicles in 5G and beyond: A Survey. *Veh. Commun.* **2023**, *39*, 100551.
98. Chen, S.; Dong, J.; Ha, P.; Li, Y.; Labi, S. Graph neural network and reinforcement learning for multi-agent cooperative control of connected autonomous vehicles. *Comput.-Aided Civ. Infrastruct. Eng.* **2021**, *36*, 838–857.
99. Ebbesson, E.; Lund, J.; Smith, R.C. Dynamics of sustained co-design in Urban Living Labs. *CoDesign* **2024**, *20*, 422–439.
100. Benyahya, M.; Kechagia, S.; Collen, A.; Nijdam, N.A. The interface of privacy and data security in automated city shuttles: The GDPR analysis. *Appl. Sci.* **2022**, *12*, 4413.
101. Gruyer, D.; Orfila, O.; Glaser, S.; Hedhli, A.; Hautière, N.; Rakotonirainy, A. Are connected and automated vehicles the silver bullet for future transportation challenges? Benefits and weaknesses on safety, consumption, and traffic congestion. *Front. Sustain. Cities* **2021**, *2*, 607054.
102. Pucihar, A.; Zajc, I.; Sernec, R.; Lenart, G. Living lab as an ecosystem for development, demonstration and assessment of autonomous mobility solutions. *Sustainability* **2019**, *11*, 4095.
103. González-González, E.; Nogués, S.; Stead, D. Parking futures: Preparing European cities for the advent of automated vehicles. *Land Use Policy* **2020**, *91*, 104010.

MDPI AG
Grosspeteranlage 5
4052 Basel
Switzerland
Tel.: +41 61 683 77 34

Energies Editorial Office
E-mail: energies@mdpi.com
www.mdpi.com/journal/energies

www.ingramcontent.com/pod-product-compliance
Lightning Source LLC
Chambersburg PA
CBHW041359210926
51850CB00013B/3124